Conflict and Collective Action

Conflict and Collective Action

The Sardar Sarovar Project in India

by Ranjit Dwivedi

Routledge
Taylor & Francis Group

LONDON AND NEW YORK

First published 2006 by Routledge

2 Park Square, Milton Park, Abingdon, Oxfordshire OX14 4RN
52 Vanderbilt Avenue, New York, NY 10017

Routledge is an imprint of the Taylor & Francis Group, an informa business

First issued in paperback 2019

Transferred to Digital Printing 2006

Typeset in S.R. Enterprises, Nirmal Puri, Lajpat Nagar-IV,
New Delhi-110024

British Library Cataloguing in Publication Data
A catalogue record for this book is available from the British Library

ISBN 978-0-415-40116-6 (hbk)
ISBN 978-0-367-17601-3 (pbk)

It is not possible to know to whom Ranjit might have dedicated his work—since he cannot say, we have decided to dedicate this work to him

Ranjit Dwivedi, 1965–2002

Contents

Impact

Reform

List of Tables

List of Figures

Foreword

Amita Baviskar

Controversy has dogged the Sardar Sarovar Project for more than two decades. Protest against the dam first focused on the issues of displacement and environmental harm in the Narmada Valley, and then swelled into a worldwide campaign led by the Narmada Bachao Andolan (Save the Narmada Movement). The Narmada struggle now symbolises far more than opposition to a large dam with questionable benefits and enormous costs. It has compelled the public conscience to recognise the flaws intrinsic to dominant models of development. It has fomented a debate about citizenship and democratic decision-making, technological choice, regional and class-based inequalities in resource use, and the place of cultural minorities within the nation. The multiplicity and range of issues raised by the project and the Andolan defy easy encapsulation. That Ranjit Dwivedi presents a comprehensive and detailed account of these complex questions is a remarkable achievement.

I first came to know of Ranjit when he published a critical, yet eminently fair, review of my book on adivasi politics in the Narmada Valley. It is that same careful, even-handed appraisal that distinguishes this work. Meticulous in its examination of substantive issues, and measured in tone, this is a provocative book precisely because it refuses to take sides in a debate marked by passionately held, highly polarised positions. Whether one agrees or disagrees with its stance or analysis, the intellectual integrity of this work shines through.

The genesis of the Narmada Valley dams can be traced to the early days of Independence when electricity and irrigation were critical constraints in the national project of achieving self-reliance through intensive industrialisation. Industrially manufactured agricultural inputs, coupled with irrigation, were deemed essential for rapidly increasing food production in the context of widespread scarcity. The Narmada dams were intended to promote national security not only in terms of food, but also geo-politics. Assessing the likely gains from the Navagam dam, planners in the 1960s pointed out that it could be used to reclaim and populate saline marshlands in Kutch, so that 'retired soldiers and sturdy peasants' could be a demographic deterrent to Pakistani incursions from across the border. The popular political slogan of that time, *jai jawan, jai kisan* (Victory to the Soldier, Victory to the Farmer) indicates the two dimensions, external and internal, along which threats were perceived. Dams continue to be part of a larger landscape of insecurity, a re-mapped riverine sacred geography. The conjoint fortunes of dams and the nation are asserted even 35 years later. In a speech inaugurating the resumption of work on the Sardar Sarovar dam, the Bharatiya Janata Party leader and then Indian Home Minister L. K. Advani linked the project to another triumph of Indian power—the nuclear test at Pokhran—and denounced either's opponents as 'anti-national'.[1]

Dwivedi's account of the early debates around the Narmada project shows that the necessity of large dams was a foregone conclusion for Indian planners and politicians. There was no debate about whether to build the dam, only about how to share its waters between various states. Alternative strategies for water management were not considered. Instead, the discussion was limited to the contending claims of the riparian states of Madhya Pradesh (MP), Maharashtra and Gujarat, which came to be adjudicated by the Narmada Waters Dispute Tribunal. The Tribunal's Award in 1978 allocated different shares of Narmada waters and the costs of the project (including resettlement) among the three states. Dwivedi argues that inter-state equity was given short shrift by the Tribunal: 'Notwithstanding the commitment to balanced regional development in the Indian federation' (p. 70), the Award was not based on an assessment of the relative needs of different states but on their resourcefulness in making their case. Gujarat fared better than MP, even though it was far more prosperous, because politically organised industrial and agrarian elites could galvanise the state administration into pressing their case. Dwivedi thus suggests that the Tribunal violated its mandate of promoting inter-state equity

and did so because it was swayed by the more effective representations of the Gujarat lobby. While a hegemonic Gujarati identity has
certainly been an advantage in creating a united constituency in
favour of the dam, I am not persuaded that Gujarati mobilisation
influenced the Award to MP's detriment. It should be borne in
mind that Madhya Pradesh had successfully staked a claim to
building several large dams upstream of Gujarat and was perceived as a state rich in natural resources. For the Tribunal, MP
was not a resource-poor state; it was up to its government to
develop its endowments. Protests against the Award in Madhya
Pradesh were concentrated in the submergence area of Sardar
Sarovar among people to be displaced, rather than among the
populace at large. If one takes into account all the dams proposed as part of the Narmada Valley Project, the Tribunal Award
does not seem to unduly favour Gujarat.

Apart from the fairness of the Award in terms of the claims of
different states, there is another important aspect to the equity
question that the Award fails to address. The Narmada Water
Disputes Tribunal's Award has, over time, been presented as
sacrosanct and, indeed, as the protracted inter-state wrangling
over Cauvery waters shows, there is a good case for accepting
the finality of such judicial decisions. But was the Tribunal
Award, with its primary focus on the riparian rights of different
states, a suitable judicial instrument for judging the pros and
cons of building large dams on the Narmada? A Tribunal designed
to address distributional conflicts generated by the federal structure of Indian states ignores such issues *within* states. Equity
concerns about whether the main beneficiaries from the
Narmada dam should be drought-stricken pastoralists in far-
flung hamlets in Saurashtra or well-to-do sugarcane growers in
central Gujarat, were not included in the Tribunal's terms of
reference. The differences among affected populations *between*
states were also unacknowledged; people were identified by the
state they resided in, and not their unequal situation within the
matrices of caste, class, gender and eco-locality. Thus the dilemmas of weighing the cultural losses of hill adivasis in Madhya
Pradesh and Maharashtra against the gains of cash-crop farmers
in Gujarat were glossed into a simplistic injunction to provide
land in compensation.

Dwivedi points out that, while virtually all the parameters
assumed by the Tribunal in its deliberations have changed, the

Award remains non-negotiable. Basic data about river flows and
the assured quantum of water for impoundment, and the extent
of submergence and displacement stand revised. Environmental concerns that went unconsidered until the 1980s have come
to the fore. The need to settle 'sturdy peasants' in Kutch has
been superseded by the need to provide drinking water to urban
populations, a priority that can be met in more ways than one.
Previously unavailable technological options to large dams have
now become viable. Yet, despite this vastly changed scenario,
the Tribunal's decision of a 455 ft high dam has been virtually
writ in stone (or concrete, in this case). The Madhya Pradesh
government periodically made noises about lowering the height
of the dam, but did not pursue that line of argument with any
seriousness, using it primarily as a red herring to distract the
opposition. Even the Supreme Court, though it invited all views
'uninhibited by the terms of the NWDT Award' (p. 224), chose to
accept the recommendation of a government-appointed committee that the dam be constructed to the height of 436 ft and then,
if resettlement were satisfactory, to its full height of 455 ft.

    To understand how a contingent set of technical features acquires
a stable, taken-for-granted character, a validity that transcends
its context, we must turn to the political economy of planning.
Drawing on Vijay Paranjpye's ground-breaking work on the economic appraisal of large dams, Dwivedi examines the three benefit-
cost assessments generated at different moments to justify the
Sardar Sarovar dam and points out that, despite being based on
different estimates of costs and benefits, all three emphatically
agree that the project would yield high benefits at low risk. The
Government of Gujarat, the World Bank, and Tata Economic
Consultancy Services, were unanimous in their positive appraisals of the project because their valuations were derived from a
common source, viz. the preferences of state planners. The model
was not disaggregated to consider disparate stakeholders who
might have conflicting preferences. Dwivedi goes on to argue
that it was, in fact, a foregone conclusion that economic appraisals
generated by project planners and funders would be positive.
'Comprehensive cost-benefit calculations were made only *after*
it had been decided to go ahead with the project' (p. 107, emphasis in the original).[2] That prior decision was a political one, based
on electoral equations between the Congress and its opponents

in different states and at the centre. Although Dwivedi does not examine the political considerations at work in the World Bank, his argument can be usefully extended to understand how multilateral financial institutions decide to support certain projects.[3] While the institutional pressure to fund the dam was somewhat differently structured in the case of the Bank, decision-making still bore a striking resemblance to the practices of the Indian government in terms of how political and economic imperatives overrode concerns about social and environmental impacts. Officials at the Bank were evaluated on the basis of how much money they loaned out, and a project initiated by a relatively stable and solvent government was an attractive funding proposition.

Dwivedi's delineation of the post-facto character of planning provokes several questions. If the bottomline for the project boils down to an electoral calculus, a concern in the last instance with appeasing resource-hungry elites, should we take planning seriously at all? If plans only serve to justify political decisions with retrospective effect, are they merely cosmetic exercises? As James Ferguson (1990) has persuasively shown, development planning is a powerful tool for legitimising state projects because it has the discursive effect of detaching and concealing the particular interests that inform project initiatives. Planning privileges forms of technocratic knowledge that mute the discordant strains of conflicting voices to a neutral murmur of rational appraisal. Planners must engage in constant misrecognition, not only of the political context in which projects are embedded, but of the vast parallel economy of corruption nourished by large projects. Turning a blind eye to commissions and kick-backs that may comprise as much as a quarter of the entire project cost is essential for upholding the fiction that large dams serve the public good. While the pervasive influence of corruption is widely acknowledged in the public domain, academics have generally hesitated to incorporate it into their analysis, perhaps because of the difficulties in documenting and proving corrupt practices. We are more at ease discussing discursive effects than the grimy details of money changing hands, preferring to politely describe corruption as a form of 'rent-seeking'. Yet, there is an unwritten benefit-cost analysis that informs political decisions that planners must dress up respectably, that needs to be better articulated in our critical analysis of planning.

The discursive effects of planning have material consequences. Purged of politics, planning apparently invites discussion and debate; its neutrality seems to offer room for 'stakeholder participation'. Conflicts get domesticated through such consultative processes. As dam opponents have found out, it is possible to develop a *critique* of a benefit-cost assessment, but immeasurably harder to come up with an *alternate* appraisal on the same terms. The kind of data and the computational techniques deployed in such assessments tend to be the tools of specialists, and governments have a decided advantage in mobilising technical expertise. Lured into the labyrinths of technical disputation, it is easy for critics to get trapped. For a social movement struggling for legitimacy, taking on planners' claims is an imperative that entails difficult choices in terms of using scarce organisational resources.

The career of the Sardar Sarovar project within the larger trajectory of national development in India has been accompanied by the emergence of an environmental technocracy at the centre, reflecting a growing concern with managing ecological issues in the public sphere. Dwivedi relates the formation of what eventually became the Ministry of Environment and Forests to the bureaucratic impediments faced by the dam project. Before the Supreme Court temporarily halted the project pending its decision on the writ petition filed by the Narmada Bachao Andolan, the only government authority to oppose the dam was the environment ministry which refused to sanction the project on grounds of inadequate research into its environmental impacts. This was a fledgling form of eco-governmentality, to extend Foucault's (1991) concept of the exercise of power as 'the conduct of conduct', that has since become more established, thanks to vigorous support from the Supreme Court. Simultaneously, project planners have learnt to negotiate these regulations about environmental conduct by generating better technical reports and, at a pinch, mobilising political backing to leapfrog over environmental concerns. Though the pastoral function of governmentality—leading and guiding the conduct of subjects— is still only fitfully performed, this is an area of environmental activism that needs to be studied in greater detail.

Against the backdrop of the project, Dwivedi traces the emergence of multiple strands of opposition that eventually culminated in the emergence of the Narmada Bachao Andolan as the main voice of those threatened with displacement. This is a valuable discussion of differences that have largely gone undocumented, highlighting the difficult choices that social movements must make under conditions of great adversity and uncertainty. Although the Andolan is now associated with a stand that opposes large dams in toto, it is useful to remember that this was not always so, and to recall the exigencies that led the Andolan to this position after exploring other alternatives. Dwivedi also discusses the strategies adopted by other groups working on displacement, especially ARCH-Vahini in Gujarat, which has had a tense relationship with the Andolan, at one point accusing the Andolan of not practising 'responsible activism' (p. 142), and failing to agitate for a better rehabilitation policy. In describing the multi-level articulations of protest, Dwivedi also mentions the significance of Northern environmental and human rights NGOs, especially their role in compelling the World Bank to appoint the Independent Review Mission that ultimately led to the revocation of the loan agreements between India and the Bank. Whether the political values and priorities of Northern groups influenced the Andolan's strategic choices, and vice versa, remains unanswered. For instance, was the International Rivers Network's world-wide campaign against large dams a factor in the Andolan leadership's decision to adopt a hard-line anti-dam stand? Comparable analyses of indigenous activism in the Amazon rainforest by Beth Conklin and Laura Graham (1995) suggest that Northern supporters and audiences shaped representational strategies in decisive ways.

While the contradictions and compromises that are an intrinsic part of alliances between unequally situated groups need to be acknowledged more fully so that localised movements have greater leverage, transparency is not necessarily an unalloyed good. Public scrutiny of such questions can be a strategic liability for movements that must constantly defend themselves against charges of being 'anti-national' and 'foreign-funded'. The more nuanced the delineation of the contours of political coalition-building, the greater its potential to harm a beleaguered movement. The documentary impulse that academics hone, to 'tell it like it is', can be a political liability. At every turn, scholars and supporters of progressive movements struggle with forms of self-imposed censorship, their choices praised by some and condemned by others. The ethical and political dilemmas that all academic practice must negotiate and resolve acquire greater intensity in this case.

Dwivedi situates anti-dam protests in India within a larger environmental movement that subscribes to the ideology of eco-logical Marxism, challenging patterns of resource control and demanding greater control for those who labour to produce value. While in sympathy with this political project, Dwivedi seems less willing to support the Andolan's wider critique of develop-ment as embodied in large dams. In his view, if its distribution of benefits and costs could somehow become fair, and its envi-ronmental problems addressed, the dam should be built. I believe that ecological Marxism offers a more thorough-going critique of large dams than Dwivedi allows. There is more to an ecological Marxist manifesto than redistribution; there is also a trenchant critique of commodity relations under capitalism, a critique that lies at the heart of the Andolan's opposition to large dams and to a globalised economy.[4] Large dams emerge as 'solutions' in a world dominated by capitalism, where rivers must be exploited to maximise profit for capitalists and the state. To arrive at dif-ferent priorities and different arrangements, both social and technological, the life-world needs to be radically transformed, and that is the political project against development. Redistrib-uting power will not necessarily bring about a more sustain-able, less consumerist world. In fact, those who gain power tend to plunge into 'me too' modes of accelerated material consump-tion. The hegemonic discourse of development works through transforming subjectivities; its techniques of persuasion are deployed in the education of desire. As desiring subjects, those who are dominated want what they have been denied—the power to consume.

The unfulfilled promise of prosperity for all, embodied in the discourse of national development, sparked off challenges that have taken the form of identity politics—caste and religious af-filiations that were folded into the fabric of a nation premised on 'unity in diversity' and a polity where electoral mobilisation relies on invoking communitarian loyalties. That the ecological Marxist critique of development and its vision of change have been eclipsed within this historical configuration does not negate their cogency. Movements like the Narmada Bachao Andolan ask impossible questions that can never be adequately addressed in the mainstream political process, precisely because they probe the very fundament of our social arrangements, our conveniences and accommodations. The failure to deal with them only points to the poverty of our political discourse. Rather like Mahatma Gandhi, whose political brilliance was grounded in a spiritual quest to reform the self, the Andolan has achieved recognition, even a measure of success, without generating a major engagement with its larger critique. Yet this critique remains alive, slipping in and out of everyday acts of resistance and reflection, so that even mundane logistical tasks like organising transport for a rally can become inspired by a transcendental faith. The power of the Andolan's ideology lies in its ability to signify many interrelated, sometimes contradictory, things all at once—rich meanings that cannot always be attributed to specific groups of 'stakeholders' but that take on a life of their own. The multi-level mobilisation accomplished by the Andolan moves across ideological scales, 'radicalizing rationality' (p. 197) and, I would argue, also going beyond it to re-enchant the world by creating a secular sacred. Its generative power lies in bringing utopian visions to life by connecting them to tangible projects for social justice.

There are two major themes in this work that, in my view, deserve serious consideration by sociologists of social movements. The first concerns truth-claims—the discursive production of authoritative 'facts' by contending social groups. Dwivedi argues that a critical analysis of a project must do more than simply evaluate the merits of opposing claims. Beyond examining their validity, it must trace the 'conflicting interests and experiences, power and knowledge struggles' that underlie different positions. Dwivedi deftly performs a double move in the study, analysing competing arguments for internal consistency, and also interpreting them as rhetorical strategies that seek to achieve certain political effects. Thus he shows that the cost-benefit analyses conducted at different points in time are not only economic estimates, but also ritualised exercises that seek legitimacy for the project. While his investigation of such facts and figures shows a relish for the cut and thrust of technical argumentation, Dwivedi is more circumspect in his treatment of judicial authority in the form of the commissions and courts charged with examining the case of the dam. The truth-claims they produce command our attention; it becomes impossible to wholly challenge their authority. This is the double-bind that the Andolan could not escape. It was compelled to demand a comprehensive review of the project from the state even as it knew that no such review would ever be on its own terms. The terrain of struggle has throughout been tilted in favour of those in power. And still the struggle went on, against odds that over-whelmed, in the hope of a temporary reprieve, some unexpected conjuncture, the idiosyncrasy of a Supreme Court judge—on such tenuous threads hung the fate of thousands. The Andolan had to submit to the knowledge that some truth-claims carry more authority than others, knowing that the judicial writ would be binding whether they agreed with it or not. Given their finality, I believe that judicial truth-claims demand the greatest scru-tiny of all. They are too important to be shielded by the majesty of the law, or by fear of contempt of court.

No scholar can critically scrutinise all the truth-claims about the project because the analyst is as much influenced by hegemonic ideologies as anyone else. The production of a meta-narrative about truth-claims requires that academic statements, such as this study, also be similarly examined in terms of their political effects. Perceptions of 'facts', and the relative value one places on different facts, are shaped by social position and constitute what Donna Haraway calls *situated knowledge*. Whether pro-duced by judges, government committees, or scholars, views from above may seem all-seeing, but they too are partial, their hori-zons bounded by the blinkers that accompany every subject-position. The journey beyond this limited field of vision requires empathy and a willingness to recognise and engage with other realities. But the arrangements of power mean that it is not incumbent upon engineers, economists or others who sit in judge-ment to enter into the world of poor adivasi farmers whose fate they decide. All truth-claims are not equal. I would argue that what gains credence and currency in the world, and what remains a voice in the wilderness, cannot be treated on par. Situating different kinds of truth-claims, made using different vocabular-ies, within the same field of politics, requires a feat of transla-tion that challenges the material and imaginative resources of most social movements.

Dwivedi poses contrasting truth-claims against the backdrop of a world increasingly ruled by risk, the second major theme in this book. Uncertainty is inherent in large, complex projects such as the Sardar Sarovar dam. Dwivedi highlights how the Andolan has transformed public perceptions of the social and environmental risks of large dams. He points to a lesser known aspect of the struggle—its effective mobilisation of technical arguments that challenge the economic, financial, social, and environmental parameters of the project. Indeed, the Andolan's core of urban, highly-qualified activists and supporters was able to take on government experts on their own terms, matching analysis for analysis, technical report for technical report. The two sets of studies arrived at diametrically opposing conclusions about the likely outcomes of the same project, a deep interpretive divide. Their divergent understandings were based on different underlying assumptions, conservative in one case and liberal in another. What made the prognostications of one set of studies seem more persuasive than the other? How did readers make up their minds about what to believe? To my mind, it is far easier to be sanguine about a project's future if one is in the privileged position of benefiting from it and is protected from its nastier aspects. Even the Supreme Court, charged with upholding the constitutional right to life of those threatened by displacement, accepted the wildly optimistic claims and assurances offered by the government at face value. Whether it is the Bhopal gas disaster or Enron power project in Maharashtra, project promoters and defenders have always sought to downplay risk, often with catastrophic consequences for subaltern groups or tax payers. The principles of a more socially responsible process of decision-making that focuses on reducing risk for the most vulnerable social groups have been cogently presented in the final report of the World Commission on Dams (2000). Hostile responses from dam-builders indicate that the report is a potential weapon in the hands of those mobilising against displacement.

Dwivedi's discussion of the politics of technical disputation in an environment marked by uncertainty illuminates a neglected facet of the Narmada struggle. At the same time, it is also important to remember that the hegemony of technical truth-claims in the public sphere is continually under siege from other experiential realities. The political economy of planning

and projects is constantly reworked and resisted by the very
subjects that it seeks to produce. Their presence in public protests is a powerful reminder of all that cannot be calculated and
computed in rupees and dollars. The visual impact of these
ground truths—the lives of Nimari farmers and hill adivasis
inhabiting a landscape that they have made meaningful, the
charismatic integrity of Medha Patkar—has been far more important than the slew of booklets and articles written on the
dam.[5] The vocabulary employed by the displaced cuts through
paper claims and technical truths to insist that justice should
not be denied. I believe that this is the greatest achievement of
the Andolan—that it stripped away the seeming neutrality of
numbers, the facticity of figures and formulae. In so doing, it
has enlarged the public sphere so that people hitherto unconsidered and uncounted could speak about what mattered to
them, their lives and their beliefs, and their collective eloquence
had the power to make the world pause and listen.

In 2000, the Supreme Court gave its final judgement in the
Sardar Sarovar case, allowing the dam to be built since 'dams are
in the national interest'. The decision has been a major setback
for the anti-dam movement, which has valiantly continued to
struggle against other dams on the Narmada. It has been a death-
blow for the people in the submergence zone who resisted the
project for 20 years. In retrospect, others, many of them victims
of 'resistance fatigue', who had earlier accepted even inadequate
resettlement as making the best of a bad deal, seem to have
played their cards better. As they sort through their complex
feelings of anger, defeat and fear, those who chose to fight till the
end continue to struggle for survival, seeking some land from a
government that knows full well the insecurity of their position.

In many ways, the climate for a sympathetic consideration of
the plight of the displaced no longer exists. The project of national
development has been transformed by neo-liberal 'reforms', and
the emergence of dominant classes ruthless in their indifference to the circumstances of their poorer fellow-citizens. These
classes demand electricity and water at *any* cost—the Tehri dam,
hydro-projects in Nepal, natural gas from Bangladesh and the
former Soviet republics, all possibilities are vigorously explored
regardless of ecological and social hazards. At the same time,
widespread farmer suicides in the semi-arid Deccan region point

to a worsening agricultural crisis around credit, seeds, pesticides, but also *water*. Those who can, move their investment out of agriculture; small land-owners are left high and dry, destitution and migration their only future. In the context of neo-liberalism, resource-intensive development has created intense inequities which cannot be managed within the older discourse of 'national development'. While the current debate around large dams addresses issues emerging from the new water and energy policies—the role of private capital in 'public projects', power-purchase agreements that overwhelmingly favour private power generation units, water pricing, and lay-offs in state water and power utilities, it faces a daunting new challenge in the way naked exploitation has become normalised around the world.

In reading this work, I was perturbed by Dwivedi's studied distance from both the pro- and anti-dam positions. According to him, most studies 'are limited to the objective of providing corroborative evidence of the badness or goodness of the SSP and other such large dam projects. The debate generated by these studies gradually froze with the periodic recycling of arguments for and against the projects' (p. 295). Putting aside the question of whether that is an accurate characterisation of the literature—and Richard White, Robert Klingensmith, David Gilmartin, Rohan D'Souza rank among several others who eschew crudely evaluative conceptual frames—I think there is a case to be made for evaluative studies. At the end of the day, we are left to confront that scary question: What is to be done? For after analysing who says what, and who will gain or lose, and demarcating areas of uncertainty and ignorance, we still have to make *judgements*. We have to arrive at provisional understandings of goodness and badness in order to orient action. This is a hazardous process, always 'subject to revision', but essential nevertheless. As someone who strives to practice what Michael Burawoy has called 'public sociology', trying to be a partisan yet critical interlocutor on issues of social justice, I found myself irked by Dwivedi's quietism. The Hindi word for detached—*tatasth* (literally, on the riverbank, unmoved by the current)—seems all too appropriate for this study. Though much of his analysis points in that direction, why didn't Dwivedi come out and say that the dam is a disaster? The provocation is Ranjit's; sadly, the pleasure of a vigorous argument with him is not to be mine.

Conflict and Collective Action

I end on a personal note with an account from August 2004 of events in Anjanvara, the Bhilala adivasi village where I first lived in 1990 when I started research in the Narmada Valley. The struggle in Anjanvara remains for me a touchstone for understanding the politics of large dams and of development:

Khajan telephoned from Sondwa on the night of August 14. He had come to the town from his village Anjanvara to collect 'relief' at the Block headquarters. For the first time ever, his village and others adjoining it on the banks of the Narmada saw their fields submerge in the rising waters of the reservoir of the Sardar Sarovar dam. Their standing crops—jowar and *bajra*, maize and *tuvar*—disappeared under water. With the land was lost their supply of food and income for this year and for the years to come.

By now, this has become a familiar story. All of us know of the tragedy of the Narmada dams. I knew that Anjanvara would be submerged during this year's monsoons. Since the Supreme Court order of 2000, we had seen the river rise slowly, irreversibly altering the landscape. We knew this was imminent. But when it finally happened, it was still awful beyond words.

On the phone, Khajan's voice is hesitant, blurred with tears. He manages a couple of halting sentences: 'Amita, *maari zingi doob gayi...Maara kheton ma boot phirtali*' [My life has drowned...a boat now sails over my fields]. Binda, his wife, has been crying ever since. They are farmers, this land was their life. They haven't received any land or equivalent monetary compensation from the government, their legal entitlement. Last year, they had rebuilt their house further up the hillside, so they were at least spared the additional trauma of losing everything they owned. But next year, they will have to move out of the area and they will have nowhere to go. Their struggle for survival—fighting hunger, disease and exploitation, trying to secure a better life for themselves and for their children has just become immeasurably harder.

Khajan is the *sarpanch* (elected head of local government) of the Sakarja panchayat. He had come to Sondwa, the Block HQ, to collect the wheat and rice sanctioned by the District Collector as 'relief'. The meagre amounts allotted per household will barely last them a couple of months. Beyond that, if the government thinks of them at all, it probably expects them to fight starvation in the time-honoured way—migration.

Khajan says that government officials and shopkeepers in Sondwa regard him as a figure of fun. '*Maadarchod mere upar hanste hain*' [The mother-fuckers laugh at me]. The *bazaarias* have always despised and feared adivasis like Khajan who were key members of the Andolan and Sangath, and whose collective defiance over a twenty-year period forced them to change their practices—offer respect and fairer treatment, better prices and wages. These adivasis are now once again at the *bazaarias*' mercy. There is ignominy in pleading for a few quintals of grain from a state you once challenged and officials lose no opportunity to make the experience as humiliating as possible.

Further into the conversation, as he repeats the same disjointed sentences, it occurs to me that Khajan is slightly drunk. But the comforting numbness that liquor should bring is denied to him. The pain pierces through and he returns again and again to the immense grief of a living death: *maari zingi doob gayi*.

The drowning of Anjanvara is one reality. A different world is to be found in Ahmedabad, where water from the Narmada now surges through the desiccated bed of the Sabarmati and people rejoice that a river runs once again through their city. One river flows, the other dies. Resolving this paradox requires a sustained critique of the political economy of water. Ranjit Dwivedi's book is a valuable contribution in that endeavour.

Delhi
February 2005

Notes

1. *The Hindu*. 1 November 2000. Online edition.
2. That dam construction is a complex, contingent process on which planners ineffectually try to impose discursive control is brilliantly illustrated by Timothy Mitchell (2002) in his account of the Aswan high dam in Egypt.

3. For a conceptually path-breaking ethnography of how aid practices
   are mutually constituted from the head office to the field, see David
   Mosse (2005).
4. See the introductory essay in Richard Peet and Michael Watts (2004).
5. The role of the media in shaping public perceptions of the Narmada
   issue and other social movements merits closer examination. There
   are significant differences between the coverage in English, Gujarati
   and Hindi papers (sometimes within the same state) and the extent
   of reporting has fluctuated over the career of the project. The media
   tendency to represent complex issues by focusing on key personali-
   ties and photogenic events has sometimes led to a trivialisation of
   the movement.

Preface

The present book is a slightly revised version of the doctoral thesis, *Resource Conflict and Collective Action: The Sardar Sarovar Project in India*, that Ranjit Dwivedi successfully defended in 2001 at the Institute of Social Studies, The Hague, The Netherlands, and for which he was awarded a distinction. Its subject matter, namely a detailed analysis of the Sardar Sarovar Project (SSP) and the conflicts generated by it, is one of extraordinary complexity, requiring the coordination and integration of multi-level and multidisciplinary enquiry into a single framework of analysis. The study indeed represents a highly imaginative, competent and commendable analysis of the complex policy debate and collective action(s) mobilised in opposition to and in support of the mega Sardar Sarovar dam project. This project itself involves numerous layers of preconditions and dimensions—technical, social and economic—that need to be analysed in order to gain a proper understanding of its overall implications. Beyond this, the political and juridical field of contestation that had emerged around the project over the years since its inception, had itself developed into an inordinately complex web of claims and counter-claims, strategies and understandings. The field comprises issues concerning displacement and resettlement; threatened livelihoods and promises of security; environmental hazards and uncertainties; anticipations of differential economic benefits; and the politics and policies of state and national governments, multilateral financial institutions, infrastructure corporations, as well as interventions by activists and social movements.

Ranjit Dwivedi's primary aim was not to arrive simply at an overall assessment of the pros and cons of the SSP, though that in itself would have constituted a sufficiently ambitious undertaking. While he dealt extensively with the project's contested expected outcomes through a careful and critical re-analysis of all available materials, in his research this constituted the necessary groundwork for a more informed understanding of the contestation around large dams in general and the SSP in particular.

Undertaking such a research project requires more than the usual conceptual and analytical instruments. It sets particularly high demands in terms of methodology and its application, the ability to mix and synthesise a range of diverse research methods, and the sophistication to bring this together with proper analytical distance, clarity of argument and considered assessment. With respect to each of these requirements this analysis scores very high indeed: the work is original, of high academic calibre, methodologically challenging if not exemplary, innovative as well as creative.

Ranjit Dwivedi, who hailed from Orissa in India, had been a Lecturer in Sociology in Delhi University before joining the Ph.D. programme at the Institute of Social Studies. His was but the second thesis, since the inception of the Ph.D programme at ISS in the 1980s, to be awarded a distinction. Upon leaving the ISS in 2001 he took up the position of Lecturer in Development Studies at The Open University in Milton Keynes, England. In this new position he immersed himself with all his characteristic enthusiasm and energy and, predictably for those that knew him, in the short time he was there, became a highly valued colleague of the Development Policy and Practice Group. In short, Ranjit seemed on course for an exceptionally promising if not brilliant academic career.[1]

While he was in the process of completing his thesis, actually after he had handed in his finalised draft but before his public defence, Ranjit suffered acute renal failure. He showed remarkable courage and humour throughout the process of recovery after a transplant operation. Tragically, five months after taking up his post at the Open University, Ranjit suddenly passed away while on a field trip with OU students.

As the supervisors of his doctoral research we had become increasingly convinced that in Ranjit's study a perspective was being developed which should be heard and read more widely. The present appearance of Ranjit's work as a book is a tribute to his scholarship which merits recognition by the wider research community. As such the book represents the culminating phase of his work that, under normal circumstances, he himself would surely have wanted to see through successfully. The publication of the book thus gives a sense of satisfaction not just to those of us who have been indirectly engaged in the research project, but also to the extended ISS community, and no doubt to Ranjit's family.

Quite a lot of water has flowed down the Narmada since the time that Ranjit finalised his dissertation, necessitating some changes to the book version, while at the same time retaining the intellectual integrity of his work. As such, we were aware of the need and utility of an appropriate updated contextualisation. With this in mind, we invited an acknowledged expert in the field, Dr Amita Baviskar, a leading environmental sociologist who has made extensive, sustained contributions to these debates, to provide an independent introductory essay which would provide such a contextualisation, as well as a critical, reflective commentary on the work and its significance. In fact, Dr Baviskar was aware of some of Ranjit's work before the tragedy came to pass. Kindly, she agreed to contribute such an essay, and we are most grateful to be able to include it herewith. In addition we have dropped the first two chapters from the thesis and also made some other deletions and substitutions, exploiting also the text of the statement made by Ranjit at his public defence.

We vividly recall that Ranjit Dwivedi came to ISS a passionate, articulate opponent of the SSP. At inception, he conceived his research project essentially as an intellectual elaboration of the case against the dam. As the research process unfolded the elemental hard position seemed partially to dissolve, and the initial linearity of the argument to find a rather more complex grammar and vocabulary. By the time the thesis was fully drafted, Ranjit's final views on the dam came to be dispersed and lodged at different levels and in various parts of the structure of his argumentation. It comes as no surprise that Dr Amita Baviskar expresses a sense of loss at the impossibility of being able to pursue this question directly with him. What could have been an enlightening conversation between two creative and committed scholars was not fated to take place. Instead, all we can offer is an edited excerpt from the transcript of his public defence, where Ranjit was challenged by one of us with such a question. His nuanced and reflective response is revealing and relevant.

To Resist or not to Resist

Question posed by Professor Ashwani Saith, Co-Promotor and Examiner:

It has been a while; much water has flowed down the Narmada from the point of five or more years ago when you walked into my room, Ranjit. There is a metamorphosis between that time and now—from a self-righteous activist who was relatively clear about which side of the line or what bank of the river he stood, to an academic who has managed to write all this dense stuff and still left us wondering which bank one should really be on! My question revolves around what you have said is the central issue—to resist or not to resist. And I would like to preface it with a small comment on perceptions. Perceptions are formed on the basis of imperfect information. You never have access to anything like perfect information—that is a notional concept. Everybody acts on the basis of perceptions—as an activist you move in certain ways; as an academic you tie your understandings and perceptions to another basket of items. After having taken a dip in the Narmada waters, indeed having gone underwater and held your breath for five years, when you come up, I want to ask: have your perceptions changed, and if so, what is the basis for that change? Eminent activists or prize-winning writers are very clear about the rights and the wrongs of it all. I was recently in Ahmedabad. There was a big banner just outside the airport: 'Thank you Supreme Court for saving our lives with this water'. I think there is a lot of truth, or one real truth, behind this. Asked to make a judgement, the Supreme Court put it away several times but then had to say something. Is the academic to be the only person who will not say something? To resist or not to resist? How do you read it now? What is the net value added of your five years of tapasya *in terms of the outcome—the way you perceive it, and how you would then act on the basis of that perception?*

Response by Ranjit Dwivedi:

I would resist—perhaps because of a conviction that I have, that some of the basic things that one expects in the Sardar Sarovar Project, I suspect, have not been done properly. That would include issues like a proper cost-benefit analysis about which I have a chapter and about which I have reflected. We have three cost-benefit analyses before the project started and all three of them have got different internal rates of return, and the question that I pose is: which one to believe in such a situation? Then there is also the criticism that one has about displacement and rehabilitation. We are throwing out a lot of affected people from the valley but without a proper rehabilitation scheme. If you go to the rehabilitation centres you will find that there have been successful cases, but there have also been extremely unsuccessful ones. And while you feel happy about those that have been settled, you also try and question how they have got what they have got, and then suspicion starts building in your head. But that's about it: it's just suspicion that you walk out with. But when you go and see people who are living in tin sheds and complain to you that it is extremely hot in summer and extremely cold in winter and full of water in the rainy season, you start wondering where this is all going. So probably on those two scores—and that's what I call the basic elements—I would resist. I would certainly resist. But I would not go to the extent to which a section of the activists have gone, to whom the criticism is much more than that—to whom the dam should not be built. So their list of criticisms and conflicts, and questions that they have are enormously large and it isn't possible for policy-makers of a nation-state to be able to even think of those kinds of expectations. Professor Willem van Schendel mentioned that because of its links with global networks, the movement is in a position to mobilise these kind of resources very quickly. But the Government of India is not in a position to counter this by mobilising what are its global networks. So at some point while trying to accommodate some of the criticisms from the movement, or the radical version of the movement, at some point it says 'no more, this is where we stop'. The problem for me is that they are stopping too soon. And that is where I will resist. But I think if the displacement and rehabilitation aspect is taken seriously, and social and environmental impacts and rights issues are integrated with the cost-benefit analysis: then I am all for the dam.

[Excerpts from the Public Defence]

We would like to acknowledge the sustained encouragement and support provided by Sharada Srinivasan in this endeavour; and the highly efficient professional advice and inputs of Omita Goyal and Esha Beteille. Special thanks are also due to the Board of the Institute of Social Studies, and in particular to Dr David Dunham, Deputy Rector, for their unequivocal institutional and financial support throughout the process; to the family of Ranjit Dwivedi for agreeing to this publication; and to Dr Amita Baviskar for her open willingness to participate so thoughtfully in this unusual, but valuable, enterprise.

Martin Doornbos
Ashwani Saith

Institute of Social Studies
The Hague
February 2005

Notes

Articles and Chapters in Books: 'Action, practice and "Best" practice in development', in Part 2 of Course TU871, Development: Context and Practice, The Open University, Milton Keynes, 2002; 'Models and methods in development-induced displacement', *Development and Change*, 33 (4): 709–32, 2002; 'Environmental movements in the global south: issues of livelihood and beyond', *International Sociology*, 16 (1), 2001; 'Environmental movements in the global south: an outline of the critique of the livelihood approach', in Henri Luistiger-Thaler, Hamel Pierre, Jan Nederveen Pieterse and Sasha Roseneil (eds) *Globalisation and Collective Action*. Macmillan and St. Martin Press, 2001; 'Displacement risks and resistance: local

perceptions and actions on the Sardar Sarovar (Gujarat), India',
Development and Change, 30 (1): 43–78, 1999; 'Resisting dams and
"development": contemporary significance of the campaign against
the Narmada projects in India', *The European Journal of Development Research*, 10 (2): 135–83, 1998; 'protected areas, resource conflicts and action groups: some observations on current trends in
conservation advocacy', *Sociological Bulletin,* 46 (2), 1997; 'Why some
people resist and others do not: risks, interests and collective action
at the local level in the Sardar Sarovar project', Working Paper Series No. 265, Institute of Social Studies, The Hague, The Netherlands, 1997; 'Peoples movements in environmental politics: a critical analysis of the Save Narmada Movement in India', Working Paper Series No. 242, Institute of Social Studies, The Hague, The Netherlands, 1997; 'Parks, people and protest: some observations on the
mediating role of grassroots environmental action groups in resource
conflicts', Working Paper Series No. 228, Institute of Social Studies,
The Hague, The Netherlands, 1996.

Book Reviews: Tom Brass (ed.) (1995) 'New farmers' movements
in India', Essex, Frank Cass, in *Development and Change* 28 (2),
1997; Madhav Gadgil and Ramachandra Guha (1995) 'Ecology and
equity: the use and abuse of nature in contemporary India', New
York and London, Routledge, in *Development and Change* 28 (1),
1997; Amita Baviskar, 'In the belly of the river: tribal conflicts over
development in the Narmada Valley', Delhi, Oxford University Press,
1995, in *Sociological Bulletin* 45 (2)., 1996; Andre Beteille, '*Society
and Politics in India: Essays in Comparative Perspectives*', Delhi,
Oxford University Press, 1992, in *The Sunday Observer,* 10–16, January 1993; Dipankar Gupta (ed.), *Social Stratification*, Delhi, Oxford
University Press, 1992, in *The Sunday Observer*, 12–19, October
1992; Ramachandra Guha, '*The Unquiet Woods: Ecological Changes
and Peasant Resistance in the Himalayas*', Delhi, Oxford University
Press, 1989, in *Sociological Bulletin* 38 (2), 1989.

Other Articles: 'Gulf War and the environment', *The Middle East:
Perspectives from India*, 1 (1) pp. 60–1, 1993; 'Listening to the Red-
Green Concert', *The Sunday Observer*, 27 September–3 October,
1992; 'Socialism with a human face', *The Sunday Observer*, November
22–28, 1992; 'The politics of environment', *The Deccan Chronicle,*
22 September, 1992; 'Reservation policy: a political camouflage for
progress', *Link,* 27 August, 1989.

Acknowledgements

This research project was first conceived in the year 1989 when the struggle in the Narmada Valley was at its nascent stage. However, the offer of a faculty position in Sri Venkateswara College, Delhi University, led to the postponement of the project at that stage. As the Narmada movement gained in strength in the 1990s, my interest in the research project intensified. The offer of a fellowship by the Institute of Social Studies could not have come at a more opportune time. As I joined the course, I realised soon enough that development studies was a vast and all encompassing field and quickly settled down to the idea that PhD at ISS would be a difficult but challenging venture. As the project reaches completion, I have no hesitation in mentioning that I found the ISS environment an extremely stimulating one. In the course of working on this thesis, I have learnt a lot from the large network of people who supported the venture. I take this opportunity to mention a few here.

My sincere thanks and gratitude goes to my promoters Prof. Martin Doornbos and Prof. Ashwani Saith under whose scholarly guidance I have learnt the nuts and bolts of doctoral research. To me, they have always been a rich source of ideas and arguments. Even on the personal front (especially when life has tended to get very difficult) I have found them strongly supportive and most reassuring. Prof. Ben White showed a keen interest in my work and being and I am grateful to him for having found time to comment on several chapters of the thesis. My interactions with him have been extremely enjoyable and fruitful.

In India several people contributed to my research. Ashish Kothari at the Indian Institute of Public Administration provided me with an office and allowed me to make use of his impressive data base on SSP. He also facilitated my trip with the Jungle Jivan Bachao Yatra. Shripad Dharamadhikari of the Narmada Bachao

Andolan (NBA) facilitated my access to the rich documentation of NBA on SSP. Amit Bhatnagar and Jayshree of the Khedut Mazdoor Chetna Sangh (KMCS) provided stimulating updates on SSP and the struggle around it. They were also kind enough to guide part of my village visits. From them, I was also to learn first hand the trials and tribulations of social activism. D.T. Buch of the Sardar Sarovar Narmada Nigam Ltd. provided me with some vital government reports on the SSP. Prakash Deo Singh, my classmate from Jawaharlal Nehru University and now a faculty member there accompanied me on field visits in Madhya Pradesh and Gujarat. His insights and ideas on the subject were extremely useful. Interactions with NBA activists have contributed to my understanding of the movement's perspective on SSP and related issues. Numerous people in the submergence villages in Madhya Pradesh and the resettlement sites in Gujarat have eagerly shared their experiences, views and feelings that have influenced my understanding on displacement and people's struggles.

At ISS, I have gained from discussions and critical comments of several academic staff members. Some of them found time to go through the thesis as the chapters progressed. I wish to particularly mention Jan Nederveen Pieterse, Peter Waterman, Bridget O'Laughlin, Des Gasper, Eric Ross, Jos Mooij and Haroon Akram Lodhi. Cris Kay, Erhard Berner, Mohammed Salih, Shanti George and Ajay Tankha, whose comments and insights on my work were highly rewarding. I have also benefited from my interactions with Prof. Thayer Scudder of the California Institute of Technology and Dr. Michael Cernea of the World Bank.

Ank v/d Berg has been a constant source of support both in good times and bad. Dita Walenkamp, Maureen Koster, Martin Blok, Paula Bownas, Ton Rimmelzwaan, Henry Robbemond, and Ricardo Mahesh were always willing to help whenever I approached them. Jean Sanders, the editor of this thesis, with her keen eye for details, has heightened its readability.

PhD colleagues in ISS provided for camaraderie, friendship and readily available intellectual stimulants. Foremost among them were Gebru Mersha, Edwin Croes, Daniel Kotzer, Maha Mahfouz Abdelrehman, Purnendu Kavoori, Ashesh Ambasta, Yasuke Kubo, Edsel Sajor, Karen Gabriel, P.K. Vijayan, Abebe Haile Gabriel, Richmond Tiemoko, Gabriel Rugalema, Daniel Charvez, N.C. Narayan, Imani Tafari Ama, Vijay Gudavarthy,

Mahmoud El Zain Hamid, Asiimwe Godfrey and Merra Tegene Malesse. The numerous hours of exchange provided inspiration to write besides clarity of ideas, answers to queries and challenging questions to pursue. Many participants of the MA and diploma programmes shared valuable time to make life at ISS comfortable and enjoyable and helped me overcome much of the loneliness of PhD research.

My family and friends have remained in regular touch enquiring about my progress and wellbeing. My brothers Biswajeet and Satyajit constantly fed me with the latest jokes going around which was most rejuvenating. My parents Dipty and R.P. Dwivedi shared my achievements and apprehensions. Their counselling and guidance provided me with a lot of strength especially during times of crises. M.N. Panini, Wicky Meynen, Anurag Srivastava, Asha and J.P. Das, Rekha Wazir, Themba Tshbalala, Morabo Morojele, Khaleed Hamid Farooqi and Mukta Srivastava remain special.

My son Suyash and Sharada Srinivasan have been the greatest source of inspiration for this work. Sharada was an avowed critic of my work. She has been a steady companion through thick and thin and has been inspirational in holding life together. Suyash has been away from me during most of the research period. The times we have spent together have been most joyful and high spirited. In many ways he has borne the cost in the making of this thesis.

Ranjit Dwivedi

Introduction[1]

S   ince the 1980s and 1990s large infrastructure develop-
    ment projects have become the subject of major conflicts
    and controversies around the world. In the developing
countries infrastructure projects, particularly in the water and
power sectors, have generated political resistance over displacement and environmental despoliation as well as intellectual criticisms over the ways resources are used and meanings produced.
This book focuses on a popular case, the Sardar Sarovar 'Narmada
project' in India to show when, how and why development action
by the state generated conflicts and change. It subscribes to the
view that a deeper understanding of conflicts and change is possible when particularising development in terms of actors, context
and events.

Large infrastructure projects appear to mirror what can be
called the modernisation crisis. The established assumption of
modernisation—that social change can be orderly and predictably manipulated through planned investments—is increasingly
subverted by unintended consequences, side effects, and the
systemic production of risks, suffering and despoliation. Project
planners are finding it increasingly difficult to justify these outcomes in terms of 'externalities'. The consequence in recent years
has been a dramatic fall in the acceptability count of large development projects. The state, the highest political organ that
usually involves itself directly in matters of large projects, finds
its authority eroded. Its decisions no longer go unchallenged.
The means with which the state sets out to develop, distribute
and value resources appear to be politically contested. The view
is gaining ground that the state increasingly relies on power
rather than authority to push these projects. In this sense, the
modernisation crisis is also becoming a legitimisation crisis.

In more than one way, the crisis that confronts large development projects concerns perceived limitations in the state's vision

and strategies of change. The size and scale of development in-terventions, the knowledge base for planning and implementing such projects, and the authority and mandate of the state to carry through decisions—all aspects over which the state's prerogative was unquestioned—form the core of the crisis. The opinion is gaining ground that large projects imply big and un-acceptable risks and costs with little comparable benefits. They are top-down, decided and controlled by development mandarins, thriving on social exclusion rather than participation. The deci-sion-making process is not transparent nor are decision-makers accountable. This perception is backed politically by popular collective action that increasingly confronts large projects at multiple levels: local, national and global. Advocacy, lobbying, protest and resistance actions by NGOs, social and environmen-tal movements, citizen groups and community organisations have multiplied dramatically. Perhaps more importantly, such collec-tive actions are said to signify not just limitations to the state's agenda of change but also visions and strategies with which to overcome those limitations.

These perceived limitations of state-led change form the basis of an emerging 'critical discourse', a systematic organisation of knowledge of, and experience with, large development inter-ventions that conflict with truth-claims by the state. Within this discourse, meanings and actions illuminate the crisis along mul-tiple dimensions. In causal terms, and perhaps rather obviously, the crisis is development-induced. The development order—institutions, projects and actors—is the crisis producer. What is not so obvious is that in the language of mainstream develop-ment, the crisis is treated as if it were outside the order of projects, external to the stated objectives and costs and benefits. In expe-riential terms, the crisis is conflict-based. It is the accumulation of conflicting experiences and interests which are difficult to integrate and negotiate. In political terms, the crisis unfolds in power struggles and collective action. State interests expressed in terms such as 'collective good' and 'public purpose' are chal-lenged through protest and resistance. Development action becomes an arena of struggle.

Understood in this way, the crisis questions the definition of development as expanding people's choices for their own benefits and giving them the freedom to exercise that choice. In experi-ences with large projects, development has more accurately meant costs as well as benefits, risks as well as opportunities, curtailment as well as expansion of choices, suffering as well as freedom, impoverishment as well as enrichment, and despoliation as well as creation. In short, large projects illustrate the profoundly contradictory phenomenon that development is. The crisis seems to us to stem from the difficulty increasingly faced by planners of large development projects in integrating these conflicting experiences and outcomes.

In our view scholarship in development studies has yet to address this problem in its multidimensionality. Of course, the fact that development entails benefits and costs and that it generates side effects is not a revelation. For instance, development economics acknowledges externality as cost or benefit arising from developmental action which does not accrue to the firm or individual carrying out the action. It also assumes that in developing economies it is difficult if not impossible to prohibit actions that will cause external diseconomies. Having acknowledged these externalities, however, it leaves them out of the purview of evaluating benefits and costs, noting that project activities that generate such externalities may require interventions (legal and policy obligations, information-generation and motivation) and internalisation on grounds of efficiency, equity or sustainability.

There is a more sensitive understanding of side effects and externalities in the sociology of development action.[2] For the most part, scholarship devoted to the social-environmental impacts of development projects has generated awareness of the adverse nature of externalities and unintended consequences. Yet one cannot but observe that much of this sociology remains positivist, applied and evaluative. It is still confined to measuring externalities in specific project situations in order to judge a project's overall worth, 'that is whether the project had indeed accomplished its overall objective(s) and triggered the desired consequences or some unanticipated ones' (Cernea 1985: 7).

In valuing development, the sociology of development action at least makes an effort to understand externalities, side effects, suffering, despoliation and unintended consequences. Yet this understanding is limited. It is ultimately geared towards assessing the worth of development (project) as good or bad on the basis of a cost-benefit balance sheet. It is limited in its ability to explain how 'good' projects become 'bad' and vice versa, how the criteria of judging goodness and badness change over time, how a project can be good for some and bad for others, and how goodness and badness are socially and politically constructed attributes.

This limited understanding cannot address the core of the crisis. In our view, the sociology of development action must acquire the capability of not merely demonstrating whether development projects are good or bad, albeit with a methodology that is inclusive. Rather, it should generate explanatory tools with which to understand how development constitutes a complex web of contradictory actions. The crisis is better understood, not by generating a single truth, i.e. an objective, scientifically demonstrated value of any given project, but by unravelling the actual process of development in which various truth-claims are advanced. Underlying these conflicting truth-claims are conflicting interests and experiences, power and knowledge struggles. These hold the key to deeper understanding of the complexities that underlie the crisis marking large development projects.

This book situates claims and experiences using a heuristic device to distinguish between three levels of conflicts in development: over interests, norms and values. Instead of evaluating the Sardar Sarovar Project (SSP) as good or bad, the objective was to unravel the process in which 'truth claims' were advanced, and how interests and experiences, power and knowledge struggles underlay conflicting truth claims. The purpose was to link the causes, the experiences and the politics around these claims. The issues were how claims emerged, how experiences and interests fuelled them, and how actors mobilised power and resources to sustain them.

The Sardar Sarovar Project (SSP) is a large dam project in the western state of Gujarat in India. It forms part of a larger investment on the Narmada river called the Narmada Valley Development Project involving the construction of 30 large, 135 medium and 3,000 small dams spanning the states of Madhya Pradesh and Gujarat. The SSP is the terminal dam and the largest investment in the Narmada basin. It is intended to achieve the storage and regulation of Narmada river flows to irrigate 1.8 million hectares of land, supply drinking water to 40 million people, and to have installed capacity of 1,450 MW of power over the next 30 years.

Given these figures, it is not difficult to imagine that this project is one of the largest in India. Since its inception, however, it has been the subject of serious controversies. Whereas the government of Gujarat considered it the 'real life line of the state', grassroots groups dubbed SSP as one of the 'greatest planned environmental disaster'. On the one hand, the SSP promised to change the shape of water-short and drought-affected regions in Gujarat, and on the other, it was criticised as a white elephant that drained resources and affected the cultural and biological diversity of the region. In fact, the Narmada Bachao Andolan (or Save the Narmada Movement, hereafter NBA), which led the collective action against displacement and despoliation in the valley, is one of the severest critics of the SSP.

Rather than positioning oneself in the bipolar world of proponents and opponents, it was more prudent to ask, what conflicts and competitions marked the making of the SSP, how did the SSP affect lives and livelihoods, and how were claims in the project politically represented. The complexity marking the study subscribed it to no single method and it drew from project analysis, environmental sociology, political economy, conflict and social movement studies. The data used were both qualitative and quantitative. The study uses both descriptive and explanatory tools of analysis with which to generate understanding of the crisis. Such tools aim at showing 'what' is happening in the SSP, 'how' and 'why'. In social science research, it is difficult to draw clean boundaries between description and explanation (Sayer 1992; Dey 1993; Miles & Huberman 1994). We describe the context of actions in the Narmada, the intentions and meanings that organise state and non-state actions, and the development process in which actions are embedded. We also attempt to explain why the crisis unfurls in the given context. That is to say, we look at 'the power or liability' (Sayer 1992:104) of events and actions that cause, produce, or influence the crisis in the SSP.

The study has been conducted at many levels. The reason for a multiple-level analysis stems from the nature of the phenomenon under study. A micro-level anthropological study focused on a village or a people appeared inadequate and inappropriate. While a macro-level survey of large development projects would have generated comparable data and insights, it would have required enormous resources and time for a comprehensive coverage of different aspects of the projects. It therefore seemed more feasible to focus intensively on a popular case and to follow the actions, events and actors at various levels: local, national and global.

This research concerns processes, actions, relations and episodes of events (Sayer 1992: 242). The main units of analysis, however, are actors (individuals and collectives) embedded in the process of development in the Narmada and acting in the making of its crisis. The main database therefore consists of their perceptions and actions.

Field visits were conducted in two phases, in 1995 and 1996. Data collection techniques included the collection of documentary material, participant observation, unstructured interviews, and group interviews.

The research followed roughly a 'project cycle' model and focused on design, appraisal, construction, evaluation, review, impact and alternatives in the SSP. However, in each of these phases, significant actions influencing the course of events and the life of the project were introduced. Actions of governments, multilateral institutions, NGOs, evaluators, and affected people, at different levels, were important to understand the deployment of claims, interests, power and resources in the conflicts. The research showed that the 'pro- and anti- SSP divide' was not as it is academically projected pitching state vis-à-vis civil society. Rather it highlighted collision between provincial governments, disagreements among affected people, clashes among NGOs and differences among evaluators.

The main results of the study concerned the nature of conflicts in the SSP and in particular, the collective action led by the NBA against the project. The conflicts were intense. Whether it was the riparian struggle between two states, downstream and upstream, or the struggles of the affected people with the governments, use and access to land, water and forest resources assumed centrality. The latter brought issues of rights and entitlements to the forefront. The conflict turned environmental when the NBA began playing a leading role, broadening the struggle to include questions of resources, values and visions. The fight was, as the NBA put it, for redefining the means and goals of development. The movement indicted the underlying water and power policies with which projects such as the SSP are justified. In the process, it generated a critique of large infrastructure projects in general and particularly of large dam projects.

The NBA transformed the struggle from one of local farmers to a broad citizens coalition seeking the demonstration of public purpose in the SSP and similar other projects. Collective actions over information and knowledge questioned the SSP's desirability and even viability as a development project. At stake were underlying principles, rules, decision tools, procedures and processes. Sufficient evidence from the politics of the NBA existed to argue that the project was planned on a 'minimal information, maximal uncertainty' dictum. Given the parameters of the SSP, the legitimacy of a large number of stakes remained unrecognised, liabilities and compensation were not properly negotiated and information on entitlements did not flow to the stakeholders.

The significance of the results is to be seen in the contribution the study makes to three research issues. The first pertains to the theoretical leaning of focusing on actually existing development; second, to a methodological query concerning critical and crisis analyses; and the third, to the substantive issue of understanding the NBA and its collective action. The focus on development as actually practised is drawn from the 'third approach' of William Fisher's edited volume '*Towards Sustainable Development? Struggling over India's Narmada Waters*'. Fisher argued for a theoretical tool of analysis that went beyond opponents and proponents of the project to look at what is actually happening in development. We believe that what was recognised in Fisher has been realised in our study. Our understanding is that the SSP crisis is an outcome of a political process that cannot accommodate and negotiate the conflicts that it has generated.

Related to this theoretical tool is a methodological question— whether to distinguish crisis from critical analysis. In our opinion, critical analysis of development action involves judging the validity of claims, arguments and statements made in relation to the action. It involves methods with which to evaluate whether claims are true or false, accurate or inaccurate. Crisis relates to whether a sphere of action, or for that matter a system, ought to continue or not. It is caused when criticisms can no longer be accommodated or assimilated. Whereas the bulk of research on Narmada and other such struggles has subscribed to critical analysis (for example to argue that the SSP is bad or good), we argue that crisis analysis can take us a step further into looking at claims and counter-claims and their mutually exclusive but valid characteristics.

And third, is the understanding of politics and practices of the NBA. The study indicates that the NBA is best understood as a multi-level network with which it mobilises resources and discourses in its campaign against the SSP. In particular, it deploys what is called 'risk politics' in contemporary environmental movements. By highlighting the fudges and tensions in information, facts and figures concerning the dam, the movement's environmentalism rests on radicalising and democratising scientific rationality. In other words, the NBA was new in two distinct ways. First, it was neither a peasant organisation nor a working class configuration but a multi-class outfit. Second, it made development more accountable while it wanted it to be replaced by alternatives.

Given the crisis and conflicts in the SSP, it is difficult to imagine possible ways forward. However, we use the three-dimensional conflict matrix—interest, norms and values—to project potential conflict-resolution mechanisms. The first and easiest option is to seriously address political status of interests, particularly those of affected people in the Valley, over displacement and rehabilitation. The second option concerning interpretation of norms and rules is perhaps more difficult in an era of globalisation, but can make transparent and democratised decision-making tools. It could also generate political impetus for the third option concerning control over values and visions in which weights are attached to the alternative viewpoints of development that could practice democratic techniques and local methods of change.

Notes

1. This chapter is a slightly revised version of the text of the author's public defence presentation delivered on 29 January 2001.
2. In sociology of action, unintended consequences have long engaged scholars both at the macro and micro levels. For most part, however, a theory of functionalism has guided such scholarship in which unintended outcomes have been seen in functional terms (see Merton 1957).

1

Development-induced Conflicts: Concept and Analysis

Introduction

At the outset, it is necessary to outline, the conceptual and analytical elements which provide theoretical frame work understand the empirical phenomenon that is the subject of this book: the Sardar Sarovar Project (SSP). Three questions are addressed here: (*a*) what approaches to development enable us to identify factors that cause and influence conflict in project situations? (*b*) how should we analyse different aspects of conflicts pertaining to interests, experiences and claims? and (*c*) how should we approach and assess outcomes and actions manifested in struggles and movements? The discussion around these questions is based on theoretical literature regarding project management, political economy, environmental sociology, social movement and conflict analyses.

Development as Projects

Development is conceived differently in discourses, ideologies, models, processes and practices. Yet the different images are deeply interlinked. It is difficult to imagine a discourse without its practice or a model without its actions, and vice-versa. Moreover, it is highly debatable whether development in fact constitutes a monolithic set of discourses and practices or perhaps a widely diverse range.

For instance, Escobar (1995: 5–6) views development as a 'historically produced discourse', arguing that 'thinking of development in terms of discourse makes it possible to maintain the focus on domination...and ...to explore the most pervasive effects of development'. The 'focus on domination' leads Escobar to see that 'development discourse...has created an extremely efficient apparatus for producing knowledge about and the exercise of power over the Third World [by the First World] (ibid.: 9). In contrast, Hobart (1993) speaks of several discourses on development, notably those of developers, the local people and the national government. Analysing these conceptualisations, Grillo and Stirrat (1997: 21–2) argue against viewing development as a monolithic enterprise. Rather, they stress that 'within development there is and has always been a multiplicity of voices...even if some are more powerful than others'. In that context the authors even identify diversity of discourses among developers, local people and government. Preston (1994) distinguishes three different discourses on development: those of planners, free-marketers and the public sphere. This conceptualisation is somewhat different from that of Peet and Watts (1996: 20) who speak of development discourse historically and normatively as a 'western discursive formation' constantly oscillating between state, market and civil society. In other words, they appear to be talking of one development discourse containing several competing theories, ideologies and models. Likewise, for Moore and Schmitz (1995), the development discourse is a composite comprising symbols, actions and institutions, subsuming both theory and practice.

Rather than viewing development as a unified overarching discourse, we believe it is more useful to think of a plurality of development discourses. Nonetheless, our interest in development action leads us to focus more on development practice. To visualise development as practice (or a set of practices) is to focus on its strategic, operative dimension. That is to say, we conceptualise development as 'activity'. It is, as Hulme and Turner (1990: 99) state, 'a conscious action needed to bring about desired transformation'. Given the manner in which the practice of development has been reflected upon, synthesised and systematised, it has come to mean the conglomerate of policies, plans, programmes and projects that protagonists of development pursue in order to achieve their goals. In other words, we are referring to development apparatuses—the institutions and instruments—through which development as a goal is achieved (see Moore 1995). Development practice encompasses decisions, choices, actions and targets (Porter et al. 1991). Development policies and practices closely reflect overarching discourses and models, and in that sense can be as diverse as discourses. Yet no matter how development is conceptualised as a discourse, in practice its delivery has relied on defined ways and means. Perhaps it would not be wrong to say that there is less diversity in 'doing' development than in thinking about it.

One of these defined ways of doing development is through projects. In the modernisation discourse (of developers and planners), development projects are the 'privileged particles of the development process' (Hirschman 1967). They are investments of human, physical and financial capital in a time-bound intervention to create productive assets. According to Hirschman, development projects are special kinds of investments: they imply purposefulness, specific locations and the introduction of something qualitatively new. Over a specific period they create assets, systems, schemes or institutions which continue in operation and yield a flow of benefits after the projects have been completed.

Two broad categories of development projects are conceivable. 'Hard' projects, consisting of large infrastructure and industrial investments, which may be distinguished from 'soft' projects, usually in the social and human development sectors of health and education. Hard development projects involve heavy investments, being said to represent 'if not the engine then the wheels of economic activity' of any country (WB 1994b:14). In the infrastructure sector examples of such projects include public utilities such as power, telecommunications, piped water supply and piped gas; public works such as roads, major dams and canal works; and other transport sectors such as railways, ports and waterways. Considered to be in the 'public interest', the planning of such projects usually involves the highest political authorities, especially when in the public sector.

Hard projects often involve systematic and careful advance planning. They are guided by the 'blueprint approach', implying that they go through clearly defined and time-bound stages in a project cycle involving identification, preparation, design, appraisal, implementation, operation and evaluation. The dominant view in project planning is that the structural rigidities of blueprint approaches suit hard infrastructure projects in view of the fact that infrastructure investments are capital-intensive and involve major physical assets and large financial resources. The sequence of activities is logically ordered in a fixed, predetermined path (see Bridger & Winpenny 1983), so that design, planning, appraisal, implementation and evaluation leave little room for manoeuvre. Each of these stages in the project cycle is sustained and supported by technical analyses 'concerned with the cost effective achievement of well defined goals' (Hulme 1988: 272).

The orthodoxy of the blueprint approach in the project planning discourse has drawn criticism both within and elsewhere. Alternative approaches to development projects stress political factors, such as the role played by actors and interests, competition and conflicts, bargaining and negotiation, that ensue from actors pursuing their own interests and the potential impact which they have over the outcome (see Long & van der Ploeg 1994; Schaffer 1984). Hulme (1988) attributes the analytical focus of political approaches to the interaction of different sets of interests that commonly conflict and where outcomes emerge from a complex network of power struggles, in which a variety of resources are manipulated. In this approach, the project cycle functions through agenda setting, manipulation, bargaining and tradeoffs.[1]

Others tend to shift the analytical emphasis to the wider 'environment' in which projects are enmeshed. Thus, Korten (1989) and Uphoff (1992) adopt a 'process approach' to projects which stresses their flexible nature and changeable procedures. Instead of treating projects as closed, controllable and unchanging systems, the process approach acknowledges that 'connections between inputs and outputs are not linear, responses to inputs are often non-proportional, actions generate unpredictable effects and the same inputs under similar conditions do not always produce the same results' (Mosse 1998: 5). Concern with the process arises from the realisation that 'development action is undeniably complex, often unpredictable and locally variable in its effects and significantly influenced by political, institutional and social realms' (ibid.: 6). In a different vein, Barnett (1981), following a Marxian political economy persuasion, views projects as a series of socio-economic relationships existing through time. Here, development projects manifest the politics and ideology of dominant class interests. Their outcomes necessarily advance the interests of the dominant classes in society, whereas the processes set up by the changes may well be extremely unfortunate for the vast majority of people.

Approaching Development Projects

The case at hand and the corresponding research questions cause the weight of analysis to be placed on political approaches to development projects. Orthodox planning approaches allow little scope for analysing crises and conflicts in development projects. Of course, scholars have attempted to strengthen planning and managerial approaches by infusing them with sociological and political understanding. Thus Cernea (1985: 7) argues for a sociological input at every stage of the project cycle and not just in the ex-post evaluation stage. This to him is the 'wrong end of the cycle—too late to affect the social process'. The plea is to integrate sociological models, procedures and professional skills into the development project cycle. While the application and applicability of social science knowledge in the project cycle is indeed crucial, Cernea's overall concerns, which seem to facilitate and guide development project action with few obstacles, suggest a persuasion that is closer to the orthodox approach than to political approaches.

Our view is that sociological and political understanding should not be restricted to the status of 'inputs' in the project cycle, and that they are crucial to an understanding of the actual process of development. If our central concern is to generate a deeper understanding of developmental failures and crises, then sociological and political tools must be deployed in the analysis of why and how such crises and conflicts plague development projects. We need to apply socio-political understanding to show how structures and actors influence developmental outcomes and how developmental crises and conflicts are caused, experienced and represented in the political arena. An overtly political approach aiming to unravel these aspects in development projects can provide a deeper understanding of particular cases,

while simultaneously contributing to questions of theory and
(its) application.

Structure: Actor Linkages

A central question regarding political approaches to development
projects is the relative explanatory power of structures and actors. Some political approaches focus on macro-level structures,
systems and arrangements. Understanding of the micro level is
deduced from larger structures. Micro-level encounters are subject to, and determined by, the constraining character of larger
structures. In Marxian political economy, for example, a classical perspective that highlights the power of structures, particular social formations are approached and explained through modes
of production.[2] Other approaches focus on actor encounters that
contribute to the construction of social situations. These approaches emphasise interpretations of actions and meanings.
Table 1.1 differentiates between political approaches that are
structure-oriented and those that are actor-oriented.

**Table 1.1 Political Approaches to Development Projects:
Macro–Micro Linkages**

Approaches	Main Focus	Analytical Framework
Structural Approaches	Macro-level arrangements	Systemic reproduction of socio-economic relationship; serving the interests of dominant classes and elite.
Interaction Approaches	Micro-level practices	Situational encounters between actors differently endowed with power, resources and agency.

This divide between structural and actor-oriented approaches
may be better understood if it is set against the backdrop of
extreme theoretical positions that accentuate it on the one hand
while noting attempts that want to bridge it on the other.

Theories such as Marxian political economy and rational-choice
accentuate the structure:actor divide. In Marxian political
economy, the economy that is structured by a dominant mode of
production creates appropriate forms of politics, consciousness
and ideology. Contradictions emerge from periodic structural
crises in the economy—e.g. between capital and labour or between

capital and nature. These contradictions create opportunities for transforming unequal production relations (predominantly class relations) that govern the means of production. Actors viewed in class terms engage in conflict as collective agents, i.e. as class for itself, and struggle for social transformation directed at controlling the state and ownership of the means of production. For most Marxian political economists structures are irreducible ensembles of social relations, such that individual entities are what they are by virtue of their location within them. As stable sets of relations (of production), structures generate individuals as social agents and in so doing acquire explanatory significance.[3]

At the other end of the spectrum is rational-choice theory, in which actors (both individuals and collectives; see Bates 1981, 1989) pursue their interests in a rational manner by calculating their available resources and the possible gains to be made in a given situation. The rational-choice approach assumes that competition and conflicts of interest are built into institutionalised economic and political power relationships. Conflicts occur when individuals and groups pursue their disparate interests. The focus is on (rational) actors with competing interests, their probability-utility calculus, their preference structures, the information at their disposal, the availability of alternatives and the consequences of choices (Elster 1986).[4]

Both theories have their limitations. There is some dissatisfaction with the rigid formulation of structured actors in Marxian political economy arising from its production logic. Scholars have been critical of the formula-oriented approach to conflicts and conflicting actions which automatically emerge from material contradictions,[5] of the limited analytical usefulness of class as actor and of the neglect of other terrains of struggle apart from class struggles in Marxian political economy (see Slater 1992). The theory provides critical insights into the understanding of the macro environment—the trajectory of development as a historical project, the investment patterns in productive means—i.e. technology and infrastructure and issues pertaining to distribution and distributive conflicts. Its heavy reliance on structural categories, however, has aroused suspicion, especially in 'post-Marxist' perspectives. The central argument here is that structural categories—endowed with objectified, external and constraining characteristics—fail to capture the dynamics of interaction and political practice. The preferences in such perspectives are for more interactive frameworks of analysis. Thus, for Laclau, 'institutions do not constitute closed structural frameworks but loosely integrated complexes requiring the constant intervention of articulatory practices' (Laclau 1990: 223 cited in Slater 1992: 296). Neither can individuals be fixed as subjects occupying specific positions in the structural constellation around them, nor can their identity be determined through structural categories (Laclau 1985).[6]

Dissatisfaction with the rational-choice approach may be considered as follows. First, the approach does not explain clearly how individuals overcome their self-interest and the free-rider problem and become involved in collective action (Peet & Watts 1996: 29). Second, there is the ambiguity of whether to assume that actors share interests, norms and values, or to demonstrate the construction of solidarity and sharing in the process of collective action (Eder 1995: 32). Third, and most important, is its methodological privileging of *intentional-utilitarian* agents in social explanation, ignoring the specifics of context and circumstances (Long and van der Ploeg 1994).[7]

The fact that the 'structure versus action' issue is central but far from settled in development theory can be supported by referring to the decade-long debate in development sociology (see Booth 1985, 1994; Corbridge 1990; Mouzelis 1988). That debate generated friction between those insisting on the retention of structural forces as the development problematic and those arguing for an interactive and constructivist approach to explain the diversities of outcomes of social processes at the local level.[8] For Schuurman (1994: 23) the structure-agency debate in development theory has been about 'an unwarranted reductionism of systemic features to the practices of certain agents on the one hand and the reductionism of practices of certain agents to systemic characteristics on the other.'

The two approaches have their respective advantages and limitations. At the outset it may be argued that whereas structural approaches neglect the capabilities of human actions, interaction approaches show little sensitivity to power differentials among actors. Neither approach is sufficient in itself for an understanding of the public action sphere in development projects. Particular conditions as well as broader forces that shape and are shaped by local circumstances need to be integrated for a proper understanding of the encounters (see Booth 1994:10).

Attempts to integrate structural and action-oriented approaches in development theory have been greatly influenced by Giddens' contributions (1984). His concept of 'duality of structures' explains that, insofar as actors draw on rules in the enactment of social practice, they are endowed with the capability to resist, modify or change those rules. The extent to which this is possible depends on the resource constraints (e.g. asymmetry of power and domination) that actors face in a given setting. According to Giddens, the 'setting of interaction', while facilitating certain forms of action, can equally be constraining. Settings or locales enter into the very fabric of interaction in multiple ways. They involve structured rules, e.g. what one might or might not do in a place, as well as provide resources that actors can draw upon in the enactment of social practices. Locales are also constitutive of an asymmetry of resources and knowledge available to each actor (Cassell 1993). It is the possession of resources and their mobilisation that makes the exercise of power possible.

> Social practices involve actions, which make a difference to the world in some way, no matter how small. But agents,[9] those who are able to effect change, must possess appropriate resources in order to do so.... The transformations of nature, and the deployment of persons that accompanies it, are inconceivable without human access to power and the resources that must facilitate it (Cassell 1993:11).

Following Giddens, the question of whether one is looking at structures or actors appears to be redundant. Simply put, structures, rather than being viewed as determining actions, are seen as being reproduced in social practice. In the specific context of development theory, however, Giddens' handling of the structure-agency problematic raises a few questions. As Schuurman (1994: 32) implies, if development projects represent structured rules, then it becomes important to ask: rules made by whom and for the benefit of whom? It is also important to probe the question of resource allocation, i.e. who gets what at whose cost? These questions make us bear in mind the 'unequal distributive dimension of the production of and control over rules and resources' (ibid.).

Applied in the context of development projects (that imply organised social change), Giddens' theory appears inadequate to seek out power differentials as well as differences in orientation among the various actors involved at different levels.

In their studies on agrarian change and development, Long and van der Ploeg (1994) draw from Giddens (1984), arguing that patterns of social relations embodied in state institutions, markets or technology are not relevant as 'determinants'. Rather, it is more useful to explore whether they are considered as self-evident limits beyond which action is judged to be inconceivable or as boundaries that are targets of negotiation, reconsideration, sabotage or change, i.e. as barriers that are to be removed (ibid.: 77). Agency, they argue, is not merely an attribute of the individual actor, but is the result of possessing cognitive abilities and persuasive powers.

> Agency, which we recognise when particular actions make a difference to a pre-existing state of affairs or course of events entails social relations and can become effective only through them.... Effective agency requires the strategic generation/manipulation of a network of social relations and the channelling of specific items (such as claims, goods and information) through certain nodal points of interaction (ibid.: 66).

By adopting this interactive framework, Long and van der Ploeg dismiss as 'simplistic' the characterisation of planned project interventions by state agencies as structural expressions of class relationships, state-peasant relationship or the logic of accumulation. Planned intervention to them is an ongoing, socially constructed and negotiated process. Pleading for a focus on intervention practices rather on intervention models, they argue that:

> the notion of intervention practices allows one to focus on the emergent forms of interaction, procedures, practical strategies, and types of discourses and cultural categories present in specific contexts. The central problem for analysis becomes that of understanding the processes by which the external interventions enter the life-worlds of the individuals

and groups affected and thus come to form part of the resources and constraints of the social strategies they develop. In this way external factors become internalised and often mean quite different things to different interest groups...(ibid.: 78).

The plea for 'deconstructing' planned development interventions instead of 'objectifying' them is based on the following logic. Through the deconstruction of any given development project—a 'project' defined by Long and van der Ploeg as a model or blueprint of actions—one comes across a series of 'actor projects' that conflict and negotiate with each other and, in the process, change the shape of the development project itself.

The implication that may be derived from all this is that political approaches to development projects must necessarily and adequately interconnect structures and actors in the analysis. In structural terms, development projects—as models of action—are historical-social projects. Critical theory holds that what ruling interests—whether as state actors or as dominant economic classes—in a given society intend to do with human beings and things is projected through them. In that respect, development projects constitute acknowledgeable political domination. Unlike assumptions in managerial approaches, planned interventions are not neutral terrain. Behind the organisation of means and choices between alternatives that make development interventions appear as models of purposive-rational action, lie 'hidden and therefore unacknowledged social frameworks of interests in which strategies are chosen, technologies applied and systems established' (Habermas 1978).

A structurally sensitive framework also enables us to analyse the particular kinds of power, control and legitimacy that development planners seek and exercise vis-à-vis emerging societal interests and demands. Development projects bear implicit assumptions about what development is and whom it is for. That is, justifications of projects are influenced by how they correspond to the goals and visions of development. The political-economic context in which development gains meaning, i.e. who says what it is, and power, i.e. who directs it, form crucial constituents in understanding specific interventions. In that sense, a development project represents institutional power and knowledge; it represents larger political-economic structures.

Simultaneously, it has to be borne in mind that development projects are socially constructed processes in which actors, albeit with varied degrees of resources and power, exercise agency and influence outcomes. Despite overarching structural settings, actors reproduce, challenge and change development interventions. The public sphere of a development project involves competition, bargaining and negotiation among numerous actors. The analytical framework must therefore adequately address the questions of how competition and struggles among various actors, interests and values manifest in particular ways and at particular times, and why conflicts manifest more acutely in some development projects than in others.

Development Projects as Fields of Action

This specific research does not aim to *establish* that planned development interventions are reified forms of institutional power. Instead, having taken due cognisance of this view, we are inclined to focus on actor encounters and their impact on the shape of development. Our purpose is to identify the range of interactions among and between project actors and actors in the wider societal setting or 'environment'. Their pursuit of interests and values influence the planning, evaluation and implementation of a project in complex ways. In our view, there is a need to develop a richer and nuanced understanding of the complexities and interconnections of social actions in development projects.

In this pursuit of actor interactions, some preliminary remarks appear in place. In the literature on development politics, actor encounters are often conceptualised in terms of opposing categories such as planners' discourse versus peasants' discourse, or simply the 'state' versus 'people' (Escobar 1995; Kothari 1990; Parajuli 1991; Shiva 1991). To us these juxtapositions underpin a populist ideology and close complex processes of cooperation and solidarity, conflicts, contests and negotiations among actors within each category. While there may be much moral force in a makeup that pitches people/civil society/community against the state, it leaves out crucial issues pertaining to interest politics, including the making of actors, their stakes in the development process, and their differential power and resources.

A more open mapping of actors in development projects is needed for the analytical framework to successfully interconnect macro-level settings with the coming and becoming of actors at the micro level. In this regard it is useful to draw tools from political economy and environmental sociology in order to anticipate the possible range of actors in the action field. In political economy terms, power relations govern the process of resource distribution and, in turn, the distribution process maintains relations of domination and subordination. In development projects, the power relation is reproduced in a zero-sum game involving a set of losers and winners in the process of resource distribution. Although scholarship on political economy attaches considerable significance to the question of who gains and who loses, it tends to forestall the answer by identifying dominant industrial, agrarian and administrative class interests as beneficiaries, and marginal industrial and agrarian classes as losers. As discussed later, a simplistic alignment of actors as losers and winners along class lines may not be adequate for an analysis of the complex web of actors and actions underlying planned development interventions. Even more, such a classification yields a typical structural explanation of political actions of winners supporting, and losers resisting, development projects.

Analytical tools from the emerging field of environmental sociology add new dimensions to development action. Development projects are conceptualised as 'modified or built environments' (Humphrey & Buttel 1982: 4) that create risks and opportunities for different categories of people. Unlike in managerial approaches where unwanted and uncertain outcomes are explained as externalities and unintended consequences, environmental sociology approaches speak of the systemic production of suffering, risks and despoliation. Beck (1992) observes that the risks and sufferings of built or modified environments are unevenly distributed; some bear more risks than others. Moreover, risks are perceived and defined differently as each social group seeks to interpret risks in terms of its own interests and values (Heimer 1988). Thus, risks in modified environments are to be viewed both as systemically produced and socially constructed. One can then anticipate the field of actions to be influenced by the acts of those experiencing and perceiving risks.

A second set of analytical issues pertains to the process of social construction of risks, particularly the awareness and consciousness of environmental risks and despoliation generated through political actions of social movements and environmental action groups. Environmental sociology concerns itself deeply with the role played by social movements in risk definitions and in framing choices before actors in the political field (Beck 1995; Hannigan 1995). In the context of modified and built environments, where risks are systemically and socially produced, environmental movements can be expected to play major roles in representing and redefining risks (see Renn 1992). As potential players in the action field, environmental action groups add new dimensions to the action field and introduce new questions in terms of analysing the political representation of risks. How do environmental despoliation, risk and suffering find political representation? In other words, how do movements represent these risks? These questions pertaining to movement politics require analytical tools in their own right and are addressed later in this chapter.

What needs to be stressed here is that while the actions of risk bearers and their political representatives may seem to dominate the action field of development projects, it is necessary to recognise and acknowledge other actors and interests that constitute the action field. Palmlund's (1992) suggestion to classify actors according to six generic roles is helpful here. Apart from risk bearers and risk advocates, the social arenas of risk include risk generators, risk researchers, risk arbitrators and risk informers. This classification anticipates the participation of numerous actors in the action field: project proponents (governments, bureaucracies, donor agencies), research organisations (governmental, non-governmental and semi-governmental) and evaluation bodies (courts, professional review groups, monitoring and evaluation units), apart from social groups which hold and define stakes.

As with the complex constellation of actors, development projects as action fields include a complex set of interactions—competition, negotiation, bargaining, arbitration, conflicts and resistance. We have mentioned earlier that the modernisation discourse on development projects views them primarily as state actions for social transformation. In our opinion, viewing projects as action fields not only opens up the characterisation—state action—to closer scrutiny but simultaneously allows us to accommodate the entire gamut of actions generated in and around a development project. To capture this set of actions the term 'public actions' is useful in that it constitutes both adversarial and collaborative actions. Dreze and Sen (1995: 89) consider public actions as being of profound significance to the successes and failures of economic and social change in general and development efforts in particular. They single out actions of citizens' groups, community organisations and social movements, scrutinising the accountability and morality of state action and its impact on public life and rights. By conceptualising development projects as public action fields, we hope to have incorporated this sensitivity into the analytical framework. Development projects in this respect become not just state-acted change, but arenas of conflictive and communicative actions.

Two aspects of the analytical framework need to be reiterated in the analysis of development projects as action fields. First, that (hard) development projects are not merely 'blueprints' that are put into concrete shape with cement and steel, but are shaped and altered by the choices, interests, interactions and interrelations of a wide range of actors. The diversity of actions disallows any classification of actors as passive winners and losers; rather, they possess the ability to initiate, facilitate, resist and alter development projects. To the extent that they exercise their agency, the shape of the development project will be dependent upon the interactions among the initiators, facilitators and resisters. The diversity of social outcomes encountered in development projects is due to the interplay of a variety of social actors and the qualitatively new properties emerging from such interactions (Giddens 1990; Uphoff 1992).

Second, and conversely, actions and outcomes in development projects are constituted by circumstances or settings which influence options and choices available to different actors. That is, different actors are endowed with or have access to different resources at a given point in space and time.[10] Circumstances, while constraining, also provide opportunities and resources, albeit in different measures and strengths to different actors. They shape experiences and perceptions, political agency and orientation. They are, however, not to be viewed as 'determinants' but as 'influences' on the field of action. It is on the basis of resources drawn from the circumstances that actors define their interests, as in the perception of opportunities, gains, risks and losses. The resultant action, whether resistance, conflict, bargaining or negotiation, pervades multiple levels and generates a complex web of interconnections with significant impact on the shape of change itself. Our aim is precisely to unravel this complexity in the context of the Sardar Sarovar Project.

Conflicts in the Action Field

If development projects are fields of action, conflicts dominate the action fields. Large resource transformation, reallocation and redistribution through projects induces competing and conflicting claims and interests with regard to access, use and control of those resources. Thus, development projects are like the planting of mines. Of course, resource conflicts have been part and parcel of human history. As a historically specific process, however, 'development' in contemporary societies marks qualitative shifts in scale, intensity, and modes of resource transformation, deployment and use. It generates higher stakes and interests compared to other modes of resource use and, by implication, more intense conflicts.

Suitable analytical tools are required to enable an analysis of conflicts in the action fields. Structural approaches, as in Marxian political economy, help us to anticipate some aspects of conflict: the major gainers and losers, their class basis, and the resource struggles. Since we adopt an interactive approach, however, we need to develop a more comprehensive understanding of conflicts in terms of the actors involved and their stakes. Explaining the multidimensionality of conflict in the action field is a fundamental step towards understanding crises in development projects and a prerequisite to developing crisis/conflict management proposals.

Following Touraine (1985: 750–1), conflict is seen as constituted by clearly defined opponents or competing actors and the resources they are fighting for or negotiating to take control of. For Touraine, conflicts[11] refer to organised actors and to goals that are valued by all competitors or adversaries. He identifies three elements in conflicts: the *identity* (*i*) of the actor, the definition of *opponent* (*o*) or adversary, and the *stakes* (*s*) or goals. *Identity* is the self-definition of the actor of what it is, on behalf of whom it speaks; *opponent* is the actor's principal enemy as explicitly identified by it; and *stakes* refer to the vision of order or organisation the actor envisages in the long term.

Touraine distinguishes three levels or dimensions of conflict. The first, *organisational processes,* generate conflictive actions over organisational status and change; here, conflict is over the competitive pursuit of collective interests. At the *political-institutional* level conflict takes the form of political force that challenges the rules of the game, e.g. in the decision-making process (and not just the distribution of advantages).[12] Finally, at the level of *cultural orientation*, conflicts concern the control of cultural patterns in a given society manifested in struggles between the elite deployment of cultural patterns and resources, knowledge, investment, and ethical principles and the masses, over representations of truth, production and morality.

The organisational, institutional and cultural levels of conflict are distinguished by the degree of integration among the three elements: identity, opponent and stakes. In organisational conflicts the elements are loosely interrelated, with self-centred actors competing in a field that exists independent of them; in political-institutional conflicts there is limited integration among actors, opponents and stakes, whereas in cultural conflicts the components are integrated and interdependent.[13]

Touraine's model is helpful in analytically segregating the different dimensions of conflict in development interventions. That is to say, using the model one can separate actions oriented towards competition over interests from those that challenge decision-making and/or policy-making devices and procedures, and those that seek control over cultural patterns, i.e. meanings, knowledge and values. Furthermore, such actions may be seen as succeeding one another in any given context as (in a social process where) 'lower-level' conflicts give rise to higher-level opposition. That is, struggles over the political status of needs and interests may lead to struggles over interpretations of institutional norms and finally to struggles to control 'truth'. The three levels of conflict constitute actions that are capable of affecting the action field in different ways (Table 1.2).

Table 1.2 outlines a model that captures the conflict dynamics in planned development interventions at different levels. Lower-level conflicts may be said to involve the expression of

Table 1.2 Elements and Dimensions of Conflicts in Development Projects

Types	Levels	Interaction	Stakes	Problem Areas
Reactive	'Lower' Organisational	i, o, s	Gains & losses in resource distribution including livelihoods	Conflicts over political status of interests
Reform	'Middle' Political	i-o, o-s, i-s	Decision-making process & information control	Conflicts over interpretations of norms & rules
Radical	'Higher' Cultural- Social	i-o-s	Redefining meanings & challenging discourses	Conflicts over control of values & knowledge

Source: Developed from Touraine (1985).

disapproval by actors in competing pursuit of their respective interests. They constitute interest-driven acts of protests and resistance and demands for negotiation and bargaining. In contrast, actors engaged in radical conflicts contest established norms, institutions and practice, and express what Offe (1985) calls 'non-negotiable demands', with very little scope for political exchange and reformist practices.[14] Compared to lower-level conflicts, radical conflicts exude conscious constructions of solidarity and identity, the very form of their antagonism transmitting societal messages (Melucci 1989: 56). On the institutional level, the struggle is to control 'zones of uncertainty' (Touraine 1985: 753). The argument then is, other than in (a) lower-level defensive conflict where interests are at stake, and (b) in radical conflicts where cultural patterns are at stake, the stakes in middle-level conflicts imply control over information, rules and decision-making strategies.

Our suggestion is to view the conflict model as a heuristic device that helps in identifying different dimensions of conflicts in the SSP. It allows us to see conflicts in development projects over interests, norms and values, enabling us to situate claims and experiences of different actors. How different actors are situated in different positions in the action field will of course depend on the specific empirical conditions. For example, local community groups can potentially engage in reactive, reform and radical conflicts. Similarly, governments may be involved in conflicts at the organisational, institutional and cultural levels. Yet to us the device is useful to identify various dimensions of conflicts and to locate different actors along those dimensions in terms of their stakes and their practice. More importantly, it helps us analytically to link the action field of development projects with social and environmental movements.

Collective Action in the Action Field: New Social Movements

As already seen, movements enter the action field as actors and advocates publicly counting and scrutinising the risks and impacts of development projects. Here we shall deploy analytical tools in the assessment and understanding of the practices and politics in social-environmental movements. This is both a delicate and intricate task since the relevant movement literature is extensive and rich, and marked by diverse perspectives and methods. The term social movements, according to Diani and Eyerman (1992), is primarily an evocative label attributed to a series of empirical phenomena easily analysed under the rubric of social conflict, collective action or political protest. In a minimalist sense, however, social movements may be defined as collective action: (a) expressing social conflicts (Touraine 1985; Lindberg 1995); (b) based on large mobilisations (Cohen 1985); and (c) challenging or breaking the limits of the system in which action occurs (Melucci 1989).

Touraine uses social movements to refer to conflicts around the social control of cultural patterns. This, he admits, is an 'arbitrary semantic decision' (1985: 760) as his use of the term to refer only to conflicts over cultural patterns is not backed by strong arguments. He advances the hypothesis, however, that social movements are higher-order conflicts and that 'some component of social movement must be found in all social conflicts', i.e. in conflicts around organisational processes and political decision-making which, to him, are 'non-integrated and lower-level social movements' (ibid.: 761–62).

To Touraine, social movement becomes the end product of a process of collective mobilisation where actors come to share a particular identity, clearly and commonly define their opponents and construct visions of altering established structures of relations. Melluci (1989: 29) endorses this formulation; for him, a social movement breaks the limits of compatibility of a system by violating boundaries, by pushing the system beyond the range of variations that it can tolerate without altering its structure. He sees social movement as different from interest conflicts that take place within a system of shared rules and where competing interests respect the limits of a given social order. Moreover, since social movement is a struggle over cultural pattern, i.e. visions and imagination, knowledge and meanings, it is unfair to endow it with instrumentality (as in interest conflicts) and to evaluate its effectiveness in terms of whether or not it has achieved its professed goal.

The meaning of social movement in the works of scholars such as Touraine and Melucci, however, needs to be placed in its context. Both these scholars, as well as Offe (1985), belong to the European 'new social movement' (NSM) school, unravelling structural conditions that gave rise to social movements in postwar Europe (especially since the late 1960s). NSM scholarship has argued that social movements are culturally pervasive but historically specific public actions which change form as societies themselves change. Thus, social movements of the industrial era are quite different to those in post-industrial societies. In the post-industrial era, social movements increasingly represent conflicts over symbolic goods (meanings, lifestyles) which are different from struggles over material goods (resources). NSM theorists consider struggles over identity and autonomy in contemporary movements around environment, gender, and peace and disarmament as distinct from 'old social movements' based on class and distributive politics. Studies in Europe and elsewhere show that the core members of these movements belong to the so-called middle class living in material conditions that facilitate their relative neglect of material, economic and redistributive demands (Eder 1995; Offe 1985).

Scholarship devoted to the rise of 'green consciousness' through environmental movements in Europe echo NSM sentiments. Thus, Cotgrove (1982) argues that goals pursued in the environmental movements are post-material, ushering in a moral debate over the nature of ecologically responsible society.

Others such as Kriesi (1989) associate environmental conscious-ness with the rise of the new middle class, the major constituent of NSM. Even scholars such as Beck (1992), who link the rise of environmental consciousness to risks to 'life chances' (and not just seeking new lifestyles), speak of the consciousness emanat-ing in post-industrial society.

Although (as is evident in Table 1.2) we consider struggles over symbols (which we term radical action) as an essential ele-ment in the action field of development, our use of the term 'social movement' requires important theoretical and method-ological qualifications. Our use of the term is tempered by the realities of the so-called 'scarcity society' (Beck 1992). The theo-retical underpinnings of NSMs in developing countries are more nuanced, the core of NSMs constituting a multiplicity of action groups: urban squatter movements, neighbourhood councils, coalitions for the defence of indigenous traditions and regional interests. NSMs are judged to be economical and cultural, de-noting struggles over meanings and over resources (Castells 1983; Escobar 1992). Their 'newness' is attributed to their being (a) non-party political formations (Kothari 1984; Seth 1983), outside the realm of institutional party politics on the one hand and trade unions and lower-class peasant politics on the other, and (b) forces heralding a new vision of development by 'redefining politics and articulating alternative forms of governance' (Sethi 1993a).

These NSMs are considered to emanate from the structural failures of the modernisation project largely practised and con-trolled by post-colonial developmentalist states that have so far exercised hegemony over power and knowledge (Evers 1985; Laclau 1985; Parajuli 1991). Such struggles are often in opposi-tion to the state and other socio-political institutions which serve the interests and values of dominant national elite and attempt 'to regulate both accumulation and legitimisation as well as capi-talism and democracy' (Parajuli 1991: 175). NSMs here are deemed to be counter-hegemonic formations resisting domina-tion, albeit in different degrees and forms, as well as 'providing some basis for a developmental and democratic alternative to the system as it now works' (Wignaraja 1993: 5).

This conceptualisation of NSMs in scarcity societies reflect-ing a developmental crisis, manifesting struggles over resources and meanings and a pluralistic paradigm of development and governance, is echoed in theories on 'third world' environmentalism and environmental movements (Bryant 1992; Friedman & Rangan 1993; Gadgil & Guha 1995; Guha & Martinez-Alier 1997; Shiva 1991). The rise of environmental consciousness and movements is attributed primarily to the socio-ecological impact of a narrowly conceived development model based on short-term commercial criteria of control and exploitation of environmental resources, and serving almost exclusively the needs and interests of the rich minority. Unlike in the so-called post-industrial risk society, environmental conflicts and movements in the scarcity society have a strong material basis emanating from state hegemony over water, forests and land resources which serve the subsistence and livelihood needs of the majority.

The theoretical qualifications posed here suggest that development projects *are* an action field for social and environmental movements. They also underline the fact that, in scarcity societies, it is difficult to disentangle conflicts over resources from those over risks or, for that matter, struggles over resources from those over meanings. This virtual collapse of lower, middle and higher conflicts into one phenomenon—social movement—has a methodological cost. We surely understand the 'why' of the movement, but not the 'how' (Cohen 1985; Dwivedi 1998).

This brings us to the methodological qualification. Structural explanations on social movements attribute them to crises and contradictions in the larger system, be it post-material or material. Focused as they are on the 'why' of movements, they are viewed as political reactions to (new) forms of (structural) domination and exploitation. Such methodological approaches, however, tend to conflate conflicts and movements. Perhaps more importantly, the tendency is to valorise the transformation potential of such movements (Salman 1990:102).

To us, it is relevant for several reasons to explain the 'how' of movements. Our main concern is to locate social movements within the purview of an action field within which we wish to see how collective action is socially constituted, how social movements mediate in conflicts, how a movement gains actor-hood, how a movement represents interests, how a movement generates awareness on social impacts and environmental risks, and, above all, how it affects and/or alters the action field. This concern with movement 'practices' is to gain a deeper understanding of political actions in the action field. It is also based on the methodological assumption that there is an autonomous dimension of politics which is not exclusively a reaction or response to structural conditions and crises.

Mindful of the dialogical character of social movements, i.e. interacting systems and actors, we draw some analytical tools from contemporary scholarship on social movement formation and development. First, that the action field comprised of opponents and allies provides the movement with a 'system of constraints and opportunities' (Melucci 1988: 332). As Tarrow (1994) suggests, the action field generates opportunities through cleavages among elites, shifts in ruling alignments and the availability of influential allies. Likewise, constraints arise from the growth of counter-campaigns with rivalling claims, state repression as well as co-optation, and policy innovations of governments. Second, that the movement, though not a unified empirical datum, strives to construct a collective identity. This is mainly achieved through an 'articulatory practice' (Laclau and Mauffe 1985) that aims to produce alliances and networks at multiple levels. It is in this practice that the movement acquires a definition for itself (Castells 1997: 69–70). Third, social and environmental movements engage in 'cognitive practice' (Eyerman & Jamison 1991). They are producers of cognitive constructs, capable of inserting new meanings, knowledge and information into the action field. It is through this cognitive practice that movements generate awareness on social impacts and environmental risks, and link them to issues of democratisation, rights and accountability.

Conclusion

It is clear then, that the project under study (the SSP) bears complex interconnections of structures and actions, as a result of which an analysis of the case must proceed with the aim of unravelling these connections if a deeper understanding is to be gained of the crises and conflicts that mark development projects. We have proposed that the project be approached as a field of action and tools be developed for analysing conflicts and collective actions in this field. This approach is the most appropriate

for interrogating how development conflicts and crises are produced and how they are experienced and challenged by actors in
the field.

We also feel, however, that approaching a development project
as an action field requires it to be located clearly within a wider
setting. Understanding this setting, its spatial and temporal
dimensions, its changing characteristics and attributes, provides
a contextual framework necessary for anticipating likely influences on the action field. In their book *Liberation Ecology*, Peet
and Watts (1996: 16) use the concept of 'regional discursive formation' to suggest that 'certain modes of thoughts, logics, themes,
styles of expression ... run through the discursive history of a
region'. They argue that 'regional discursive formations originate
in and display the effects of, certain physical, political-economic
and institutional setting'. They emphasise that development
ideas and discourses are always regionalised. If modernisation
to them is a western discursive formation, then the genesis and
character of dependency theory has deep links with Latin American experiences, whereas discourses on 'state planning and some
traditions of Marxist theorising have a distinct Indian or south
Asian character' (ibid.: 24). We think this concept, despite sound-
ing big, usefully captures the idea that development discourses
and practices tend to have a regional flavour. An exploration of
development action would therefore require familiarity with the
distinctiveness of the regional discourse.

In the next chapter we review ideas and debates around Indian
development experiences in order to map the evolution of, and tensions in, this regional discursive formation. It will be obvious to
the reader that there is no unanimous view on the attributes
and operations of the setting. By bringing these debates to the
fore, however, we hope to open up some links between the causal,
experiential and political aspects of development crises which
will similarly aid the understanding of the case in hand.

Notes

1. Hulme (1988) restricts this analysis to the design phase of the project,
   but acknowledges that this understanding can be extended to the
   entire project cycle. For political approaches to project appraisal
   and evaluation see Gasper (1987), Barnett (1981), Stewart (1975).

2. Similarly, for structural anthropologists it is the stable (if not permanent) structures that explain and even determine motivations, practices and actions.

3. Reflecting on reductionism in developmental Marxism (distinctly broader in scope than Marxian political economy), Mouzelis (1990) notes four types: (*a*) agent>agent—e.g. economic agents exercise power over political agents; (*b*) agent>structure—e.g. economic elites reproduce favourable frameworks of institutions and policies; (*c*) structure>structure—e.g. laws of motion of capital determine institutional forms; and (*d*) structure>agent—e.g economic structures determine political actions.

4. Knight (1994:15) uses a rational-actor approach to analyse social conflicts and institutional development and change in society. The assumptions underlying 'methodological individualism' in the theory are clearly spelt out. 'Individuals act in pursuit of various goals and interests and choose their action in order to satisfy those interests efficiently. Rational-choice approaches are usually associated with self-interested desire and goals. But it is more important to assert that individuals act intentionally and optimally towards some specific goals. Actors choose their strategy under various circumstances. More often than not they are confronted by situations characterised by an interdependence between other actors and themselves.'

5. As Maheu (1995: 2) explains in the context of relationships between social movements and social classes: 'it is more difficult than ever to make a causal or even purposeful link between dissatisfied groups and their participation in social movements.... The theory (of deprivation) does not explain, when mobilisation indeed takes place and why its principal actors turn out to be those least exposed to the actual experience of deprivation.'

6. Notwithstanding this suspicion for structural categories, Laclau considers macro processes such as bureaucratisation and commodification as evidence of 'relative structuration' impinging on the life worlds of subjects (Slater 1992: 297).

7. Here one notes the efforts to bring together these opposing theories (in amalgamation) through what has been called 'analytical Marxism'. The attempt here has been to integrate macro-level analysis of classical (functional) Marxism with micro-level interactions. The specific plea is for a game-theoretic rather than functional treatment of such issues as internal divisions among classes and interests of the state so as to strengthen the micro foundations taking the behaviour and preferences of different agents (see Roemer 1986). While conceding to the purist's critique that analytical Marxism is only a variance of rational-choice theory, we make

use of some of its insights to analyse how groups engaged in resource struggles in a development encounter adopt different strategies
and exercise different options in a common situation, while making
no claims to have adhered to its methods of analysis.

8. The debate on the 'impasse' was by and large among British devel-
    opment sociologists. I thank Ben White for drawing my attention
    to this aspect of the debate.

9. Giddens makes a subtle distinction between 'actor' and 'agent'. The
    latter engages in intended action for change. As a majority of schol-
    ars use the two terms interchangeably, a stronger emphasis on the
    distinction is to be welcomed by adding representation as a specific
    characteristic of agency. That is to say, agents act 'on behalf of'.

10. Structural categories are useful here to capture this inequality in
    actor endowments.

11. Touraine differentiates between social and historical conflicts: the
    latter attempt to control the process of historical change and can
    be both offensive and negative.

12. The term 'political' implies the level of policy and is not opposed to
    Laclau's argument of it being a dimension. Struggles at all levels
    are nothing but political.

13. The integration of the three elements in the three conflict types
    has been symbolically expressed as follows. The '*i*'dentity and
    '*o*'pponent and '*s*'ocial goals are strongly defined and integrated in
    struggles over cultural pattern as i-o-s. In struggles around political
    processes, the relationship is of a lesser order integration repre-
    sented as i-o, o-s, i-s. At the organisational level the i,o and s are
    autonomously placed.

14. For Offe (1985: 830–1) the non-negotiability of demands results
    from the unwillingness as well as the incapability of actors. Un-
    willingness to negotiate stems from the perception of actors that
    no part of their central concern (values of 'survival' and 'identity')
    can be compromised 'without negating the concern itself'. Incapa-
    bility arises because (*a*) actors have nothing to offer in return for
    concessions and (*b*) they lack both internal coherence as in formal
    organisations, and a set of ideological principles from which an
    image of desirable arrangement of society could be deduced. While
    we accept 'unwillingness' as a central explanation of non-negotia-
    bility, in our view the 'incapability' factor, because of ideological
    and organisational incoherence, needs to be tested empirically as
    actors in radical formations often invest resources in constructing
    ideological coherence as well as solidarity bases.

2

Development, Crisis and Conflicts: The Indian Experience

Introduction

Large development projects do not exist in a vacuum. Wider, systemic forces and relationships overarching the action field influence actions and interactions. An understanding of this wider setting helps in contextualising the action field, offering vital clues towards linking conceptual and analytical tools with the specificity of the case. For instance, development projects cannot be viewed in isolation from how development itself is perceived as an ideal. Conflicts and tensions emanating from project situations have a bearing on how development strategies are seen and how costs and risks are appraised, rationalised and justified in the larger environment. Likewise, movements derive their meanings from this environment, which holds significant evidence concerning what the movement is all about.

This chapter culls out some characteristic and distinctive features in the larger setting which are specific to development in India. But before detailing how these aspects unfurl in the Indian setting it is necessary to explain their significance in an understanding of the SSP. First, development projects echo and enshrine ideologies and derive sustenance from them. Actors who make and design the projects foresee developmental goals and visions. Assessing those goals and visions and the struggles that they may have generated provide clues to an understanding of conflicts in the action fields. Second, development projects feature as part and parcel of broader development strategies of resource allocation and distribution which introduce us to the rationality of the process, i.e. why certain means are preferred over others. Third, an overview of developmental impacts and outcomes is an essential guide with which to mark limits and bottlenecks in the preferred strategies, particularly towards goals of growth, equity and balanced regional development. Such an overview also serves to assess the intended and unintended consequences and outcomes of change and development. Finally, political actions that respond to these developments provide clues to constraints and opportunities in the larger setting. They also indicate the relative power of actors, the intensity of power struggles, and the impact that such struggles can have on development thinking and practice. It is these four aspects of the wider setting that are explored in this chapter.

Development as Ideology: Large Projects as 'Temples'

The post-colonial Indian development model rested on the simultaneous pursuit of four basic goals: (a) the national integration of a diverse social structure; (b) economic development for raising the standards of living of a large majority of the poor; (c) social equality in an inegalitarian social order, and (d) political democracy in a culture that had valued authority based on status and power concentrated in the hands of a minority elite (Kothari 1990:24). Set in the context of anti-colonial nationalism, these goals were essentially political, 'imposed' by state authorities on the society, with the implicit assumption that the pursuit of such goals would benefit society as a whole.[1]

As the cornerstone of these basic goals, economic development implied modernisation, accelerated industrialisation and catching up with the West. The 'Nehruvian model' of development meant an economy aimed at accelerating growth through capital-intensive industrialisation and the state-ownership of strategic sectors. The state commanded the economy through an enlightened bureaucracy, central planning and a large public sector within which heavy industries and huge irrigation and power projects were assigned crucial transformative roles. Advanced centres for research and training, elite science and technology institutes, and universities, featured prominently in the model to generate skilled human resources for development.

The Nehruvian model's subscription to the modernisation ideology sharply contrasted with the development visions and ideals of Gandhi. In the Gandhian world-view, development assumed a completely different import to its commonly understood meaning as expansion of choices.[2] Gandhi's vision of post-colonial India rested on decentralised village communities and collectives, small-scale strategies and solutions through craft production and simple technology, and voluntary limits to consumption.[3] He believed that industrial development would crowd out traditional cultures, values, knowledge and technologies. In his view, one advanced with the eclipse of the other. This zero-sum process is clearly espoused in one of his oft-quoted statements:

> God forbid that India should take to industrialisation after the manner of the West. The economic imperialism of a single tiny island kingdom is today keeping the island in chains. If an entire nation of 300 million took to similar economic exploitation, it would strip the world bare like locusts (Gandhi 1951:31; cited in Guha 1989; *see also* Baviskar 1995; Omvedt 1993; Shiva 1991)

As Chaterjee (1984: 160) demonstrates, in the Gandhian ideology,

> any kind of industrialisation on a large-scale would have to be based on certain determinate exchange relations between town and country with the balance tipping against the latter whenever the pace of industrialisation quickens.

In Gandhi's view, 'industrialisation on a mass scale will necessarily lead to passive or active exploitation of the villages as the problems of competition and marketing com[e] in' (cited in Chaterjee 1984: 160). The Nehruvian panacea of 'socialist industries', what Peet and Watts (1996) term the complementary and contradictory rationalities of the centralised state's attachments to industrial gigantism and heavy goods on the one hand,

and the reciprocity and networks at the enterprise level on the other, did not convince Gandhi. In an interview with Francis G. Hickman in 1940, Gandhi states:

> Pandit Nehru wants industrialisation because he thinks that if it is socialised, it would be free from the evils of capitalism. My own view is that evils are inherent in industrialism, and no amount of socialisation can eradicate them (cited in Chaterjee 1984:160).

The Gandhian vision of a small-scale, needs-based and survival guaranteeing moral structure of the traditional (village) economy was deemed an impractical set of solutions to the needs of post-colonial Indian society. It survived merely as a feeble populist critique to a developmental ideology backed by modern large-scale technology, bureaucratic authority and capitalist economic activity in industry and agriculture. Political legitimacy accorded to modern technology intertwined with a nationalist discourse on self-reliance. Although the objective was to catch up with the West, the ground rules for the race included the building of an industrial infrastructure, mobilising to the extent possible resources from within the country.[4] Villages and rural areas were conceived as intellectually and socially backward, whereas industrial actors, engineers and technocrats became the revolutionary vanguards of change and transformation. Their deification became apparent in Nehru's frequent eulogies describing them as builders of modern India and, more specifically, in his characterisation of large development projects as 'modern temples of India'. In his semantic elevation of development as a religion, development projects as temples and engineers and technocrats as high priests propagating the faith,[5] Nehru contributed not only to the power of developmental ideology but made it the only option available. To him, 'no country can be politically and economically independent, even within the framework of international interdependence, unless it is highly industrialised and has developed its power resources to the utmost' (Nehru 1956: 414).

The Developmental State: Manager-cum-Player in Resource Allocation

The role allocated to the State in the Nehruvian model in matters pertaining to resource allocation and distribution first needs to be seen in the context of a colonial legacy. While colonialism aimed primarily at appropriating surplus for 'non-developmental use' (Kohli:1989: 24), it initiated its own demise by generating new political forces, i.e. a nationalist elite committed to the ideology of nationalism that could build on the rudiments of a modern state structure for socio-economic transformation and politically directed development. Thus, the strategy of development was not only oriented heavily towards the state; it was nationalistic and oriented towards economic self-reliance, weakening if not breaking links with international capitalism. The Indian State committed itself to rapid industrialisation and achieved remarkable success in this sphere as compared to other post-colonial economies. Kaviraj (1997: 236) notes the imperative of keeping at bay neo-colonial pressures that caused the Indian state to prioritise industrialisation, in particular the development of the capital goods sector, over the objectives of social equality, and acknowledges the positive aspects of the strategy:

> In retrospect...it...appears that the mixed economy—not surrendering the entire productive and distributive mechanisms of the society either to the unrestricted rapacity of private business or to the chronic rheumatism of the state bureaucracy—introduced a long term check and balance relation in the political-economic system and increased room for manoeuvre.

In the earlier stages of development, however, the resource requirements of industrialisation were not a matter of choosing among different strategies. In many ways the leading role of the state was a forced one. An important factor that set the basis for state-led development and for much of developmental planning, was the acute deficiency of material capital and other structural constraints which prevented the introduction of productive technologies. At the time of independence industrial and commercial groups were incapable of carrying the 'burden' of industrialisation on their own. They supported and endorsed planned state initiatives and modernising activities and advocated protectionism against foreign competition.[6] Second, a society marked with diversity, inequalities and regional imbalances required a strong central agent to be responsible for resource allocation and distribution. Although the leading productive role of the state was soon to be marked by inefficiency, this was a relatively small price to pay to counter the distributive irrationalities of private (foreign) industrial forces.

The redistribution of assets and income and the underlying objective of social equality were assigned secondary importance, although the sustenance of a democratic polity implied that redistributive goals were held to be of legitimate political concern. Chakravarty (1987) argues that planners tolerated income inequality provided it was not excessive and could be seen to result in a higher rate of growth. The assumption was to let the national cake grow before attempting any substantial redistributive exercise, such as land reform and redistribution. Juridical changes in the status of productive assets were largely ignored. Thus, the major problems of the economy during the first decades of independence were 'how much to save, where to invest and in what forms' (Chakravarty 1987: 10). Public investments hence were concentrated in the infrastructural sector as well as in the modern, capital-intensive industrial sector. While investment in heavy industries was state-led, the private industrial sector was confined to 'relatively labour-intensive and light consumer goods' (ibid.: 15). Agriculture was used as a 'bargain sector', a sector that could produce the required surplus in a relatively short time with some infrastructure investments. Large irrigation investments relieved the long-standing stagnation in the agriculture sector, but it nevertheless remained vulnerable and monsoon-dependent. When the green revolution package was introduced through the 'new agricultural policy' in 1969, 'it certainly broke the stagnation... but did so at the price of increased polarisation in the countryside' (ibid.: 27; see Byres, Crow and Ho 1983; Dhanagare 1987).

Impacts and Outcomes: Uneven Development

At the macro level, and viewed from a perspective of enhancing the 'public good', the performance of the Indian economy has been impressive but uneven. India's development record as a nation-state with a population of a billion people, includes self-sufficiency in food, a diversified industrial infrastructure, skilled human resources, growing commercial and urban centres, access to amenities such as transport and communication, power and water supply, education and health care. In growth terms, the GDP has shown an increase over the years, as have national and per capita income. Savings and investments have doubled, literacy rates have improved significantly, and death and birth rates have shown a decline. Further, India is self-sufficient in foodgrain production, the industrial production index has increased, and achievements in the power and irrigation sectors are impressive.

These indicators of the overall performance of the economic modernisation project have to be seen in conjunction with the simultaneous persistence of poverty and unemployment, economic inefficiency and environmental degradation. More than 30% of the population continues to live below the poverty line, concentrated mainly in the central and eastern regions of the country (Srivastava 1995). At the beginning of the Eighth Five-Year plan an estimated 23 million people were unemployed of whom 7 million were in the educated unemployed category. The fastest deceleration in the rate of employment growth has been in the agriculture sector, which accounts for about two thirds of employment. This includes the high output growth regions like Punjab, Haryana and western Uttar Pradesh. Child labour is conservatively estimated at around 30 million, coexisting with forms of unfree and forced labour in different parts of the country. Industrial growth has gone hand-in-hand with industrial sickness; estimates of the number of sick units range from 200 thousand to 400 thousand (ASG 1994). Public sector undertakings, though fulfilling objectives of employment generation and regional balance, have been the seat of technological, economic and managerial inefficiency. As far as the state of environment is concerned, forest cover has rapidly dwindled, the rate of industrial–urban pollution is among the worst in the world, and ground water depletion, soil erosion and desertification have reached worrying proportions (see CSE 1982; 1985).

The uneven performance of the Indian economy therefore implies that economic growth and development have existed side-by-side with poverty and unemployment. Such outcomes could be explained as characteristic of 'uneven development' (O'Connor 1989)[7] where resources allotment is lopsided and their distribution unequal, and where productive social relations governing these resources remain hierarchical. The appropriate question to raise in this context would be: benefits for whom and at what cost?

The Class of Gainers

As far back as 1929, Nehru asked: for whose benefit must the industries run and the land produce food?[8] Over the next several decades, this question intensified and encompassed major conflicts and contradictions in resource distribution. In political economy studies the focus has mainly been on the power of the dominant social classes to determine the nature of the development process and to corner its benefits. In assessing developmental outcomes and, more specifically, towards understanding the dynamics of resource allocation, Bardhan (1984) argues that conflicts over scarce resources involve three dominant groups. First, the industrial capitalists who have profited from import substitution policies and have learned to turn the industrial licensing system to their advantage. Second, the rich farmers who have benefited from the green revolution, government subsidies and support prices. And third, the professional bureaucrats who have gained considerable corrupt income from their administration of programmes for the benefit of farmers and their control over the investment decisions of industrialists and the business class.[9]

> When diverse elements of the loose and uneasy coalition of the dominant proprietary classes pull in different directions and when none of them is individually strong enough to dominate the process of resource allocation, one predictable outcome is the proliferation of subsidies and grants to placate all of them, with the consequent reduction in available surplus for public capital formation (Bardhan 1984: 218).

Rudolph & Rudolph (1987: 51) argue in similar vein when they identify the dominant class interests and benefits of development underlying what they call the 'Nehruvian settlement'.

While private sector industrialists and management and administrative professionals were senior partners in the settlement—benefiting from import substitution, self-reliance and an expanding public sector—the junior partners were rural, large land-owners who controlled state governments and allocated resources in ways that benefited their rural interests.[10]

The commanding role of the state in development implied that employment opportunities were created for public sector industries and their administration. 'Bureaucrats were needed to supervise, account, keep in touch with and control the state sector; any expansion of this sector economically meant a proportional increase in its supervisory ministerial bureaucracy' (Kaviraj 1997: 239). This implied not only a dramatic expansion of the bureaucratic class but also the concentration of administrative power in its hands. The latter was derived from the regulatory (and redistributive) functions associated with development. As Frankel (1990: 498) observes:

> By the 1980s, the bureaucrats manning [*sic*] the public sector were far more powerful than their counterparts in the large private business houses. They presided over the commanding heights of the economy in the organised industrial sector and administered a formidable regulatory apparatus for the licensing and expansion of public enterprises, import and export of capital enterprises, allocation of foreign exchange and clearances to raise capital from the public. Between 1969 and 1973 alone, their powers were expanded by nationalisation of banks, general insurance and the coal-mining industry; the Monopoly and Restrictive Trade Practices Act (MRTP) devised to limit investment by the large industrial houses to the heavy investment core sector and the Foreign Exchange Act (FERA), restricting the role of foreign capital in Indian industry.

The bureaucratic spread manifested itself in the power to define development, a proactive but nonetheless top-down approach by central planners to decide what is good for the people in the countryside. It was also manifested in a reactive and obstructive manner which involved the selective distribution of patronage and benefits derived from the regulatory regime, the use of public resources for private purposes, in the process undermining the very goals of development that it set. Noting the 'rent-seeking' power of the bureaucratic class, Bardhan (1984:58) states:

> The majority of businessmen who do not have the clout or the money power of the conglomerates have to approach these [bureaucratic] dispensers of permits and licenses essentially as supplicants.... The bureaucracy impinges somewhat less on the interests of rich farmers, even though in matters of administered prices, procurement, restrictions on grain movements and trade, and distribution of credits and fertilisers, it has numerous powers to exercise and favours to dispense (cited in Toye 1993: 168).

In the private agrarian sector, with the abolition of the zamindari system but in the absence of substantive land reforms, the agrarian social structure came to be dominated by a powerful intermediary class. The politics of this class ranged from brutal oppression of the underprivileged on the one hand, to the formation of political alliances demanding state subsidies in agricultural inputs and higher prices for farm output on the other (Brass 1995; Patnaik & Hasan 1995). Their power derived from caste and regional–ethnic identities, and they could control political regimes and political parties at the regional level. It can be argued that this class operated as a powerful deterrent and undermined the ability of the state to undertake significant redistributive decisions.

Uneven Development in a Federal and Democratic Polity

Development in India cannot be understood adequately without referring to the federal polity and its implications for planning and decision-making. The central government, including the ministries, departments attached to these ministries, and the Planning Commission, the state governments and their ministries, departments and boards, all feature in a multi-layered decision-making process. Mechanisms such as the National Development Council, consisting of the prime minister and chief ministers of the states, offer scope for negotiating centre-state relationships. However, the centre-state relationship has witnessed periodic demands for decentralisation and autonomy on the part of the states.

Provincial states have asserted themselves in the politics of development planning. With the waning of nationalism and a competitive political structure in place, struggles for limited resources led to a form of pronounced regionalism with different regional interest groups staking claims and competing for resources. The assertiveness of provinces by no means implied breaking free of the centre, but they did become incubators of conflicts. The struggles of various interest groups—political parties, caste associations, bureaucracies, rich-farmer lobbies—within and among themselves, implied not just competition for the access to and control over resources, but also brought rules, procedures, policy measures and practices into the arena of conflict. Development planning became 'a process of competition and struggle over resources, goals, strategies and patronage without an overall design' (Brass 1990: 253), and regional differentiation in agriculture and industry became more pronounced (Bharadwaj 1995: 204–10). Regionally powerful lobbies which could influence investment decisions stood to gain disproportionately from the location of development projects, their shape and magnitude, as well as the redistribution that ensued from them.

Development planning in India is thus not a simple, one-way flow of directives and decisions but a complex process of allocating and reallocating resources. Its implementation has involved the dynamic interplay of, and competition among, numerous actors representing national, sectoral, regional, class and primordial interests, which have variably influenced the autonomy of state planners from societal interests.

The conception of the wider setting as dominated by class forces and relationships has to contend with scholarship that considers political democracy as a major variable in influencing state autonomy on the one hand, and developmental outcomes on the other. For Harris (1994) democratic politics rather than systemic class control and the usurpation of benefits is of significance in assessing the Indian developmental setting.

The [developmental] regime...is one in which no single class exerts economic or political dominance and neither is there a state organisation which is powerful in relation to society. In a context of parliamentary electoral politics the maintenance of [state] power had come to depend on a precarious trading of concessions to different social groups including the rural poor as well as...surplus appropriators (ibid.: 179).

To consider the rural poor as 'beneficiaries' of development may not be a correct way of looking at their situation. Yet forces of parliamentary democracy have yielded centrally-sponsored welfare schemes of poverty alleviation and employment generation for the rural poor. These programmes—Integrated Rural Development Programme (IRDP), National Rural Employment Programme (NREP), Drought Prone Area Programme (DPAP), to mention a few, have had largely divergent impacts with reasonable success stories in some states co-existing with significant failures in others.[11]

In most parts of the country, the radical programme of land reforms has long been abandoned save for the abolition of zamindari systems and other forms of intermediary rights to the land.[12] This has prompted the thesis that reformist welfare schemes of poverty alleviation and employment generation are compensations that conform to the system. Politically, these programmes do not in any way challenge the system and its dominating classes, unlike, say, the land reform programme; economically they are sustained because in various ways they tend to benefit these very dominant classes through patronage and pilferage.[13]

It is significant to note that, like the welfare programmes for the rural poor, land reforms have by and large been successful in some states. This leads Kohli (1984) to pursue the question of why such redistribution programmes (land reforms, small farmers schemes and wage and employment generation projects) were more successful in some states than in others. He goes on to demonstrate that, despite similar socio-structural conditions, differences in 'regime types' are of considerable consequence for the effectiveness of state-initiated redistribution.[14] His approach brings to the fore the significance of (going beyond the class model of) politics and political agency in distributive conflicts.[15] According to him:

A highly interventionist state tends to politicise its society.... As the state reaches everywhere, ubiquity characterises its politics.... An interventionist democratic state typically encourages considerable politically oriented activism.... Widespread politicisation however does put a high premium on a polity's institutional capacity simultaneously to accommodate the resultant demands and to promote socio-economic development. Well-organised political parties therefore become especially crucial (Kohli 1990: 30).

In this book, we are more inclined to assess the role of 'non-party political formations' rather than political parties per se. However, Kohli's infusion of political variables helps us to understand how state–society relations influence development outcomes. He believes that state autonomy in much of the Third World should be understood not so much for its capacity to act against the interests of propertied classes (1989: 28), but rather for the ability of state actors to mould societal preferences to minimise later opposition to political initiatives or to take the lead in initiating socio-economic changes that may not be actively demanded or opposed by any group in society. Thus, development in this perspective implies a politically-defined objective on the part of state authorities and not a set of demands that is societal. Concomitantly, in an open democratic polity like India, characteristics such as demand-oriented political participation by various interest groups, the periodic needs of political regimes to secure legitimacy and thus to make policy concessions to competing social groups, and the penetration of state institutions by the more powerful in society, constrain state autonomy in relation to the society (ibid.: 31). Thus, societal forces have a constraining and not determining impact on the development agenda. In similar vein, state agencies, by trying to accommodate diverse social interests, come to reflect major societal cleavages.

Crises, Experiences and Responses: Locating Interests and Agents 'From Below'

The crises and contradictions in uneven development entail costs and risks for a large segment of the population. Marxian political economy explains systemic contradictions in terms of unequal and exploitative exchange relationships between capital and labour. In his attempts to incorporate environmental crises as part and parcel of capitalist contradictions, O'Connor (1992) highlights the 'second contradiction' between capital and nature, where capital has greater need for organised and rational access to nature as the supply of external nature (as production condition) shrinks. Materialist theory holds that experiences with these contradictions call into existence different political agencies of change—organised labour movements and the environmental movement.

There is no unanimity among scholars in terms of analysing experiences and political responses to systemic contradictions in the Indian context. In fact, sharp differences over what mode(s) of production typify the Indian political economy—capitalist, feudal, semi-feudal, colonial, etc.—influence the analysis of political interest and agency. Thus, for example, the industrial working class (as part of the organised sector) is seen by some as an exploited class whereas others consider it as a beneficiary class of development (see Gadgil & Guha 1995; Kaviraj 1997). However, there seems to be unanimity regarding the experiences and costs of industrialisation and economic development borne by the agricultural workforce, i.e. poor peasants and agricultural labourers.

According to Patnaik (1992), in conditions of extreme land concentration, terms of trade, whether *for* or *against* agriculture, does not make any difference to the agricultural workforce. In the former (favourable prices, higher subsidies), the cost is borne by the workforce (together with industrial workers and the urban petty bourgeois[16]) because of a fall in real wage rates; in the latter, the rural landed class passes on the costs to its workforce. He suggests that the latter has been the trend in India from the early 1970s, a form of semi-feudal capitalism. Interestingly, he points out that 'a squeeze on the agricultural workforce in such a situation may generate a feudal revival under the guise of reactionary populism' in the countryside, because 'where revolutionary movement is weak, reactionary populism can gain ground easily' (ibid.: 203).

Marxist scholars in India tend to associate reactionary populism with a political discourse that emerged with agrarian mobilisation in those pockets where the industrialisation of agriculture had been accomplished. Led by the so-called landlord/

capitalist class farmers, this mobilisation had its base mainly in the north (Punjab, Haryana, western UP), west (Maharashtra and south Gujarat) and some areas in the south of the country (see Banaji 1995; Brass 1995; Byres 1995; Dhanagare 1995; Hassan 1995). The demands for better terms of trade for agriculture, i.e. fair prices and related issues and a subsuming abhorrence for land reforms, land ceilings and higher wages, have led scholars to conclude that the politics of the farmers' movement exude a reactionary character.

Dissatisfaction with the Marxist projection of the modernisation crisis and political responses led Omvedt (1993) to explore contradictions between relations of production and conditions of production—following O'Connor. In Omvedt's opinion, Marxist scholars overlook 'relations of exploitation and surplus extraction between toilers (land-owning small peasants) and those controlling conditions of production, markets and state' (1993: xiii–xiv). She argues that non-wage labourers, subsistence producers, petty commodity producers, and collectors of minor forest produce are as much at the receiving end of extractive and exploitative production relations in industrial capital as wage labour.[17] Omvedt (1993:112) supports the urban bias thesis of Michael Lipton,[18] arguing that the primary contradiction exists between 'Bharat' (primarily the villages but also the informal, unorganised urban sector workers who are 'refugees from Bharat' in the cities) and 'India'[19] (the westernised industrial bureaucratic elite, inheritors of colonial exploitation).[20]

The political implications of Omvedt's arguments are ambiguous. Her analysis of systemic contradictions suggests the inclusion of *small* land-owning peasantry along with poor peasants and workers. Yet she seeks a 'broad-based all-inclusive farmers—wage workers alliance' in the countryside and views positively the politics of the farmers' movement. This ambiguity notwithstanding, in extending understanding of the modernisation crises to include production conditions, Omvedt (1992: 52) generates an environmental sensibility that is by and large missing in Marxist analysis:

'Development' or growth of production relied on the extraction of resources (use of energy) from outside the production system itself, such that the costs were borne by the external areas. These 'external areas' included both the peasants and tribals...whose labour was exploited (directly or in the form of extraction of their surpluses) and 'nature', the planetary system which had created the stored energy being extracted.... The costs of this can be measured both in terms of the exploitation and poverty of toilers, and in terms of the land and soil degradation and the various forms of environmental crisis we see today.

Extension of the political–economic analysis to the realm of nature and environment adds a new dimension to the cognitive mapping of costs and benefits of modernisation and conflicts therein. The result is that environmentalists differ substantially from Marxists in calling for a broad-based industrialisation that is agriculture-led (albeit with an attack on landlordism), with necessary investments in energy, irrigation and infrastructure, and the intensive use of local resources.[21] For environmentalists, resource-intensive development cannot achieve an improved quality of life for all its members as it inherently generates economic polarisation and ecological destruction. Intensified industrial resource use not only impoverishes nature, but marginalises subsistence farmers, rural artisans, herders, fisherfolk, ethnic minorities and women by dislocating cultural and customary forms of resource management (Agarwal 1997; Gadgil & Guha 1995; Shiva 1988). Thus, as with the skewed distribution of benefits of development, its costs are borne unequally by different groups in society. The distribution problematic in development becomes two-pronged—the outcome of development is cornered by the dominant groups, whereas the process of development tends to marginalise those already deprived or left out and who lose habitats, livelihood and systems of life support. The costs borne in the process include ruptures in long-established social fabrics and institutional practices, loss of biological and cultural diversities through homogenisation, and the sustenance of an exclusionary and depriving resource regime (Kothari 1984; Parajuli 1991; Seth 1983). According to Shiva (1991: 19):

This process has been characterised by huge expansion of energy and resource intensive industrial activity and major projects like big dams, forest exploitation, mining, energy intensive agriculture, etc. The resource demand of development has led to the narrowing down of the natural resource base for the survival of the economically poor and powerless, either by direct transfer of resources away from basic needs or by destruction of the essential ecological processes that ensure the renewability of the life-supporting natural resources.

Shiva criticises the linear theory of modernisation and endorses the dependency model of analysis to highlight the exploitative exchange relationships in the process of development.[22] In her opinion, just as the invisible cost of modernisation in the West was borne by its erstwhile colonies, likewise the costs of economic development will be passed on to 'internal colonies' and hinterlands within a country. 'Where colonialism ended, the slogan of economic development stepped in' (Shiva & Bandyopadhyay 1989: 38).[23] According to Shiva:

The ideological universalisation and enclavisation of the process of growth and development is the reason for the simultaneous existence of underdevelopment alongside economic growth in newly independent countries like India, which accepted rapid industrialisation as the path towards development. Like the erstwhile colonies, interior and resource rich areas of the country are bearing the costs of resource diversion and destruction to run the resource intensive process of development. As a result communities living in these interior regions are facing a serious threat to their survival (1991: 27).

Shiva's analysis is echoed in much of the contemporary research on environmental resource struggles. For Gadgil & Guha (1995: 15), the process of development has come to be equated with the channelling of an ever more intense volume of resources, through the intervention of the state apparatus and at the cost of the state exchequer, to serve the interests of the urban and rural elite. Material, energy and informational resources are used and consumed in India within a 'high-cost low-quality economy plagued by environmental degradation and social inequality' (ibid.: 50). As states extend their control over resources such as land, water, forest and minerals for industrialisation and modernisation, communities dependent on those resources are adversely affected and become deprived of (traditional/customary) access and use. Thus, conflicts occur on a variety of scales, relate to many different natural resources and mainly pit 'haves' against 'have-nots': trawler owners against artisanal fisherfolk, tribals against paper mills, peasants against irrigation authorities (ibid.: 63). However, as 'resource-based conflicts have fissured Indian society along many, many axes', they 'are the sharpest where there are rich possibilities of cornering state-sponsored subsidies, as with river valley projects' (ibid.: 97), ensuring struggles among the elite.

The political expression of these conflicts is what the Indian environmental movement is all about. Gadgil and Guha demonstrate its driving force to be 'victims of environmental degradation', who are organised by social action groups geared towards 'halting ecologically destructive economic practices', 'promotion of sustainable use of natural resources and bringing about environmental restoration' (ibid.: 98–99). Although the movement (containing many different strands) has a significant presence of sections of the elite (as leaders, sympathisers and ideologues), it is essentially 'environmentalism of the poor'.

At stake here are local needs and use of resources, community rights and social justice. The 'environmentalism of the poor' is first and foremost expressive of disenchantment with state-led development strategy. Scholars such as Shiva, however, hold that these movements create political space for envisioning alternatives to the development model. Unlike the Marxian political-economic approach in which alternatives rest on land reform and a concomitant removal of the fetters in the agricultural sector for its accelerated industrialisation, the environmental alternative rests on a partially but substantially recovered Gandhian discourse based on decentralisation, the local control of resources, and small-scale technology.

In fact, the environmental alternative is seen as part of a larger discourse on alternative development gaining ground in which issues pertaining to rights, identities, local knowledge and democratisation are raised by grassroots NGO movements. For Kothari (1990: 66) this movement for alternative development consists of a series of obvious and inevitable strands, of struggles against existing hegemonies, of organised resistance, of mainstream protest, of civil liberties and democratic rights. It is an effort to open up new spaces in state and civil society. However,

Kothari believes that the alternative impetus comes from a new class of people, 'activists' drawn from the conscious, enlightened and troubled middle class engaged in the voluntary sector, NGOs and the popular movement sector, who facilitate the 'convergence of new grassroots politics and new grassroots thinking':

> The environment, the rights and the role of women, health, food and nutrition, education, shelter and housing, dispensation of justice, communications and dissemination of information, culture and lifestyle...none of these were subject matter for politics, at any rate, not for domestic politics, and certainly not for mass politics in which ordinary people were involved. This has now changed. (ibid.: 67)

Insofar as these issues feature prominently in grassroots politics, the movement for alternative development is deemed to change state–society relationships by creating an awareness of ecological responsibility and democratic renewal.

Conclusion

What emerges from the preceding discussion on the debates surrounding development discourses and practices in India, is the realisation that although democracy conditions relations and actions in development in India, the tendency of dominant societal actors who influence development is strong and cannot be denied. One could therefore argue that, in this typical setting, development action is subjected to conflicts and struggles emanating from democratic conditions as well as dominant economic and power relations. Thus, crisis formation is by and large the by-product of the politics of democracy and the economics of dominance.

From this contextual framework we go on to understand the crisis in the SSP, the way it unfurls, how it is experienced and what political responses it generates.

Notes

1. The term imposition can have different meanings in different texts. For some scholars (Shiva 1988; Gadgil & Guha 1995) the term implies

non-participation, top-down and forced. While these aspects of the development model are dealt with later in the chapter, here the term is used following Kohli (1989: 28), who argues that the term suggests that developmental goals were state-directed and not society-directed, but nonetheless finds the term 'impose' too strong 'in the absence of any clear evidence of mass opposition to the state's developmental goals'.

2. For a more detailed discussion on their different views see Gopal (1980).
3. It has been convincingly argued by Parekh (1995) that towards the end of his tenure, Nehru turned increasingly towards Gandhi, realising that modernism as an ideology and its technological solutions clearly proved insufficient to address the problems of poverty, unemployment and regional imbalance. For example, and as will be noted in the next chapter, Nehru increasingly aired his views on the benefits of small development projects as opposed to large ones that he once hailed as the temples of modern India. On a different plane, although Gandhian design did not tolerate some large-scale interventions such as railways (see Parekh 1989), on at least two occasions Gandhi dissociated himself from local struggles against large power projects which resulted in the collapse of these movements and the completion of these projects. These two projects—Mulshi and Hirakud—are referred to in Chapter 5.
4. The Nehru–Mahalanobis strategy made Indian economic development dependent on foreign capital, but this was clearly with the objective of making the country self-reliant in the long term (see Brass 1990: 249).
5. See several of Nehru's speeches to Associations of Engineers and Scientists (CBIP: 1989). Some extracts feature in the next chapter. Here it is relevant to refer to one made in 1948. Referring to the 'magnificent cathedrals of Europe and to the great temples and mosques of India' as products of the 'faith of the builder', Nehru mentions that the new types of public works 'should also be fine and beautiful because of that faith' (CBIP 1989: 2e). Nehru's typification of a static society with administrators and the dynamic society with engineers is very much akin to Lenin's distinction between bureaucrats and technocrats. Lenin considered technocrats to be a major force in socialist reconstruction and on occasions exhorted the collective bodies to guard technocrats and specialists 'as the apple of their eyes' (Lenin 1965: 194).
6. The Bombay Plan of 1944, formulated by leading industrialists of that time, wanted the state to invest in the infrastructure sector, in the expansion of industries such as steel which required massive investments, and implicitly, if not explicitly, to offer them protection against foreign capital. In short, the state was to bear a large pro-

portion of the costs and risks of development but take much less of the surplus (Chaudhuri 1995: 97). Kohli (1989:56) uses this fact to dismiss the Marxist interpretation that the national movement was essentially a bourgeois movement as the latter would have meant free trade and minimal state intervention. However, Chatopadhay (1992: 147) argues that the Indian bourgeoisie 'fully aware ... of the enormous backwardness of the state of capitalism in the country...envisaged ...the active intervention by the state in the economy in the interests of the capitalist class as a whole.'

7. According to O'Connor (1989: 1–2), uneven development is usually defined in political, economic and sociological terms referring to uneven spatial distribution of industry, commerce, wealth, consumption, and labour relations. It could also imply polarities between industrial zones and those supplying raw materials, rich and poor countries, town and country.

8. Nehru's Presidential address to the Lahore Congress in December 1929. For details see Gopal (1975: 133–37).

9. For a critique of this three-fold classification of proprietary classes see Brass (1990: 272–76).

10. Industrial development also led to the growth of a working class. Inward-looking, the politics of this class, mobilised in a centralised unionised form, were directed more towards wresting concessions and benefits from the state due to 'their intuitive sense of being beneficiaries of development' (Kaviraj 1997).

11. For a political-economic analysis of the anti-poverty programmes see Vaidyanathan (1995). For an exhaustive analysis of the IRDP which takes into account various factors such as the scale of the operation, the targeting principles, replicability and participation, see Saith (1992: 93-97).

12. Some scholars attribute the failure of land reforms to provincial state-level politics and local leadership of the Congress Party whose support base was drawn from the rural landed gentry and which frustrated all 'radical moves' of the Indian state (see Brass 1990; Kaviraj 1997). Others such as Patnaik (1992: 186) argue that the 'bourgeois leadership of the freedom movement' sought to promote capitalist development 'on the basis of an agrarian structure which witnessed no significant reduction in the extent of land concentration.' Patnaik argues that the transformation of agriculture within the context of the capitalist development of the economy was characterised by 'landlord capitalism' in the countryside—i.e. erstwhile landlords and rich peasant tenants shifting to new methods of cultivation and making productive investment in land. For further discussion on the debates on the 'agrarian question' in India see Utsa Patnaik (1986). In the context of Asia see Byres (1991).

13. Referring to these programmes Gadgil & Guha (1995: 35) write, 'some offer straightforward patronage to the poor. Others provide subsidies for digging a well or purchasing a cow or a goat. Still others generate employment in constructing roads or digging trenches around forest plantations. A portion of the handout does reach the intended beneficiaries, but a substantial fraction is misappropriated by the political-bureaucratic machinery.'

14. For Kohli, regime type can be differentiated by leadership, ideology and organisation as well as the class basis of the regime in power. Comparing findings from three states, Kohli argues that a well organised, parliamentary communist regime in West Bengal has successfully initiated redistributive programmes. A populist leadership in Karnataka had limited success in implementing social reforms, and a faction government dominated by commercial peasant interests in Uttar Pradesh had no success in that endeavour (ibid.: 223).

15. In his later work on the crisis of governance in a democratic context, Kohli refers to this aspect. In his view, the emphasis of both structural-functional and Marxist scholars in explaining societal conflicts is on socio-economic forces; scholars in both these traditions tend to neglect the autonomous significance of political forces and consider them as dependent variables (1990: 28).

16. According to Patnaik, 'this squeezed real wages, even of the organised workers and attempted to sustain this squeeze through draconian anti-working class measures and anti-working class ordinances' (ibid.: 200).

17. Elsewhere she argues, 'the extraction of surplus has not only been in the form of extraction of surplus value from wagelabour, but also from women's unpaid domestic labour, serf labour and sheer "looting". Politically enforced pricing polices allowed the extraction of surplus from small commodity producers whose products were paid at low rates, and from the land itself through the underpricing or non-pricing of natural resources treated as "free goods". Surpluses extracted and accumulated through non-wage methods have been as important, if not more important, as surplus accumulation based on wage labour' (Omvedt 1992: 56).

18. The model rests on the power of an independent 'urban class/group' which engineers and benefits from the urban bias in development (see Lipton: 1978). For a discussion of the problems pertaining to measurability regarding 'bias', inconsistencies in defining what is 'urban class' and lack of evidence in demonstrating the political functioning of the urban class(es) as interest groups, see Toye (1993: 164–66). For a Marxian political economy critique of the model as 'neo-populism' see Byres (1979).

19. This characterisation echoes Lipton's urban–rural dichotomy. The division between 'India' and 'Bharat' implies the policy bias of the state for industry as against agriculture and for urban as against the rural sector. For an elaboration see Omvedt (1993).

20. This is the central argument of Sharad Joshi, the leader of the Shetakri Sangathana, the farmers' movement in Maharashtra. Also see Omvedt (1995).

21. At a theoretical level this tension between environmentalism and Marxism has been recognised by several Left-oriented scholars. As Ryle (1988:19) notes, 'the fundamental tension between socialist and ecological perspectives lies in the fact that while political ecology starts from the relation of humanity to nature, and with the general interest in ensuring that this relation is sustainable, socialism has concerned itself primarily with the distribution of power and wealth within societies.... The socialist analysis...discloses who benefits and who is oppressed and envisages the ending of that oppression'. The observations of Raymond Williams (1995: 508–9) are somewhat similar: '...the majority position among socialists has been that the answer to poverty, the sufficient and only answer is to increase production. [To them] Poverty had to be cured by more production as well as by the more specific policy of changing social and economic relations.'

22. Studies in the environmental genre, deeply permeated by the dependency theory and its variant world system model, highlight the class dimension in environmental resource use and distribution. Redclift (1987: 47) observes this fact in Latin American studies and cites Sunkel (1980: 47) in support: 'The surplus generated by the exploitation of nature allows an extremely favourable and pleasant artificial environment to be created for the middle and high income sectors, but for broader sections of population the results are fairly precarious.'

23. This statement is strikingly similar to one made by Sharad Joshi of the Shetkari Sangathana: 'Before independence it was the British government which took their raw materials for a song, processed the raw materials in London and Manchester and sold the finished product at enormous profit. Today, Pune, Calcutta and Bombay are the London and Manchester' (cited in Omvedt 1993: 112).

3

The Sardar Sarovar Project and State Conflict: Development on the Narmada

Introduction

From the time the idea of the Sardar Sarovar Project emerged to the time that the first bucket of concrete was poured, the making of the SSP has been marked by power struggles and conflicts. This chapter* maps the competing interests of major actors and examines the mechanisms of negotiation and conflict mediation that were activated in the context of a federal democratic polity. The conflicts among riparian states are especially important as they have implications for the shape of the SSP as well as for the socio-environmental movement around it. Covering three decades of the SSP, from 1950 to 1980, the central question guiding the analysis is: how do national and regional interests permeate the design and shape of the SSP? By mapping these issues and conflicts, this chapter anticipates some of their reverberations in the anti-dam campaign.

---

*The primary data are mainly drawn from project reports, government committee reports and recommendations, legal documents (affidavits and awards), and government communication (correspondence and gazette).

Harnessing Narmada Waters:
Preliminary Investigations

The Narmada river rises on the Amarkantak Plateau at the north-eastern apex of the Satpura mountain range at an altitude of 3,468 ft. It enters the Arabian Sea in the Gulf of Kambhat after traversing a course of 1,300 km in the states of Madhya Pradesh (MP), Maharashtra and Gujarat. During its course in MP, the Narmada has six falls[1] and passes through narrow gorges with large intervening areas of fertile plains. Its total catchment area is 98,800 sq. km of which 86,190 sq. km (87%) lie in MP, 1,538 sq. km (1.5%) in Maharashtra and 11,399 sq. km (11.5%) in Gujarat. The river forms the common boundary between MP and Maharashtra (35 km) and between Maharashtra and Gujarat (38 km).

The idea of harnessing the Narmada waters arose in colonial times. The First Irrigation Commission of 1901 refers to a proposal made in 1863 for the formation of joint stock companies for the construction of irrigation works and a dam across the Narmada in order to irrigate tracts of land between the rivers Tapti and Mahi. This was expected to be highly beneficial to the then province of Bombay and profitable to its promoters. The 1901 Commission observed that the 'question of the Nerbudda canal does not appear to have been considered...probably because it was realised that conditions were less favourable than the Tapti[2] which has been frequently under consideration' (cited in Paranjpye 1990:15).

It was not until 1947 that the first investigations on the Narmada were made by the Central Waterways, Irrigation and Navigation Commission (CWINC). Topographic and hydrological studies of the Narmada basin were carried out 'with flood control, power and extension of navigation as the general objectives in view' (NWRDC 1965:19). Based on these topographic studies seven sites were recommended—five on the Narmada and one each on the Tawa and Burhner rivers. A million kilowatts (KWs) of continuous power and an irrigated command of 1.6 million hectares (m. ha.) were expected from these projects.

The lack of human and material resources, however, necessitated investigation work in the first few years to be restricted to 'such projects as will give the maximum results in the shortest possible time' (NWRDC 1965: 20). Consequently in 1949, the Central Ministry of Works, Mines and Power cleared detailed investigations on four projects:

(*i*) Bargi project—three dams and a system of canals to command 0.7 m. ha. and generate firm power of 80,000 KWs; (*ii*) Tawa project—a 170 ft. dam to irrigate 0.5 m. ha. and 12,700 KWs of power; (*iii*) Punasa project—a 237 ft. dam for 223,000 KWs of power and for flood control in Broach district in Gujarat, Bombay province; and (*iv*) Broach project—a barrage and canal system for 0.3 m. ha. of irrigation command in Broach district based on regulated release from the Punasa project.

CWINC carried out investigations and prepared reports on these four projects. Renamed the Central Water and Power Commission (CWPC) in 1951, the erstwhile CWINC carried out studies into the hydroelectric potential of the Narmada in 1955. The report indicated that about 1.3 million KWs firm power could be generated at 16 sites on the Narmada and its tributaries, 14 of which were in MP, one (Gora) in Gujarat and one (Keli) on the common boundary between Gujarat and Maharashtra. Ten of these sites were on the main river, four on the tributaries and two on the irrigation canal from Bargi. The proposed sites on the Narmada are shown in Figure 3.1.

Starting in 1957 investigations were also carried out by the CWPC for irrigation projects at Bargi, Tawa, Punasa, Barwha and Kolar. A sixth site on the river Barna (a tributary of the Narmada) was also investigated. An estimated 1.85 m. ha. of total irrigation development was contemplated from these projects. Investigations were also undertaken into irrigation from the Hiranphal and Sitarewa projects.

The Broach Project:
The Making of the Sardar Sarovar

When the CWPC finalised its report on the Broach Project in 1956 it included the construction of a 160 ft. weir across the Narmada at Gora and a right bank canal to irrigate 0.5 m. ha. in a gross command area of 0.52 m. ha. Water requirements were estimated to be 2.8 million acre feet (MAF).[3] In 1957, the CWPC recommended that the site be shifted 1.5 miles upstream to

**Figure 3.1 Proposed Sites for Investigation in the Narmada Basin
by CWPC in 1956**

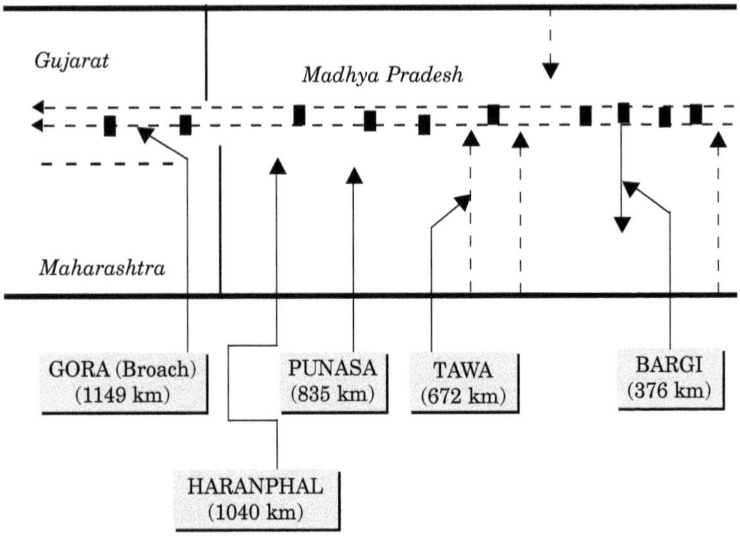

Source: CWPC 1956.

Key
Boundary between Bombay and MP (before 1960):   ▬

Boundary between Gujarat and Maharashtra (after 1960): - - - -

Narmada River: ◄ ⁻ ⁻

Tributaries: ↑

Bargi Irrigation Canal: ↓

Major sites (figures in brackets indicate distance from the ▬▬◗):

Project sites: ■ (from right to left): Burnher, Rosaria, Basania, Bargi (total three sites, two on its irrigation canal), Chinki, Sitarewa, Tawa (distance mentioned is one of confluence with the main river), Kolar, Hoshangabad, Punasa, Barwha, Hiranpal, Keli, Gora.

Navagam, with the provision of a high-level canal to irrigate a gross command of 0.3 m. ha. in Mahi and Sabarmati basins in Gujarat. This required 1.7 MAF of water.

The revised Broach Project thus contemplated irrigating a gross command area of 0.8 m. ha. utilising 4.5 MAF (2.8 + 1.7) of water. 'The implementation was contemplated in two stages. In

Stage-I, the height was restricted to 160 ft. with provision for wider foundations to enable raising the dam to 300 ft. in Stage-II. A high level canal was envisaged in Stage-II' (NWDT 1978a: 16). The revised Project with full reservoir level (FRL) of 300 ft. affected the proposed Keli dam between Hiranphal and Navagam, in view of which the idea of dam was dropped in 1959.[4]

Adding Hydropower to Irrigation Benefits

In 1959, further modifications to the revised Broach Project were suggested by the then state of Bombay and by a panel of consultants appointed by the central Ministry of Irrigation and Power. The former suggested raising the FRL from 300 to 320 ft. in Stage-II and the provision of powerhouses, one at the riverbed and one at the head of the canal. In their report submitted in April 1960 the panel recommended that the two stages be combined into one (FRL 320 ft.).

On 1 May 1960, the State of Bombay was bifurcated into the states of Maharashtra and Gujarat. The Navagam site came under the jurisdiction of Gujarat. In August 1960, the Planning Commission sanctioned Stage-I of the Broach Project, providing for the construction of a dam at Navagam with a height of 162 ft. with wider foundations for raising it to 320 ft. in Stage-II. Annual irrigation of 0.39 m. ha. was contemplated in Broach and Baroda districts with a low-level canal. The riverbed powerhouse was to be built in Stage-II, which envisaged:

> the raising of the dam to FRL 320 ft. Irrigation was proposed to be extended to an additional area of at least 9 lakh acres [0.36 million ha] pending investigations as recommended by the consultants for extending irrigation in North Gujarat including the Little Rann of Kutch by means of a high level canal off-taking with full supply level [FSL] 295 ft. Power to the extent of 625 MW at 60% load factor was also envisaged after Punasa and other upstream storage were constructed and after meeting the then anticipated needs of MP (NWDT 1978a:17).

On 5 April 1961, Jawaharlal Nehru inaugurated Stage-I of the Project. The construction of approach roads and bridges, project headquarters and staff quarters began and further investigations into dam foundations were taken up. Land in six villages in Kevadia (four completely and two partially) were acquired for construction of the project colonies and an estimated 4,000 people were displaced.

Adding Storage Benefits

Neither Stage-I (FRL 162 ft.) nor Stage-II (FRL 320 ft.) of the Project had any provision for storage capacity. Stage-I involved a barrage to provide seasonal irrigation. Stage-II involved raising the barrage to generate power and to extend irrigation to some areas beyond the Sabarmati river with a canal system 143 km long. In 1961, the Gujarat government undertook a study on how to utilise the flow of the Narmada in the catchment below Punasa in MP. Survey work related to the submergence and the command areas had been entrusted to the Survey of India in 1960. The surveys and fieldwork revealed that sizeable amounts of water could be stored in a terminal reservoir in conjunction with the Navagam dam. These studies showed that very large storage could be provided at Navagam reservoir if its height were to be raised to that of Hiranphal, to be constructed upstream of the Navagam site (see Figure 3.1). The desirability of raising the height of the dam became evident when water benefits to the drought-prone areas of Kutch, Saurashtra and north Gujarat were envisaged. As the Narmada Water Disputes Tribunal (NWDT) was to note later, 'the command area and the canal surveys as they progressed indicated larger potentiality for irrigation under the proposed high level canal. The studies indicated that a reservoir with 460 ft. would enable realisation of optimum benefits by utilising the flow below Punasa and would make it possible to extend irrigation to a further area of 20 lakh acres [0.8 m. ha.]' (NWDT 1978a: 17).

In August 1963, Gujarat proposed an FRL of 425 ft. for the Navagam dam. Instead of two canals as envisaged earlier, only a high-level canal at FSL 300 ft. was proposed. The gross water utilisation was estimated at 12 MAF (including 0.6 MAF for industrial and domestic use) and annual irrigation revised to 40 lakh acres (1.6 m. ha.).

The addition of features and benefits to the Broach Project had a direct bearing on the claims of the upstream state of MP over

Narmada waters, which had also undertaken periodic revisions in its estimates with regard to the utilisation of Narmada water. In 1961, during discussions with the CWPC on the Punasa project report, and subsequently in the preliminary report on the Barwha project, MP had estimated total water requirement at 7 MAF with a provision to irrigate 32 lakh acres (1.3 m. ha.). These figures were revised two years later in 1963: estimates of water requirements were increased to 12 MAF (including industrial and domestic uses) and the potential irrigated area in the Narmada basin in MP was increased to 46 lakh acres (1.85 m. ha.).

Federalism, Nationalism and River Valley Projects

Competing claims over the Narmada river waters by two newly formed states,[5] each ambitiously committed to developing irrigation and power infrastructure, set the conditions for a long-drawn-out inter-state dispute. Disputes between riparian states over river waters were in no way unique or recent. As the Second Irrigation Commission of 1972 was to note a few years later,

> inter-state disputes in the United States of America have kept jurists, courts and lawyers busy over the past seventy or eighty years and individual disputes have lasted forty to fifty years. [In India], the dispute between the sharing of Cauvery waters arose between the then provinces of Madras and Mysore as far back as 1909 and was ended by an agreement in 1924. The dispute relating to the Palar river between Madras and Mysore was once resolved in 1892 by an agreement only to be revived in 1927, 35 years later (RIC 1972a: 345).

In the 1960s, the states of Maharashtra, Andhra Pradesh, Karnataka and Tamil Nadu were in conflict over the sharing of waters of the river Krishna. The waters of the river Godavari were similarly contested by MP, Maharashtra, Karnataka, Andhra Pradesh and Orissa. River basin development for irrigation and power were constitutionally state subjects[6] and inter-state rivers became obvious sites of contest. Each state in the development race pushed for new river valley projects, even though old ones were incomplete and in some cases had barely begun.[7] The federal democratic structure underlying development planning enabled the playing-out of competing interests and claims. With the restructuring of federations in the late 1950s and early 1960s, expressing and creating strong 'sub-national' identities, these contentions became even stronger.

Although water was constitutionally a state subject, within the then prevalent discourse on nation-building, inter-state competition for resources and disputes over water were seen as an outcome of the pursuit of parochial and particularistic interests that went against the spirit of the nation and national good. Nationalistic rhetoric was used by none other than Nehru; his 'modern temples of India' were becoming enmeshed in particularism of a regional kind over which he did not seem to have much control. While laying the foundation stone for the Hirakud dam in Orissa in 1948, Nehru publicly indicted the 'provincialism' in river valley development:

I would like to point out that the Central government has such big schemes ready to be executed in different provinces of India. They have been prepared not on a provincial basis, but with the objective of benefiting the people of the country in general. It would be regrettable if such schemes led to provincial controversies. We should view these projects with a broad outlook and see whether these schemes benefit the people of the country in general or not (CBIP 1989:2).

When states cooperated on river valley projects, Nehru unequivocally acknowledged their efforts. Inaugurating the Parambikulam Aliyar project in 1961 he stated:

I am happy to be here for a number of reasons. One is that I am glad to be in river valley functions... [We] use the waters of our rivers for bringing good to the people.... There is nothing more important in India than the two purposes [of river valley projects]—irrigation and electric power. Therefore one of the chief functions of our governments today is production of irrigation and more and more electric power.... But above all I am happy to be here because this Parambikulam Aliyar project is the outcome of co-operation between two of our state governments and I would like to congratulate the government of Kerala and the government of Madras...on this co-operation.... Government of India functions because the states co-operate with each other and with the Government [of India]. Otherwise India would go to pieces (ibid.: 81).

Notwithstanding the Nehruvian vision of nation-building and progress in which states cooperate, pool resources and serve national interests, resource conflicts between states dominated the national agenda. The central government played the role of the 'umpire'. Its ministries facilitated negotiations and set up commissions and committees to arbitrate disputes. If these failed, resolution was sought through the Inter-State Water Disputes Act of 1956 which constitutionally provided for the formation of a tribunal by the central government to settle disputes related to water.

Negotiation and Central Government Mediation in the Narmada Project

The bone of contention between Gujarat and MP was the height of the Navagam dam. A higher dam implied a greater irrigation command for Gujarat and fewer prospects for dams downstream of Punasa for MP. It also involved the submergence of land in MP and Maharashtra. The apportionment of costs and benefits of the projects on the river contributed equally to the dispute. In November 1963, at a joint meeting mediated by the central Ministry of Irrigation and Power, the chief ministers of Gujarat and MP settled for a dam height of FRL 425 ft. for the Navagam project. The costs and power benefits from the Punasa dam set at FRL 850 ft. were to be shared by Gujarat and MP in the ratio of 1:2. Gujarat was also to provide soft-loan assistance to MP for the Bargi projects. The resolution, called the 'Bhopal Agreement', was shortlived. Immediately after the joint meeting, in a communiqué to the Minister of Irrigation and Power, the chief minister of MP contended that the height of the Navagam dam should not be constructed to more than FRL 162 ft. as that was the riverbed level at MP's border with Gujarat. In the note the chief minister reiterated MP's total claim to the Narmada within its boundaries over which it alone had the right to execute any alteration.

In September 1964, the Ministry of Irrigation and Power appointed the Narmada Water Resource Development Committee (NWRDC) headed by A.N. Khosla, then Governor of Orissa. While it may seem surprising that the central government chose to appoint the constitutional leader of a province to head a committee on Narmada waters, Khosla's antecedents dictated the choice. A renowned engineer, it was under his chairmanship that CWINC had first mooted the multipurpose river basin development of Narmada in 1946. He had also headed an ad-hoc committee in 1948 that had recommended priority investments in four projects—Punasa, Bargi, Broach and Tawa—in the Narmada basin. The terms of reference of the Khosla Committee framed by the Government of India in consultation with the three riparian states of MP, Gujarat and Maharashtra required: (*i*) designing a Master Plan for *optimum* and *integrated* development of the Narmada water; (*ii*) phasing its implementation for *maximum* development of the resources and other benefits; and (*iii*) the examination, in particular, of the Navagam and alternative projects if any, and determining the *optimum* reservoir level(s).

Large Projects: Maximisation of Gains and the Nehruvian Doubt

The felt need for maximising and optimising development gains, expressed in the semantics of the terms of reference of the Khosla Committee, dominated thinking at that time. Conditioned by resource constraints (e.g. finance, engineers), infrastructure constraints, and food and power shortages, policy objectives sought optimal and maximal outcomes from strategic choices and intervention designs. Insofar as large projects promised to fulfil these objectives, they became attractive choices.

To Nehru, in fact, large river valley projects were dream projects. He was a great believer in developing the irrigation and power sector. 'You may without knowing anything about a country blindfold state its advancement by finding out how much power it produces', he once stated. Yet it is unfair to ignore the doubts he raised over these policy objectives. He was aware of and worried about the 'disease of giganticism' that pervaded these sectors, projects which 'concentrated progress at one place' and the 'benefits of which did not reach the great majority of people' (Jawaharlal Nehru's speech at the CBIP general meeting, 1958. See CBIP 1989: 52).

> State governments are constantly pressing our government and our Planning Commission for various schemes—all big projects—and they have a right to do so. But this is all the relic of giganticism to which we have fallen prey. We have to realise that we can also meet our problems much more rapidly and efficiently by taking up a large number of small schemes especially when the time involved in a small scheme is much less and the results obtained are rapid (ibid.: 53).

His reflections on the effects of large projects, the comparative costs of big and small projects, the need to spread development across the country through small projects and by soliciting participation of the people, all seem to bring him close to Gandhian teachings. In a way, Nehru was the first critic of Nehruvian modernisation:

> I am not sure if the cost of production of power in a small project is great because the cost of a small project has to be judged after taking into account all the social upsets connected with enormous concentration of national energy... [and] upsets of the people moving out and their rehabilitation. Also it takes a long time to build a big project.... The social value of a vast number of small projects is much greater than that of one, two or three big projects.... In those small schemes you can get a good deal of what is called public co-operation...to get the public associated, to get their understanding and good-will (ibid.).

Notwithstanding these later critical reflections, Nehru accorded large river valley projects legitimacy as symbols of self-confidence and self-reliance of a newly independent and rapidly modernising country. The feeling of national pride which he evoked, particularly when such projects were completed, had a strong association with the belief that they would in due course serve national interests of industrialisation and food self-sufficiency.

The Narmada Dispute: Mapping National and State Interests

The Khosla Committee ignored Nehruvian doubts in its approach to deliberations on the Narmada waters. Instead, it shared Nehru's passion for a nationalistic framing of such projects and reiterated the need to derive maximum benefits from river valley development. 'National interest should have over-riding priority. The plan should therefore provide for maximum benefits in respect of irrigation, power generation, flood control, navigation, etc. irrespective of state boundaries' (NWRDC 1965). National interests implied food security (hence irrigation and flood control) and accelerated industrialisation (hence power generation) on the one hand, and strategic interests and national defence on the other.

NATIONAL SECURITY AND THE NAVAGAM DAM: The Khosla Committee was set up at a time when India was negotiating a period of political crises. The showdown with China in 1962 had been costly and in 1963–64 agricultural production had fallen short due to severe drought. Nehru's death in mid-1964 also generated security concerns.

During the Committee's deliberations an important factor was the provision of irrigation to arid areas around the international border with Pakistan in the Rann of Kutch (Gujarat) and in Rajasthan. Gujarat had proposed that the Navagam canal should have FSL 300 ft. at the head so that it could reclaim and irrigate land at the Little Rann of Kutch.[8] The idea was to resettle 'retired soldiers and sturdy peasants' in Kutch bordering Pakistan. The then Secretary of Irrigation, and later Chief Secretary to the government of Gujarat, Lalit Dalal, staked his claim to this idea and drew parallels with Israel's 'agricultural kibbutz manned by reserve soldiers with full equipment and sophisticated weapons to keep a watch on the activities of its unfriendly neighbours' (GoG 1991:1):

I thought that we could follow the Israel model and establish our retired soldiers...as...India cannot financially afford a standing army in Kutch and this was the most practical and economic way of guarding our border with Pakistan. I added my suggestion to our case which was presented to Khosla. [He] visited Gujarat in April 1965 and was specially impressed by the suggestion made by me for developing Kutch and establishing out-posts on the border (ibid.:1–2).

The idea of using irrigation projects for defence purposes dates back to the Mughal era. Dalal's claim of conceiving the idea is questionable. Even before Partition, proposals had been made for strategic reasons to bring water from the river Indus to the Great Rann through a feeder canal from the Hajipur Barrage across the lower Indus (RIC 1972b: 104). Dalal's claim is also unreliable because in December 1964, in drawing up the Master Plan for Narmada development, the Khosla Committee had already noted that irrigation should be extended 'in particular to the arid areas along the international border with Pakistan to encourage sturdy peasants to settle in these border areas' (cited in NWDT 1978a: 20; also NWRDC 1965: 210).

It is beyond doubt, however, that in deciding the size of the Navagam dam and the irrigation canal, the Khosla Committee accorded high priority to national security aspects. While the Gujarat government had proposed irrigating only the Little Rann, the Committee suggested that land reclamation and irrigation in the Great Rann of Kutch be included[9] (NWRDC 1965: 211). It should be mentioned here that during the deliberations of the Khosla Committee, border tensions between India and Pakistan were high and the Indo–Pakistan war in 1965 began in the Rann of Kutch.

GUJARAT'S INTERESTS AND THE NAVAGAM DAM: For Gujarat, national security interests and the proposal to take water to Kutch augured well for its claims to Narmada water and the height of the Navagam dam. In its representation to the Committee in 1965, Gujarat indicated that it would require about 13 MAF of water to irrigate 1.4 m. ha. (35 lakh acres). Following a suggestion by the Committee, it added 2.45 MAF to irrigate 0.2 m. ha. (4.5 lakh acres) in the Great Rann of Kutch. A further 1.9 MAF were added to irrigate 0.25 m. ha. (6.5 lakh acres) of land that was initially proposed to be irrigated by the Kadana canal on the river Mahi. This proposal seems to have emerged during discussions with the Committee which thought that water harvesting from Mahi by the Kadana dam should form an 'integrative' system with upcoming projects (Banswara, Anas and Baneshwar) in Rajasthan and that the costs and benefits should be shared by Gujarat and Rajasthan. Thus, by 1965 Gujarat's total claim for water from the Narmada had risen to 17.55 MAF to irrigate 1.8 m. ha. (45.81 lakh acres); in 1956 the figures had been 2.8 MAF to irrigate 0.45 m. ha. (10.97 lakh acres). Accordingly, Gujarat proposed a height of 490 sq.ft. for the Navagam dam (the site having been shifted 6 km upstream).[10]

MP's INTERESTS AND THE PROPOSED JALSINDHI PROJECT: Like Gujarat, MP had revised its estimated water requirements and irrigation acreage. In its proposals to the Committee in 1965, it mentioned an area of 3.1 m. ha. (77.5 lakh acres) for irrigation; in 1961 the figure had been 1.3 m. ha. (32.18 lakh acres). Total water requirements were similarly revised from 7 MAF in 1961 to 23.75 MAF in 1965. Of significance here is the joint proposal by MP and Maharashtra to construct a hydroelectric project at Jalsindhi on the MP–Maharashtra border between Navagam and Hiranphal, about 1,095 km from the source of the river (55 km downstream of the Hiranphal site). The Jalsindhi site had featured in MP's Master Plan, but no detailed study had been made of the project. During the deliberations of the Khosla Committee, in April 1965, the MP and Maharashtra governments agreed 'to co-operate in development of hydro-electric power at Jalsindhi sharing net costs and benefits' (GoMP 1970). This agreement provided sufficient flexibility for the scope and design of the project to be changed if the two states acquiesced.

MAHARASHTRA AND RAJASTHAN AS INTERESTED PARTIES: In the proposal submitted to the Khosla Committee by MP and Maharashtra, the height of the Jalsindhi dam was FRL 355 ft. and the Hiranphal dam height was FRL 465 ft. (NWDT 1978b: 73). Maharashtra entered the conflict as an interested party wanting power benefits (mainly from the Jalsindhi project) of at least

1,000 MW[11] as also some irrigation benefits. Maharashtra's interest in a lower dam and a lower canal at Navagam was to make the Jalsindhi dam feasible.[12] It therefore backed the MP government's proposal to keep the FSL of the Navagam irrigation canal at 185–190 ft. (and not 300 ft. as desired by Gujarat) and the height of the Navagam dam at FRL 210 ft. (and not 490 ft. as proposed by Gujarat). Although Rajasthan is not a riparian state, feasibility studies had indicated the possibility of taking water from the Narmada up to south Rajasthan. It therefore supported a higher Navagam dam seeking to irrigate 0.45 m. ha. (11.5 lakh acres) of desert area on the Rajasthan–Pakistan border.

The National Water Resources Development Committee

The Khosla Committee's recommendations submitted to the Government of India, Ministry of Irrigation and Power, were as follows.

1. *Priority for Irrigation:* Irrigation should receive the highest priority because of food shortages and the heavy drain of foreign exchange involved in importing foodgrain; in the interests of national security, it stressed the need to take water to the rainless and uninhabited areas of Kutch and Rajasthan bordering Pakistan.
2. *Allocation of Water:* Table 3.1 shows the allocation of Narmada water proposed by the Khosla Committee compared to the demands of the respective states for consumptive use.

Table 3.1 **Comparison of Demands and Allotment**

States	Irrigation (in lakh acres) demanded	Irrigation (in lakh acres) allotted	Water (in MAF) demanded	Water (in MAF) allotted
Madhya Pradesh	77.50	65.00	23.75	15.60
Gujarat	45.81	45.81	17.55	10.65[13]
Maharashtra	0.10	0.10	0.10	0.10
Rajasthan	8.50	1.00	-	0.25

Source: NWDT (1978a:19).

3. *Irrigation Canal:* The full supply level of the Navagam canal should be 300 ft. The Committee rejected Maharashtra's and

MP's claims that a lower FSL (185–190) could serve the irrigation requirements of Gujarat and Rajasthan through lift irrigation on the grounds that power required for lift irrigation (about 200 MW) was unavailable and the total cost of pumping water would be enormous (it would have to include pumping installations, pipelines, etc.). Hence, flow irrigation with canal FSL at 300 ft. was the best option.

4. *Height of the Dam:* It was suggested that the height of the Navagam dam should be FRL 500 ft., Navagam being considered the best location for the terminal dam. This height was considered the optimal choice after costs and benefits were compared with those of Jalsindhi and Hiranphal dams at similar heights. The argument was that storage capacity at Navagam would permit greater carryover capacity, increased power production, and assured optimum irrigation and flood control, and would minimise the wastage of water to the sea. The installed capacity of the riverbed power station should be 1,000 MW and canal power station 240 MW (together with stand-by units, total capacity should be 1,400 MW).

5. *Benefits:* The specific benefits from the Navagam dam were envisaged to be:
   (*i*) Irrigation—1.88 m. ha. (46.80 lakh acres) of which Gujarat's share would be 1.84 m. ha. (45.80 lakh acre) and Rajasthan's share 0.04 m. ha. (1.00 lakh acre);
   (*ii*) Power generation—951 MW to be shared by MP, Maharashtra and Gujarat in the ratio of 2.5:1:1 respectively. The cost of the power component of the project should be shared by the three states proportionately.
   (*iii*) Flood control—'The severe floods which periodically caused heavy damage in the Broach district of Gujarat will cease to be a problem' (NWRDC 1965: 210).

6. *Other Projects in MP:* The 12 other major projects in MP (Hiranphal and Jalsindhi were dropped) would together produce 1,063 MW of power at 60% load factor and irrigate 1.0 m. ha. (25 lakh acres). An additional 1.6 m. ha. (40 lakh acres) were to be brought under irrigation in MP through medium and minor projects totalling 2.6 m. ha. (65 lakh acres).

The 'National' Recommendation: A Perspective

It is not difficult to see that the Khosla Committee's recommendations matched well with Gujarat's interests. One explanation could be that national interests coincided with those of Gujarat. From a national perspective, the Committee's recommendations could be viewed as a formula for the optimal and equitable distribution of water (among contending states), taking into consideration national interests of food and power shortage as well as strategic interests. For Gujarat, a high-storage dam at Navagam implied a substantial reduction in water scarcity in the state. It could make productive use of this water with the high-level canal (almost an artificial river) running through the heart of Gujarat right up to south Rajasthan, with no burden of displacement and rehabilitation[14] as the villages to be submerged were largely in MP. Conversely, MP failed to convince the National Committee on Water Resources regarding the Jalsindhi project and a reduced height for the Navagam dam.

Notwithstanding our commitment to balanced regional development, socio-economic disparities among states in terms of resources they commanded indicate the relative power that states have vis-à-vis the national development agenda. If development indicators of the 1960s are examined in agriculture, industry and infrastructure sectors, Gujarat appears as one of the most prosperous states whereas MP one of the least favoured. Political indicators—sub-national identities and the relative power of the elite in national politics—show similar differences between the resources commanded by the two states. In terms of ranking (per capita net domestic product), Gujarat ranked 3rd in 1960–61, 3rd in 1964–65 and 2nd in 1969–70. The corresponding position for MP was 11th (1960–61), 11th (1964–65) and 12th (1969–70). Between 1960 and 1970 Gujarat featured in the first five high-income states whereas MP featured in the last five low-income states (Bharadwaj 1995: 200). The per capita income in Gujarat in 1970–71 was Rs 845, whereas in MP it was Rs 489. In Bhalla and Alagh's (1979) estimate of the growth of agricultural output and yield per hectare (period 1962–70), Gujarat ranked 5th and MP 13th among the states. Some relevant indicators are shown in Table 3.2.

Table 3.2 Comparison of Some Development Indicators in the 1960s

Indicators (1970–71 figures except where otherwise mentioned)	Madhya Pradesh	Gujarat
Total population (in millions)	41.6	26.7
Total cropped area (in thousand ha)	20001	10420*
Irrigated area as % of cropped area	8.2	14.6
Installed generation capacity (public utilities) (in MW)	727	907
Power generation (in million Kwh)	2754	4176
Energy consumption (in million Kwh)	1883	3322
Per capita energy consumption (in Kwh)	45.2	124.5
Energisation pump sets/tube wells (1968–69)	25000	42000
Installation of diesel pump sets (1968–69)	25000	215000
Tractors (per one lakh ha of cropped area) (1960–61)	12	37
Chemical fertiliser applied per ha (in kg)	0.58	3.19

*1968–69 figure.
Sources: NCAER (1967), CMIE (1992), RIC (1972b).

Underlying these differences are some socio-historical correlates that influenced the economic and political resource endowments of the two states. The cities of Surat and Ahmedabad in Gujarat are important business and trading centres, the former noted for its prosperous merchant class engaged in textile and indigo trade as early as the 17th century (see Das Gupta 1975 cited in G. Shah 1990). Ahmedabad's ascendancy as an economic centre was linked to the dramatic increase in cotton prices during the American Civil War in the 1860s. Choksey (1958) notes the steady growth in cotton cultivation and other commercial crops—tobacco and groundnut—which gradually displaced food crops such as rice and jowar. The first textile mill in Ahmedabad was set up in 1861; by 1899 the number of textile mills had risen to 26 and then to 60 in 1947 (G. Shah 1990). The rise to eminence of the Kanbi Patidars—an influential peasant caste constituting 12% of Gujarat's population—affected political and economic configurations in the region. As revenue collectors they had enjoyed some political power by exacting taxes from the peasants, and their influence increased with the growth of commercial agriculture in which they were directly engaged (A.M. Shah 1964, cited in G. Shah: 70). By the early 20th century, their influence had spread into the hinterlands of Kheda (central Gujarat), Bharuch (south Gujarat), Sabarkantha and Banaskantha (north Gujarat) through the purchase of land from adivasi and koli peasants (ibid.: 71). The Patidars constituted the new agrarian elite[15] in Gujarat, undermining the Rajputs with whom, even today, they vie for political supremacy in Gujarat politics (see Kohli 1990). The industry sector, however, was dominated by Vanias, although the industrious among the Patidars did venture into textile and agro-industries.

At Independence, therefore, an industrial class and a class of rich farmers had already emerged in Gujarat. Politically influential,[16] the demands for infrastructure development—electric power and irrigation—had been periodically put forward by this class in the 1950s and 1960s (see G. Shah 1990). Gujarat's demand for Narmada waters thus has to be seen against this political-economic backdrop, which also indicates who the major beneficiaries of the dam were likely to be. Concomitantly, given the regional variations in resource endowments and development between 'mainland' Gujarat and the Kathiawad region comprising Saurashtra and Kutch, the proposal to supply water to the latter also fostered a reconstituted Gujarati identity, a political requisite for a newly formed state.

The differences in political economy are striking when we compare MP with Gujarat. In colonial times, 'the choppy nature of its terrain, its lack of political unity and its inland location made MP comparatively uninteresting to British traders and builders' (J. Adams 1984: 37). The formation of the state in 1956 was an amalgamation of sorts: 17 districts of Maha Koshal, two of Bhopal, 16 of Madhya Bharat and eight of Vindhya Pradesh were put together to form Madhya Pradesh. On its formation, the state had a little less than 40% of its area under forest cover and a significant population, one-third of the total, of adivasis and Dalits (officially Scheduled Tribes and Scheduled Castes). The diversity that marked the population of MP was so great that a leading political activist wrote this about its character three decades later:

The Madhya Pradesh of today is an unnatural creation of state power, designed to control and administer the regional entities of Malvi, Bundeli, Bagheli and Chattisgarhi nationalities. This is why in the field of democratic process, administration, agricultural development, the distribution of agricultural produce, the development of rural industries, the proper utilisation of labour power of the people, scientific planning and all round progress, education, health and cultural development, the state has proved itself to be a conspicuous failure (Niyogi 1982: 111).

The Khosla Committee's recommendations can be seen to have been influenced by the political and economic resourcefulness of Gujarat and by the established doctrines of maximisation and optimisation that constituted planners' views on development policies and projects on the one hand, and the strategic need to take water to the Great Rann of Kutch and the border areas of Rajasthan on the other. The political-economic endowments weighed heavily in favour of Gujarat, and the likelihood of a causal connection between those endowments and the capacity to influence decision-making processes at the national level was based on the premise that political and economic 'power' is necessary if not sufficient to influence national interest.

The Struggle in the 1970s

If MP could not push through its proposals before the Khosla Committee it was able to successfully keep the controversy alive thanks to the support it received from Maharashtra. While the Gujarat government welcomed the Khosla Committee's recommendations, MP and Maharashtra rejected them. The disagreement concerned mainly those proposals regarding the development of the lower Narmada reach and the height of the Navagam dam. The comments made by the MP government on the Khosla report are worth noting:

> In all fairness, if for some sound reason Navagam dam must be built to such a height as would submerge one or more power houses proposed in MP, the latter is entitled to receive from the power to be developed at Navagam and the canal, the full quantum of power that MP would have generated in its own territory...at a cost no higher than that at which it would have generated this power in its own territory (cited in NWDT 1978a: 20).

Madhya Pradesh disagreed on the amount of water allocated
to it.[17] It also raised doubts about the hydrological estimates
(quantity of annual utilisable water flow in Narmada at different
sites) by arguing that the Khosla Committee had overestimated
the water flows. Those hydrological estimates, as we shall see
in Chapter 5, were later to become the subject of conflicting
claims. However, the major contention was about the height of
the Navagam dam and the compensation that MP expected in
the event that it would have to sacrifice the Jalsindhi and
Hiranphal projects. With no agreement on the height of the
Navagam dam or on compensation, MP and Maharashtra went
ahead with the Jalsindhi project; initial construction work including approach roads commenced on the dam site.

Gujarat and MP continued to struggle over the Navagam
project through 1966 and 1967. At least four rounds of discussion
between the chief ministers of the contending states, mediated
by the central government, failed to produce any settlement regarding the distribution, control and use of Narmada waters
and the height of the Navagam dam.[18] In July 1968 the government of Gujarat submitted a complaint to the Government of
India, asking for the appointment of a Tribunal under the Inter-
State Water Disputes Act 1956. A year later, in October 1969, the
Narmada Water Disputes Tribunal (NWDT) was formed for adjudication of the dispute.

Party Politics and the Narmada Issue

The 1970s experienced several political developments in the country, some of which directly influenced the inter-state Narmada
conflict and attempts to resolve it. The first major split of the
Indian National Congress in August 1969 into the Congress (-O)
and Congress (-R) affected the governments of Gujarat and MP.
While the Congress-O controlled the party and government in
Gujarat, the Congress-R led by Indira Gandhi controlled MP. In
May 1971, the Congress-O ministry of Hitendra Desai in Gujarat
was toppled; after 10 months of President's rule, a state election
in March 1972 led to a resounding victory for the Congress-R.
By 1972, Indira Gandhi had become the undisputed leader of the
new Congress and her party ruled the four states involved in

the struggle over Narmada waters. She also introduced a kind of politics in which the unitary traits of the Indian Constitution (as opposed to its federal traits) came into full play for the first time and on a continuous basis.

The effects of these politics on the Narmada issue were immediate. The proceedings of the NWDT were suspended through an agreement reached by the chief ministers of the four states in July 1972.[19] A section of the agreement is worth noting.

> The Chief Ministers of the States concerned feel that the development of Narmada should no longer be delayed in the best regional and national interests and therefore agree to the settlement of disputes connected with this river by mutual agreement and with the assistance of the Prime Minister of India. The quantity of water in Narmada (at 75% dependability) is assessed at 28 million acre feet. The requirements of Maharashtra and Rajasthan for use in their territories are 0.25 and 0.5 million acre feet respectively .... *The Prime Minister of India is requested to allocate the balance of water* (27.25 million acre feet) between states of Madhya Pradesh and Gujarat taking into account the various relevant features in both the states. The various view points with regard to the height of Navagam Dam would be gone into and *a suitable height may also be fixed by the Prime Minister of India* (NWDT 1979 Exhibit C-1: 60, emphasis added).

The chief ministers' agreement was shortlived. In 1973–74 the Congress-R government in Gujarat faced one crisis after another. Two Congress-R chief ministers had to quit office in quick succession. The state witnessed the Nav Nirman Andolan—an urban agitation against corruption and rising prices—as well as rural agitation led by the Patidars who opposed the Levy Act 1973[20] and the amended Land Ceiling Act 1974.[21] Congress-R rule in Gujarat lasted less than two years. Its victory in March 1972 ended abruptly with the imposition of President's rule in the state in February 1974.

The Narmada Factor in the Congress-R Debacle in Gujarat

It may seem farfetched, but the crises in Gujarat during this period were linked to the Narmada dispute. When Chimanbhai Patel replaced Ghanshambhai Oza (the signatory to the Inter-State Agreement in 1972) as Chief Minister in July 1973, he attempted to unlock the Narmada dispute from the NWDT. An economist and Patidar with a mass base among affluent farmers in Gujarat, Chimanbhai Patel 'dreamed of allocating the river's huge hydroelectric and irrigation potential for planned industrial and agricultural development in Gujarat. He was prepared to offer the MP government a package of financial compensation for submerged land and shares in water and power so generous that he did not see how they could possibly refuse' (Jones & Jones 1976: 1018).[22] To do so, however, he needed the approval and support of the central Congress-R government.

Patel thus decided to finance the Uttar Pradesh (UP) state elections scheduled for February 1974.[23] He mobilised money from the groundnut lobby—farmers, merchants and millers—promising to deregulate the sector from government control. This led to large-scale exports of groundnut oil from Gujarat. As prices rose (despite a bumper harvest of groundnut that season), the oil lobby reaped huge profits at the cost of consumers (groundnut prices doubled in October 1973), setting the stage for the Nav Nirman movement against price rises and corruption, and subsequently leading to the resignation of the Chief Minister.[24] Patel's calculations backfired although the Congress-R managed to narrowly win the UP state elections.

When fresh elections became imminent in Gujarat with the dissolution of the assembly in March 1974, resolution of the Narmada conflict lost priority. The Congress-R played safe and preferred to postpone any resolution lest opposing political parties made use of the issue in the elections. Accordingly, in July 1974, yet another agreement was signed by the four states—this time Gujarat was represented through an advisor to the Governor[25]—asking the Tribunal to resume hearings.

> That the Tribunal do allocate ... 27.25 million acre feet between MP and Gujarat ... that the height of Navagam Dam

be fixed by the Tribunal ... [and] that the level of the canal be fixed by the Tribunal after taking into consideration various contentions and submissions of the parties hereto ... (NWDT 1979 Exhibit C-1: 61).

Narmada Water Disputes Tribunal: Issues and Responses

Effectively, the NWDT began its work in 1975. Gujarat staked claim to an irrigation command area of 71 lakh acres (2.8 m. ha.) whereas MP claimed a command area of 76 lakh acres (3 m. ha.).

Between 1972 and 1977, Madhya Pradesh made major changes in the Hiranphal and Jalsindhi projects. In 1972 it had proposed an FRL of 455 ft. for Hiranphal dam[26] and FRL 355 ft. for Jalsindhi dam. Later that year, however, the height of the Hiranphal project was reduced to FRL 420 ft., 'after having considered representations and agitation in the submergence zone where land was thickly populated and very fertile' (GoMP 1975). In 1977, MP abandoned the Hiranphal project and decided to increase the height of the Jalsindhi dam to FRL 420 ft.

The arguments put before the Tribunal by MP were as follows:

(a) that Narmada was essentially an MP river and therefore the state had a right to that part of the river which flowed through its territory. This implied that Gujarat could not affect the river flowing in MP and hence had to be content with a dam height of FRL 210 ft. at Navagam;

(b) that the right to use the river's resources belonged to the people of the river basin and hence no extra-basin population should claim benefits from it. This implied that Gujarat could not undertake inter-basin transfers through its proposed canal at FSL 300 ft.;

(c) that hence, the proposed Navagam canal should be at FSL 190 ft. as any increase in the full supply level (roughly a measure of height and gradient of the canal) would entail a simultaneous increase in the height of Navagam dam beyond FRL 210 ft.; and

(d) that if the Navagam dam's height was to be taken to a height that would affect the Jalsindhi power project, there should be full restitution of power benefits forfeited at Jalsindhi. This implied that Gujarat would have to guarantee either power or monetary compensation.

Maharashtra endorsed these arguments in separate submissions and sought compensation through (d) from Gujarat. Gujarat on its part demanded that the offtake of the Navagam canal from the reservoir should be at FSL 320 ft. and the height of the Navagam dam at FRL 530 ft. Compared to its proposals to the Khosla Committee, Gujarat's representation in the NWDT entailed an increase of 20 sq. ft. for the canal (FSL) and 40 sq.ft. for the dam (FRL). It received Rajasthan's tacit support of its claims.

Considerations and Award of the NWDT

The NWDT made its final award in a detailed report in 1979. The major considerations and decisions made by the Tribunal were:

1. *Hydrology*—Following the agreement of the chief ministers in 1972 and in 1974, the NWDT accepted that the quantity of Narmada water that could be utilised annually at 75% dependability is 28 MAF, and that the share of Maharashtra and Rajasthan would be 0.25 and 0.5 MAF respectively. The task before it was thus to allocate the remaining 27.25 MAF between Gujarat and MP.[27]
2. *Irrigation Command*—The NWDT estimated the irrigation command of Gujarat to be 2 m. ha. (50 lakh acres). It therefore rejected the idea of the Khosla Committee to take water to the Great Rann, arguing that this would be four times more expensive and impractical.[28] The Tribunal also disagreed with the Khosla Committee's recommendation that the irrigation command of the Mahi–Kadana project should be included in the Navagam command (as waters from the Mahi–Kadana project were supposed to be transferred to the border areas of Rajasthan) and therefore excluded 0.6 m. ha. from the Narmada command. For MP, the NWDT estimated an irrigation command of 2.6 m. ha. (68 lakh acres). In estimating the cultivable command area, the Tribunal gave weight to drought conditions in the two states, using data

provided by the Second Irrigation Commission (1972) to this effect. The latter had estimated that in Gujarat 60 *talukas* (sub-districts) were drought-affected, covering a geographical area of 7 m. ha., compared to MP's 24 *teshsils* (sub-districts in MP) covering 4 m. ha.[29]

3. *Water for Irrigation and Other Uses*—The NWDT estimated the water requirements of the two states for irrigation as well as for industrial and municipal uses as follows: 11.7 MAF for Gujarat and 19.4 for Madhya Pradesh, totalling 31.1 MAF, or 3.85 MAF more than the figure that it had accepted for allotment (27.25 MAF). It did not, however, consider water requirements as the sole criterion for apportionment; other relevant criteria were also accorded weight.[30]

4. *Criteria for Allotment*—The Tribunal rejected MP's arguments that the principle of territorial sovereignty should be followed in the allotment of water because 'flowing water is *publici juris*'. It also rejected the contention of both MP and Maharashtra that inter-basin transfer should not be allowed, arguing that the unit of analysis should not be the river basin but the entire state, including all its inhabitants both within and outside the river basin. In other words, while rejecting the criterion of state territoriality in the first instance, the NWDT endorsed it in the second instance. Both decisions favoured Gujarat. However, the Tribunal gave weight to the drainage area, 87% of which is in MP.

5. *Allotment*—The Tribunal awarded 18.25 MAF of Narmada water to MP and 9 MAF to Gujarat. The ratio between the four states of MP, Gujarat, Rajasthan and Maharashtra worked out at 65:32:2:1 respectively. The Tribunal further directed that because 28 MAF was an agreed figure and actual utilisable flow every year would either exceed or fall short of that figure, the same ratio should be applied for sharing of water in years of excess or shortage. The Tribunal further ordered that allocation would be subject to review only after a period of 45 years (i.e. 2024).

6. *Full Supply Level of the Navagam Canal*—In fixing the FSL of the Navagam canal, the NWDT rejected the demands of MP and Maharashtra that it be no more than FSL 190 ft. Both these states held fast to their position that a higher canal offtake implied a higher reservoir, which in turn implied

a higher Navagam dam, which would affect the Jalsindhi hydropower project. They had also proposed that 0.5 MAF of Narmada water could be made available to Rajasthan with a lower-canal FSL through lift irrigation.

The Tribunal rejected lift irrigation on the grounds that it was an expensive proposition, 40% more power being required to lift water to a given height than can be generated by the same quantity of water dropped from the same height (Paranjpye 1990: 43). While a low-canal level at Navagam enabled more power generation at Jalsindhi project upstream, the power expended to pump water towards meeting Rajasthan's requirements would be much higher. *Ipso facto*, the Tribunal ordered a high-level canal of FRL 300 ft., giving primacy to flow irrigation over lift irrigation.[31]

7. *Height of the Navagam Dam*—Once the canal's FSL was fixed at 300 ft. it became obvious that a higher dam would be required than had been argued for by MP and Maharashtra. The Tribunal fixed the Navagam dam height at FRL 455 ft. Table 3.3 lists the major features associated with the height of the dam.

8. *Alternatives*—In determining the height of the Navagam dam, the Tribunal considered various alternatives and their benefits and costs. The major considerations were power benefits, irrigation benefits and the area coming under submergence. Three major alternatives are outlined here:

Alternative 1: The Tribunal compared a Navagam dam with FRL 455 ft. with a Jalsindhi dam with FRL 420 ft. and a Navagam dam with FRL 210 ft.—the proposal submitted by MP and Maharashtra. While land submergence in the second proposal was low (36,550 acres) compared to the first (91,550 acres), benefits in the second proposal were equally low and therefore less desirable. The NWDT attached little weight to submergence in MP due to the high Navagam dam (FRL 455 ft.): MP had submitted high Jalsindhi/Hiranphal projects to the Khosla Committee (FRL 465 ft.) and to the Tribunal (FRL 455 ft. and then FRL 420 ft.) which entailed the submergence of almost the same amount of land.

Alternative 2: The Tribunal also compared the Navagam dam (FRL 455 ft.) with a Jalsindhi dam of FRL 420 ft. and a Navagam dam with FRL 309 ft. (allowing the FSL of the

Table 3.3 Considerations of Dam Heights and Reservoir Levels by the NWDT

Level	Purpose	Height (in ft.)	Cumulative height	Gross Water Storage corresponding to height
Full Supply Level	Canal Offtake	300	300	–
Reservoir Dead Storage Level	Irrigation	+7 to the FSL for head regulator and losses in approach tunnel and channels	307	1.68 MAF
Minimum Draw Down Level (Dead Storage Level for Power)	Power Generation	–	363	2.97 MAF
Reservoir Level	Live Storage Capacity for irrigation (1.39 MAF) + Space for Silt Deposition in Live Storage (0.30 MAF) + Carryover Storage[32] (2.81 MAF) = 4.5 MAF	+ 129	436	6.18 MAF (1.68 dead storage + 4.50)
Reservoir Level	Power generation[33] compensating for power loss in Jalsindhi project	+ 17	453	7.47 MAF (2.97 dead storage + 4.50)
Full Reservoir Level (FRL)	Increase Carryover capacity (0.20 MAF) (also implied increase in live storage, i.e. 4.50 MAF + 0.20 MAF = 4.70 MAF)	+ 2455		7.67 MAF (7.47 gross + 0.20)
High Flood Level	Flood Cushion	+ 5460	–	

Source: FMG (1995); NWDT (1978b); Paranjpye (1990); U. Shah (1994).

Navagam canal to be at 300 ft.). It noted that while land
submergence was more or less the same in both proposals,
the second was less efficient (allowing for more water wastage to the sea) and unsatisfactory (in providing release to
the Navagam canal at FSL 300 ft.).

Alternative 3: A comparison between Navagam dam at FRL
455 ft. and Jalsindhi dam of similar height was undertaken
to calculate how much power MP and Maharashtra would
lose in a high Navagam project. Assuming the FSL of
Navagam canal at 300 ft., the Tribunal noted that this requirement would entail a reduction in the power production
potential of the Jalsindhi project as the latter would have to
be planned (*a*) with a higher tail level to enable the Navagam
canal's FSL to be at 300 ft., thereby (*b*) reducing the head
available for power production (NWDT 1978b: 74).

9. *Benefits and Costs*—'Alternative 3' considered by the Tribunal was primarily to assess how much power MP and
   Maharashtra would have to forego in the absence of the
   Jalsindhi project—the NWDT estimated the loss at about
   84% of the power generated from Navagam dam at FRL 455
   ft. Accordingly, it allotted 84% of the power from Navagam
   to MP and Maharashtra, to be shared in the proportion agreed
   upon in the Jalsindhi case, and the remaining 16% to Gujarat.
   Thus, out of the net power produced in Navagam canal head
   and riverbed powerhouses on any day, MP's share was 57%,
   Maharashtra's 27% and Gujarat's 16%.

   The Tribunal noted that the capital costs of the irrigation
   and power components of the Navagam multipurpose
   project—by now renamed the Sardar Sarovar Project—would
   be in the ratio of 43.9 (irrigation) and 56.1 (power). For the
   power component of the project (riverbed powerhouse 750
   MW and canal head powerhouse 375 MW), MP and
   Maharashtra would have to pay Gujarat 57% and 27% respectively of the capital cost as well as operation and maintenance
   costs each year (NWDT 1978b: 77). The Tribunal also noted
   that, as the irrigation and power benefits of the Navagam
   dam project depended on regulated release from upstream
   projects, particularly Punasa project, which later came to be
   known as the Narmada Sagar Project (hereafter NSP),
   Gujarat was to bear 17.63% of the cost of the NSP. Under the

Tribunal's directive, 'Madhya Pradesh shall take up and complete the construction of the Narmada Sagar dam with FRL 860 ft. concurrently with or earlier than the construction of Sardar Sarovar dam' (GoI 1979: 1430).

10. *Displacement and Rehabilitation*—The NWDT had explicit orders regarding the displacement and resettlement of oustees[34] under the SSP.

(i) Gujarat would pay all costs, charges and expenses to MP and Maharashtra that would be incurred in land acquisition, compensation, establishment charges, removal of historical and religious monuments and the like.

(ii) In deciding compensation for families, defined as 'husband, wife and minor children and other persons dependent on the head of the family e.g. widowed mother' (GoI 1979: 1,432); adult sons above the age of 18 were considered as separate families.

(iii) It estimated that 6,147 families in 158 villages would be ousted in MP and 456 families spread over 27 villages in Maharashtra.[35] Gujarat was to establish rehabilitation villages in the SSP irrigation command area for the resettlement of families willing to migrate to Gujarat. For those who were unwilling, MP and Maharashtra were to establish such villages in their respective territories, the cost, charges and expenses of which were to be borne by Gujarat.

(iv) For those willing to migrate, Gujarat had to acquire and make available, *a year in advance of submergence,* irrigable lands and housing sites for the rehabilitation of oustee families.

(v) Apart from prescribing civic amenities[36] in the resettlement villages and grants-in-aid for oustee families, the Tribunal ordered that 'every displaced family from whom more than 25% of its land holding is acquired shall be entitled to and be allotted irrigable land to the extent of land acquired from it subject to the prescribed ceiling of the state concerned and a minimum of 2 hectares (5 acres) per family. This land shall be transferred to the oustee family *if it agrees to take it*' (GoI 1979: 1,433).

(vi) In the event of any disagreement that might occur with regard to the provision of compensation, each state would appoint an arbitrator for resolution of the dispute. If the

arbitrators fail, the dispute was to be referred to an umpire appointed by the Chief Justice of India from among
sitting or ex-judges of the Supreme Court. The decision
of the umpire was final and binding.

11. *Administration and Coordination*—Anticipating the need for
coordination among the four states, the Tribunal ordered the
establishment of two administrative bodies, the Narmada
Control Authority (hereafter NCA) and the Sardar Sarovar
Construction Advisory Committee (SSCAC). The NCA,
headed by a chairperson appointed by the central government,
was charged with the power of coordinating matters regarding (*a*) storage, sharing, regulation and control of Narmada
waters; (*b*) sharing of power benefits; (*c*) regulated release
from MP; (*d*) acquisition of land for the SSP, and compensation and rehabilitation of oustees; and (*e*) sharing of costs.
The SSCAC was formed to ensure 'efficient, economical and early
execution' (GoI 1979: 1,438) of those SSP units that required
inter-state coordination.[37] The Tribunal also asked for the
formation of a 'review committee' comprised of the chief ministers of the four states and headed by the central Minister
for Irrigation and Power to review the NCA's recommendations and to arrive at decisions by consensus.

People's Protest in MP

After a decade-and-a-half of disputes and disagreements, the 'idea'
of the SSP crystallised into concrete and defined project features and parameters. The project design had to be finalised by
a constitutional judicial body whose award was legally binding
on the states of Gujarat and MP. It is interesting to note that the
NWDT award was announced when the Janata party government at the centre was headed by Morarji Desai—a political
heavyweight from Gujarat—and was welcomed unanimously by
all concerned. In MP matters took a different turn.

   The Congress party, then in opposition in the MP Legislative
Assembly as well as in Parliament[38] after its post-Emergency
electoral defeat, thought to gain mileage from the award in MP.
More than two-thirds of the adversely affected villages in MP
lay in the fertile Nimad plains. Responding spontaneously to the
final order of the NWDT, the farmers from the affected villages

in Nimad formed the Narmada Bachao, Nimad Bachao Sangharsh Samiti (save Narmada, save Nimad struggle committee). The Samiti gained instant Congress support and the agitation gathered momentum. Several rallies, *bandhs* (closures) and road blocks were organised in Bhopal, Indore, Badwani and Kukshi during August and September 1979. While the agitation aroused the general feeling that MP was the loser in the Tribunal award, renegotiating the height of the Navagam dam to save the Nimad villages from submergence was the main rallying point. The adivasi villages (one-third of the total) subject to submergence in Gujarat, Maharashtra and MP were neither a cause for worry nor were they represented in the Samiti. Amidst the agitation, the MP government issued statements that it would make every effort to discuss the issue of the height of the Navagam dam (FRL 455 ft.), even though the Tribunal award was final and binding.

Although the protest in Nimad villages was spontaneous, the organisational support of the Congress party, i.e. its state-level functionaries, student and youth wings, became its main strength. When the Congress returned to power in MP and elsewhere after the 1980 elections, it withdrew support to the agitation. Having lasted just a few months, the Nimad agitation soon collapsed.

The Making of Development in the Narmada

Having mapped the conflicts over the design of the SSP, we enter a familiar terrain regarding struggles over (river) water resources development. Competing states demand shares that exceed supply. The mismatch between demand and supply of a scarce and limited resource such as water often leads to conflict. Hydroprojects upstream influence water use downstream, and vice versa. Inter-basin transfers of water also bring to the fore conflicts over the interpretation of rights of riparian states. Cases of riparian state conflicts as witnessed in the Narmada have international and national parallels. Ohlsson (1995: 21) has attempted a concise listing of international riparian conflicts over river waters, significant among which are those between Iraq, Syria and Turkey over the Euphrates and Tigris; Egypt, Ethiopia and Sudan over the Nile; Kampuchea, Laos, Vietnam and Thailand over the

Mekong; Austria, Slovakia and Hungary over the Danube; and Argentina and Brazil over Parana waters.

Similar to the conflict over Narmada waters, each of these have their own incidental causes and conditions, trajectory and consequences. Yet all involve political authorities, i.e. states and governments, committed and competing to 'develop' their resources. Water resources are to be tapped instead of allowing the river to flow wastefully to the sea; power generation is to be maximised amidst supply shortages and increasing demand; and agriculture needs to be modernised to break stagnation and to meet the increasing food demands of a growing population. Rather obviously, political authorities and planners attach high stakes to these development projects. The commitment to, and competition for, resource development has a direct bearing on the scale and size of interventions, as is evident in the Narmada development projects.

In the Narmada conflict, the two juries—the Khosla Committee and NWDT—called upon to resolve the stalemate were guided by the objective of ensuring maximum benefits from project interventions. While the Khosla Committee recommended 13 major dam projects on the Narmada, the Tribunal on its part implicitly supported 30 major projects, 135 medium projects and 3,000 minor projects on the Narmada. The SSP ended up with a height of 455 ft., a main canal that ran 450 km and a canal network of more than 50,000 km supported by the prevalent axiom— the bigger the dam, the less the waste and higher the benefits.

Despite disagreements on the design and shape of particular projects, hardly any doubt was expressed regarding their benefits. In the 1970s, the Punjab model—the Bhakra Project, the Green Revolution and economic success—was an ideal providing empirical evidence of the benefits that mega dam projects could generate. By the late 1970s there were sufficient 'success stories' of large hydroprojects in the country and elsewhere: benefits from Hirakud, Srisailam and Aswan were perceived to be of national, public and uniform good. The political question was, which government could command and control the resources and reap the benefits? Thus, whereas political-regional interests conflicted, the meaning of water resource development was shared by all (conflicting) actors. Large dams became accepted river valley development interventions, with all the expected benefits of hydropower, irrigation, flood control and water supply.

Indeed, the Narmada conflicts clearly suggest that the anticipated benefits from the various projects were taken for granted
by all concerned. In their deliberations, the juries mainly addressed the technocratic and engineering aspects of the proposed
projects. To that end, they endorsed the prevalent planners' view
that the physical structure of a dam will yield benefits. To them
the conflict regarded who commands how much of the river water,
and who would benefit from it. There seems to have been very
little appreciation or doubt expressed about the performance of
projects. Nehru's doubts had lost ground.

The certainty with which large dam projects were associated
with development gains needs to be considered together with
the technocracy's weight behind such projects which justify
heavy investments in the short run for big 'benefits' in the long
run. Notwithstanding Nehru's critical reflections on the 'disease
of bigness', he had thrown these projects up as challenges to the
community of engineers and technocrats. What were mere challenges became established technocratic beliefs that large water
resource investments were necessarily beneficial. The technocrats were divided in terms of state loyalty. Design proposals
and projects on the one side were confronted with counter proposals and projects on the other. Ironically, however, in all these
designs and counter designs there was an assertion of complete
knowledge over how projects would behave. Conflicts were only
over which state would get what benefits.

In contrast, cost-related conflicts centred mainly around financial resources. States were divided on how financial costs
should be shared and how costs incurred (without benefits)
should be compensated. In another sense, conflicts among states
regarding developmental resources were also deemed to be 'costs'
to be borne by the nation because potential benefits from projects
had stalled due to long-drawn-out conflicts. Development (national) and politics (provincial) were conflicting realms. The one
weakened the other.

Social costs such as displacement seem to come into consideration only when the adversely affected people exert political pressure. In the Narmada Valley, for instance, the height of the (later
abandoned) Jalsindhi project was reduced in order to save some
villages from submergence after the MP government received
representation from those areas. It is only fair to acknowledge

that both the Khosla Committee and the NWDT did deliberate on the subject of submergence and displacement. In particular, the NWDT award accorded specific attention to displacement and outlined a resettlement plan whose parameters were unprecedented and later became critical yardsticks by which policies and practices of governments were evaluated.

How far can we link deliberations and decisions around the social costs of displacement in the Narmada conflicts as responding to a need to recognise adversely affected people as stakeholders in the project? On the contrary, it was commonly held that people facing submergence should accept it as their contribution to the 'national good'. In the planners' view, some people had to make sacrifices for the benefit of the majority. Despite periodic protests regarding displacement, discussed in greater detail in later chapters, the political costs of displacement were low for various reasons. Displacement impacts were borne disproportionately by people from remote regions, hilly and forest tracts who had very little political clout. Experience has shown that political pressure by these constituencies was bearable. In India, only the Bedthi project in Uttar Kannada district in Karnataka was abandoned due to pressure from the affected people, among whom were prosperous upper-caste spice farmers. Further, the very language of compensation used by project authorities tended to blunt the political voices of those affected: compensation, no matter how meagre, generated interest and impeded unity among the affected people.

It is important to note here that until the 1970s, the MP government considered submergence in its territory as politically insignificant. Its proposal for the Jalsindhi project (even after height reduction) affected more or less the same amount of land as would the SSP. In the 1980s, however, the Nimad plains in the Narmada Valley witnessed a 'pump revolution': farmers in the area had access to government subsidies and credit for motor pumps and pipes. The economics of the area changed dramatically thereafter, as did its politics. The growing political clout of Nimadi farmers and their struggle to be recognised as stakeholders in the project is discussed later in this book.

If social costs featured in inter-state conflicts over the Narmada, there was no discussion whatsoever on the environmental costs. The NWDT appears to have disregarded the environmental impact of the Narmada projects such as forest loss, wildlife loss, salinisation and waterlogging, siltation, impact on the catchment and downstream, and seismic activity. If the environmental politics of the 1970s are an indicator of general awareness on these issues, the NWDT's neglect of environmental impact seems a bit surprising. By then the country had witnessed the Chipko movement against indiscriminate logging in Garhwal and Tehri regions in the Himalayan foothills. An environmental campaign against the proposed Silent Valley hydroproject set to submerge vast tracts of rain forest in Kerala had also successfully stalled the project. These popular movements heightened state awareness of environmental issues. In 1975 and 1977, the Central Water Commission, a parastatal agency, had issued guidelines for environmental impact assessment of major irrigation and river valley projects. This was followed up with the formation of the Department of Environment and Forests under direct supervision of the Prime Minister in 1980. In that same year, Indian Parliament also passed the Forest (Conservation) Act.

Despite these developments, recognition of environmental costs in general and of hydroprojects in particular was minimal and was traded for development goals and benefits. But with growing awareness, increasing political clout of environmental lobbies and grassroots groups on the one hand, and the Nimadi farmers on the other, the Narmada stage was set for a second round of conflict, this time between state and societal actors.

Conclusion

Here we have outlined riparian state conflicts, stakeholder interests, and conflict mediation mechanisms over the SSP up to the 1970s, within the framework of a federal polity.

Clearly, the debate goes well beyond the purview of inter-state conflict over sharing of resources and benefits. In the 1980s other actors entered the picture. The World Bank became one of the major financiers, followed by NGOs and action groups. Benefits and costs came under increasing scrutiny in the Narmada Valley, as did the meaning of development. This leads us to a benefit-cost appraisal of the SSP to see how, like its 'design', its appraisal was subjected to political calls.

94   Conflict and Collective Action

Notes

1. About 8 km from its source the river drops 75 ft. at Kapildhara falls and 400 metres further downstream drops again by about 15 ft. at the Dudh Dhara falls. About 400 km from the source (near Jabalpur) the river drops at Dhuandhar falls by 50 ft. At Mandhar, 800 km from the source, and at Dardhi 45 km further downstream, the river drops by 40 ft. At a distance of 960 km from the source are the Sahashradhara falls where the river drops by 25 ft. (NWRDC: 1965).
2. By 1949–50, the Tapti (also called Tapi) river had been harnessed for irrigation through the construction of a weir at Kakrapar with a cultivable command of 0.23 m.ha. The second phase of this project involved the construction of the Ukai dam, 29 km upstream of Kakrapar weir. Initiated in 1959 the second phase of the project was completed in 1972.
3. 1 million acre feet (MAF) = 1,200 million cubic metres (Mm3) = 120 mm height of water per hectare = 0.12 million hectares (m.ha.).
4. Government of Bombay, letter no. MIP-5559-J191249, dt.16.1.1959 to Chairman CWPC (cited in NWRDC: 1965: 25 and NWDT 1978a:16).
5. The state of Madhya Pradesh was formed on 1 November 1956.
6. The Indian Constitution has three lists: Union List, Concurrent List and State List, covering the division of jurisdiction among the units of the Indian federation. Entry 17 in the State List pertains to the jurisdiction of state governments over water resources: 'Water, that is to say water supplies, irrigation and canals, drainage and embankments, water storage and water power subject to the provisions of Entry 56 in the Union List' (cited in Iyer 1994: 733). Entry 56 in the Union List states: 'Regulation and development of inter-state rivers and river valleys to the extent to which such regulation and development under the control of the Union is declared by Parliament by law to be expedient in the public interest' (ibid.). Citing relevant constitutional provisions, Iyer (ibid.) argues that whereas water is a state subject, 'it is *potentially* as much a central [union] subject ... particularly as most of our important rivers are inter-state'.
7. By the early 1960s the perspective gained ground that projects were far too numerous, with states vying for allocation of resources, thus resulting in wanton proliferation and thin spread of resources (EW 1964: 1079).
8. The Rann of Kutch—the Great Rann in the north and the Little Rann in the east of Gujarat—has been described by the Second Irrigation Commission as 'a vast desiccated plain in Gujarat, baked by the sun and blistered by the saline encrustation due to ingress of

sea water. Encircling the mainland of Kutch on the north, east and south this great salt waste covers an area of over 20,720 sq.km.... The Great Rann is of strategic importance because the international boundary between India and Pakistan runs along the north border of this area for more than 400 km' (RIC 1972b: 102).

9. A study team constituted by the Union Ministry of Food and Agriculture in 1965–66 probed into the reclaiming of land in the Rann of Kutch. It reached the conclusion that it was possible to reclaim the areas in the Great and Little Rann of Kutch if the water of the Narmada river could be made available in adequate quantity. The Irrigation Commission (1972b: 104) endorsed this viewpoint, stating that 'the future of Kutch would depend on the allocation to the State of the Narmada flows.'

10. It has been mentioned elsewhere that Gujarat subsequently revised the height of the Navagam to FRL 526 ft. (see Dalal 1991: 2). However, no mention of this is made in the report of the Khosla Committee.

11. At 60% load factor.

12. A lower FSL canal would imply a lower dam at Navagam and hence better chances of getting power from the Jalsindhi project.

13. Includes reservoir evaporation loss. If we exclude evaporation loss the amount would be higher, closer to the amount demanded by Gujarat.

14. Regarding displacement and rehabilitation, the Khosla Committee in its report concluded that: 'adequate arrangements should be made for the rehabilitation and resettlement of the oustees from areas to be submerged by the proposed reservoirs. It is desirable that such people should be settled as close as possible to their original homes and in model villages with amenities like electricity, safe water-supply, schools, roads, medical facilities, etc. and further that the lands given to them in exchange for lands to be submerged, be provided with irrigation facilities to the maximum extent possible' (NWRDC 1965: 214).

15. G. Shah (1990: 88) notes that the rich and middle farmers' lobby in Gujarat is extremely influential. Through associations and cooperatives they 'apply pressure on the government to encourage export of their products, protection of prices of their crops and greater distribution of inputs at subsidised rates to boost production.' The Khedut Samaj, formed to oppose the Land Ceiling Act in 1972, is dominated by Patidars and has since then regularly taken up issues of the farmers' lobby, scarcity of seeds and fertilisers, irregularity in power and water supply, and advocates large holdings on the grounds that small holdings are uneconomical.

16. The Patidar–Rajput alliance in the early 1960s constituted the mainstay of the Swatantra Party in Gujarat which did exceedingly well in the 1967 elections, challenging the Congress and obtaining 66 out of the 168 seats in the legislative assembly (Wood 1986).
17. It claimed more water from the Narmada by arguing that it could divert Narmada water to the Tons river basin and irrigate an additional 20 lakh acres in Setna and Rewa districts which faced drought and famine conditions in the 1960s. In terms of the area irrigated as a percentage of total cropped area Setna and Rewa (0.31% and 0.43% respectively) featured as two of the lowest five districts in MP (NCAER 1967: 62). Later, in the draft Fourth Plan (1969–74), the MP government proposed construction of the Bansagar project in river Tons to irrigate 133,500 ha of land and generate 250 MW of power at 60% LF (RIC 1972b: 210). The other issue that MP raised was the hydrological estimates (quantity of water available at Narmada at different sites) which it thought were an overestimation by the Khosla Committee.
18. Dalal (1991: 2) refers to an offer made by the then Gujarat Chief Minister, Hitendra Desai, to reduce the height of the dam to FRL 470 ft. which was not accepted by MP.
19. One factor that seems relevant here is the filing of cases in the Supreme Court by the MP and Rajasthan governments in February 1972 in the Supreme Court against the interim decision of the Tribunal not to admit Rajasthan as an interested party in 1971. The Supreme Court stopped the proceedings of the Tribunal thereafter. In 1974 the cases were withdrawn from the Supreme Court by the two states, allowing the Tribunal to continue with its proceedings. The stay order of the Tribunal proceedings may have facilitated the intervention of the Prime Minister (it may well have been an orchestrated move).
20. The 1973 Levy Act was an improvement over the 1967 Act, according to which small farmers with one acre or less land were exempted from the paddy levy. The rates were progressively higher, the idea being to charge only big and rich farmers (see G. Shah 1974).
21. The amended Land Ceiling Act 1974 implied surrendering surplus land beyond the ceiling for redistribution among the landless (see G. Shah 1974).
22. Dalal (1991) mentions that in 1973, Chief Minister Chimanbhai Patel offered to reduce the Navagam dam height to FRL 455 ft.
23. Uttar Pradesh was important to the Congress as it is the largest state in India; the Nehru family has fought elections from this state.
24. For more details on the Nav Nirman Andolan see Jones and Jones (1976) and G. Shah (1974; 1977).

25. The state was under President's rule. The chief ministers represented the other three states.
26. In the 1965 Master Plan the height of the Hiranphal project was FRL 465 ft.
27. Later in this book we show how this figure was disputed and arguments were propounded that the utilisable water in the Narmada at 75% dependability is much less than 27 MAF.
28. The 'victory' over Pakistan in the 1971 war could have been a factor encouraging economic over strategic interests.
29. See RIC 1972a (Appendix 8.1) and 1972b: 219. Madhya Pradesh and Gujarat differed on the methodology of estimating drought areas. According to the indicators used by MP 18 m. ha. were scarce areas compared to 6.5 m. ha. in Gujarat. The Irrigation Commission rejected the MP government's methodology as unreliable and the NWDT endorsed the Commission's position in this regard.
30. If water requirement had been the sole criterion then the apportionment would have been 17 MAF for MP and 10.25 MAF for Gujarat.
31. Ironically, in the following years, lift irrigation was to become an integral feature of the project's canal network.
32. A certain storage capacity in the reservoir is kept not for regular operation, but for carrying over a certain amount of water from one year to the other. This capacity is used duing shortfalls in lean years. The Tribunal estimated that a total quantity of carryover storage to the extent of 8.29 MAF had to be provided in the whole river system and that this should be provided in the reservoirs of MP and in Navagam more or less pro-rata to water use, i.e. in the ratio of 18.5 (Maharashtra and MP) to 9.5 (Gujarat and Rajasthan). On this basis they decided that carryover capacity of 5.48 MAF in the reservoirs of MP and 2.81 MAF in Navagam would be required (see FMG 1995 Part-III: 3).
33. The Minimum Draw Down Level (MDDL) of 363 ft. is related to this aspect of the project and the corresponding height of the dam at FRL 453 ft. It implies that the dead storage level for irrigation for practical purposes would be 363 ft. and not 307 ft.
34. According to the Tribunal, 'an "oustee" shall mean any person who since at least one year prior to the date of publication under section 4 of the (Land Acquisition) Act has been ordinarily residing or cultivating land or carrying on any trade, occupation or calling or working for gain in the area likely to be submerged permanently or temporarily' (GoI 1979: 1431–32).
35. Oustee families totalled about 7,360; using an average multiplier of 5, a total of about 36,800 PAPs.
36. Civic amenities included primary schools, panchayat buildings, dispensary, seed store, children's park, village pond, drinking water

facilities, tree platforms, religious place of worship, electrical distribution lines, etc. Each of these amenities was linked to the number of oustee families resettled in an area. For example, one primary school (with three rooms) was to be provided for every 100 families and one dispensary for every 500 families.

37. It may be mentioned here that while the irrigation component of the SSP was the sole responsibility of the Gujarat government, the foundation work of the dam (Unit-I) and the power component (Unit-III) required inter-state financial commitments.

38. The Janata party was in power at the centre as well as in most of the states, including Gujarat and MP.

4

The Gainers and Losers: A Cost-Benefit Analysis of the SSP

Introduction

Why was the SSP considered a worthy investment and which stakes were privileged, recognised or ignored? In other words, what was the process by which the appropriateness of the SSP was demonstrated? In analysing the appraisal process, this chapter generates an understanding of how the project is valued and by whom, and how the valuing process entails a selective accommodation of interests. In particular, the cost-benefit analyses undertaken in the SSP are scrutinised to show how the demonstration of goodness through a 'number' is based on a method that tends to hide more than it reveals.

The period covered is 1980 to 1987, during which the project features were finalised, appraisal undertaken and the project cleared by the Government of India.

The data used are drawn mainly from appraisal reports and project completion reports of the World Bank; project authorities and consultants; reports of monitoring and evaluation agencies; review groups; published documents of the Narmada Planning Group (NPG), the Narmada Control Authority (NCA) and the Sardar Sarovar Narmada Nigam Ltd. (SSNNL); and published and unpublished materials of concerned NGOs and action groups.

Major Features of the SSP

Compared to existing projects, the SSP qualifies as one of the largest interventions in water resources development in India.[1] In justifying its involvement, the World Bank noted that the project

> would be the largest Indian irrigation system planned and designed as one unit. It has been designed and planned and would be constructed to standards not hitherto used in India and thus permit higher levels of reliability and timeliness of water delivery than previously achieved. It thus represents a break with traditional modes of Indian irrigation development (WB 1985a:13).[2]

On average, the Sardar Sarovar dam will divert 9.5 MAF of water annually out of the total estimated flow of 28 MAF. The reservoir behind the dam will stretch over 200 km upstream. The main canal in the command is designed to meet a peak requirement of 40,000 cusecs to irrigate 1.8 m.ha. The scale of the intervention can be gauged from the fact that the irrigated command area covers roughly 20% of the total cultivable area of 9.25 m.ha. in Gujarat. In comparison, existing medium and major irrigation projects cover only 1.05 m.ha. (Table 4.1).

Table 4.1 Existing Irrigation Potential and the SSP

Cultivable Area in Gujarat ('000 ha)	9250.40
Existing major and medium schemes ('000 ha)	1051.28
Under the SSP ('000 ha)	1793.85
% Cultivable Area under the SSP	19.39

Source: GoG (1991).

The command area of the project will be served by a canal network of around 50,000 km. The main canal together with 31 branch canals will cover 12 districts, 62 *talukas* (partially and wholly) and 3,344 villages in Gujarat. Given the wide diversity in agro-economic conditions, the command has been regionalised into 13 agro-climatic zones (WB 1985e): 'considering the ultimate objective of (a) choosing feasible sets of crops for each region, (b) allocation of water and its efficient management, and (c) planning for conjunctive use of canal and groundwater taking into account the regional potential and constraints' (GoG 1983:174). Such factors as annual rainfall, land irrigability, class, including drainage characteristics, cropping patterns, and farmers' behavioural responses and consumption levels, are considered in the regionalisation of the command.

Under the SSP, irrigation water will be essentially supply-based (crop according to water). Following the *warabandi* system of water management in northwest India, where irrigation water is rationed strictly in proportion to farm area and supplied on a predetermined rotational schedule, water from the SSP will be supplied to a village service area (300–500 ha).[3] Farmers' cooperatives and users' associations in a service area will collectively determine their common schedule for delivery of allocated water. This system of water management replaces the demand-based *sejpali* system (water according to crop) in western India where farmers obtain official sanction for cropping patterns (seasonal and annual) and are entitled to irrigation supplies according to crop needs. While the operational procedures of the supply-based *warabandi* system are less complicated, the SSP's supply is so planned as to be more flexible to respond to variations in rainfall and to minimise uncertainties, interference and conflicts. In this regard, the SSP will operate on a 'modified supply-system' (WB 1991a).

Hydropower generated in the SSP–1,450 megawatt (MW) installed capacity—would be fed into the Western Regional Electricity Grid.[4] Power will be generated both in the riverbed powerhouse (RBPH) and the canal head powerhouse (CHPH) with an installed capacity of 1,200 and 250 MW respectively. While the bulk of power in the initial years will be generated in the RBPH, the water will be increasingly channelled through the CHPH as irrigation develops. Thus, whereas power generation in the RBPH will fall, that in the CHPH will rise. The RBPH will function in two distinct modes. In its first phase it will act as a generating station; in the second phase (with the full development of irrigation, i.e. after 45 years), it will function as a peaking facility, what is called the pump-storage mode. In other words, the power station will draw electricity from the grid in non-peak hours from the Garudeswar weir downstream and store it in the main reservoir. During peak hours, this stored water will be released through the RBPH to generate electricity. Table 4.2 outlines some salient features of the SSP including the dam, reservoir, powerhouses and canal system.

Cost-Benefit Appraisal

Planned interventions require decision rules because they use scarce resources to produce desired outcomes. Cost-benefit analysis (CBA) claims to provide an appropriate and coherent set of decision-making rules with which to distinguish between efficient and inefficient uses of scarce resources. CBA requires that the value of the benefits that flow from a project over its lifecycle should exceed the costs of its inputs. Also, in judging planned interventions meant to maximise social welfare or public good, CBA examines the distribution of costs and benefits, weighting them differently when they accrue to disadvantaged individuals/groups/regions (Kabeer 1994).[5]

In the SSP, cost-benefit appraisals were carried out by three different agencies—the Narmada Planning Group (NPG), Irrigation Department, Government of Gujarat in 1983 (see GoG 1983); Tata Economic Consultancy Services in 1983 (see TECS 1983); and the World Bank in 1985 (see WB 1985c,d). Later, reappraisals were carried out by the Sardar Sarovar Narmada Nigam Ltd. in 1989 (see GoG 1989) and by the World Bank in 1990 (WB 1991b). All these economic appraisals show a favourable rate of return and hence demonstrate the goodness of the SSP.

Estimated Project Benefits

The major benefits of the SSP estimated in the 1983–85 appraisals are irrigation, hydropower generation, and municipal and industrial water (M&I) supply. Employment generation, directly in the agriculture sector due to irrigation and during construction and indirectly as a result of forward and backward linkages, also features as a project benefit.

Table 4.2 Profile of the Sardar Sarovar Project

Structure	Feature	Operation	Objective	Likely Impact
Navagam Dam: Solid gravity type	*Height:* From the deepest riverbed level: 424 ft. (129 m). From the deepest foundation level: 509 ft. (155 m). *Length:* 3,969 ft. (1210 m) i.e. the length of the road on the top portion	Main spillway gates (23); Auxiliary spillway gates (7); Spillway discharge capacity: 87,000 cubic metres per second (3 million cusecs).	Reservoir Storage Reservoir Length: 210 km upstream from the dam site FRL: 455 ft. (138.5 m) *Live Storage Capacity:* 4.70 MAF (5,800 million cubic metres)	Submergence of 370 sq. km of land at FRL 455 ft.
Hydropower installations: An underground Riverbed Powerhouse (RBPH) in the right abutment of the dam; and a Canal Head Powerhouse (CHPH) located near the rim of the reservoir upstream of the head of the main irrigation canal.	*RBPH:* six reversible turbine generator units of 200 MW *CHPH:* five conventional turbine generator units of 50 MW *Transmission Lines:* 400 KV double circuit transmission lines. *Garudeshwar Weir:* 23 m height ungated concrete structure with a storage capacity of about 26 million cubic metres, 12 km below the main dam	Power station to be integrated with the western regional grid. RBPH will cease to supply firm power in Stage-III of the project after full development of irrigation. Reversible turbines of RBPH to function as pump-storage mode to pump back water into the main reservoir during the low power demand period from storage in the Garudeshwar weir.	Power generation and peaking facility. RBPH: 1,200 MW installed capacity. Annual Power generation: Stage-I—5,220 MU Stage-II—1,772 MU Stage-III—1,045 MU CHPH: 250 MW installed capacity. Annual Power generation Stage-I—267 MU Stage-II—955 MU Stage-III—699 MU	Generate long-term peaking capacity for the western grid.

(Table 4.2 Contd.)

(Table 4.2 Contd.)

Irrigation and Water Delivery and Drainage System: Main Canal: contour, lined. Branch Canals—31 in number (including two branch canals—Saurashtra Branch—257 km and Kutch Branch—374 km chainage.)	Canal Structures: 421 in number including railway crossings, bridges and cross drainage structures. Length of Main Canal: 438 km. FSL of Main Canal: 300 ft. Canal Network Length: 50,000 km.	Discharge Capacity at the head of main canal: 1133 cubic metres per second (40,000 cusecs). Volumetric control of water; supply to water co-operatives at delivery points.	Irrigation, Municipal and Industrial water supply.	Irrigation covering cultivable command area of 2.12 m.ha. Annual municipal and industrial water supply: 1300 million cubic metres.
Accessories: Intermediary structures between CHPH and the head regulator of the main canal.	4 Rock fill Dams; 4 Link channels; 4 Regulation Ponds;1 Saddle dam; 1 Irrigation bypass tunnel.	Saddle dam: Intake facilities to CHPH. Irrigation bypass tunnel: provision of water when CHPH is not in use. Rock fill dams and link channels: for regulation ponds.	Active storage of about 13 million cubic metres for regulation. [4 inter-linked regulation ponds between CHPH and the head regulator of the canal (distance of 7 km)]	Pond storage to be used for control of daily operational mismatches between power releases and irrigation demands.

Sources: GoG (1983); World Bank (1985a, b, c and d).

Note: 1 cubic metre per second (m3/s) = 35 cubic feet per second (cusecs) = 34.95 acre inch per hour; MU is million units energy per annum.

1. *Agricultural Production:* According to all appraisals, incremental benefits would ensue from a replacement of rainfed agriculture by irrigated agriculture. It has been estimated that irrigation from the SSP would raise the cropping intensity[6] from 105% to an estimated 130%,[7] the gross cropped area increasing from 2.2 million to 2.7 million hectares. The gross irrigated area would increase from 0.34 million to 1.79 million hectares, i.e. by 1.45 million hectares. Intensification of cultivation resulting from increased irrigation intensity will be substantial,[8] from 15.6% in the command without project to 67.5% with project (GoG 1983: 541). Yield increases will also be substantial. The incremental production of foodgrains (mainly wheat but also paddy, sorghum, millet and pulses) in the project is estimated at about 2.5 million tonnes (+300%) annually, whereas the annual incremental cash crop production[9] (cotton, oilseed, vegetables, fruit, tobacco and fodder) would be about 3 million tonnes (+100%).[10] Table 4.3 gives a comparative picture of future crop production with and without the SSP, by volume, estimated by the government of Gujarat and Tata Economic Consultancy Services.

The estimates of incremental agricultural production by the Bank (by volume, i.e. '000 tonnes) are shown in Table 4.4.

The Bank's estimates of production levels have indicated that, under a strong agricultural extension system aiming at reliable and flexible water supply and approaching the quality of service obtainable from private tubewells, yields will increase substantially (WB 1985c: 53). Yields adopted for appraisal by the Bank were lower than TECS estimates but higher than early Bank projects in Gujarat.

Agricultural production in Gujarat will be about 40–45% higher than it would be without the SSP (WB 1985b: 24). The value of agricultural output (estimated at 1984 financial prices by the Bank) would increase under the SSP by about 370% over a period of 20–25 years; without the project the increase would be less than 30% (WB 1985c: 49).[11] The increase in net value would be distributed over various regions in the command (GoG 1989:175, 544; also, WB 1985e and ORG 1982). The future value of production may well exceed estimated projections, as the latter do not consider possible major breakthroughs

Table 4.3 Estimates of Major Crop Output ('000 tonnes)

Crops	(GoG) Without Project	With Project	Increment	Percentage Increase	(TECS) Without Project	With Project	Increment	Percentage Increase
(1)	(2)	(3)	(4)	(5)	(6)	(7)	(8)	(9)
Paddy	184	902	718	390	168	902	734	437
Wheat	256	1518	1262	493	236	1518	1282	543
Jowar	146	344	198	135	135	344	209	154
Bajra	277	723	446	161	262	723	460	176
Groundnut	60	195	135	225	55	195	139	251
Cotton	410	1003	593	145	379	1003	624	165
Tobacco	77	161	84	109	69	161	92	132
Sugarcane	–	–	–	–	191	672	481	251

Sources: GoG (1983); TECS (1983).

Note: (a) Figures have been rounded off; (b) fruit, vegetables and fodder, were not considered; (c) the percentage increase calculation undertaken by GoG (1983), i.e (3)/(2)*100 has been corrected to (4)/(2)*100.

**Table 4.4 World Bank Estimates of Incremental Production
('000 tonnes)**

Crops	Present	Future without Project	Future with Project	Incremental
Rice	49	61	515	454
Wheat	181	250	1409	1159
Bajra (Pearl Millet)	230	277	653	376
Jowar (Sorghum)	115	134	487	353
Pulses	71	82	208	126
All Food Crops	646	804	3272	2468
Groundnut	56	67	310	243
Vegetables	–	–	1793	1793
Tobacco	40	55	109	54
Fruit	–	–	1269	1269
Fodder	2502	2977	4677	1700
Cotton (short)	227	265	61	–204
Cotton (medium)	135	199	1057	858

Source: WB (1985c and d).

in production technology or major shifts towards extremely high-value crops.

2. *Employment and Income Generation:* An estimated 4.5 million people (1981 census) inhabit the rural areas of the SSP command of whom about 0.6 million (153,000 households) are cultivators and an additional 0.6 million (122,400 households) agricultural labourers. Twenty-eight per cent of cultivator households and 64% of agricultural labour households fall below the official poverty line. According to the Gujarat government, the rural population in the command area—likely to double by the year 2021 to an estimated 8.3 million people—would benefit most from the SSP. Despite population growth, the percentage of people below the poverty line will fall 'although it is difficult to predict to what extent, given the demographic changes and intricate shifts in income distribution' (WB 1985c: 49).

On average (in terms of both farm size and percentage increase), income from crop production would rise by 250% with the project, compared to estimates without the project. The sum of all regional incremental incomes from farming in the SSP, inclusive of net value added, would amount to Rs 10,400 million per year after full development, whereas without the project it would be about Rs 3,100 million. Total incremental wage income would rise by more than 100% and the corresponding employment generation would be about 0.6 million person years (GoG 1983: 545). Apart from full on-farm employment that will mostly benefit landless labourers, the SSP would also provide about 125,000 seasonal jobs annually during its construction (WB 1985c: 49). Construction activities during project implementation would provide employment for over 4.9 million person years—75% unskilled, 13% semiskilled and 12% skilled—spread over 20 years (TECS 1983: 91). Around 2 million person years would be in the first five years of construction.

3. *Hydropower Benefits:* The installed generating capacity of the SSP would be 1,450 MW.[12] The firm capacity of the RBPH of the Sardar Sarovar Project would be 3,635 million units per annum (equivalent to 415 MW average continuous generation), gradually decreasing to zero after 45 years as full irrigation develops in MP and Gujarat. The CHPH's firm energy contribution would be 213 million units per annum (equivalent to 24 MW continuous) in the initial years, and will increase to 440 million units per annum (equivalent to 50 MW continuous) as full irrigation develops. Both powerhouses would also produce substantial seasonal energy (also called secondary energy or spill energy) of 1,621 million units per annum in the initial years, reducing to 865 million units per annum in the final phase of irrigation development.

The two power sources would be part of the capacity expansion programme in the western power grid.[13] To that extent, their benefits were considered least-cost options compared to other modes (thermal and gas-based plants) of expanding capacity of the western grid. The benefits of the hydroelectric component were also estimated using the cost-benefit approach. When the costs and benefits of irrigation were excluded, the power component alone yielded an economic rate of return of 13% including the full cost of the dam and reservoir (WB 1985f: 97).

4. *Benefits from M&I Water Supply:* The annual volume of M&I
   to be provided would be about 1,310 million cubic metres
   (Mm3), equivalent to 1.06 MAF. Initially, it was estimated to
   cater to 131 urban centres (1,115 Mm3 water to an estimated
   population of 18.6 million in year 2021) and 4,720 villages
   (195 Mm3 of water to an estimated population of 10.8 million
   in year 2021). The number of villages to be covered for water
   supply was subsequently revised to 8,000. The benefits of M&I
   water would accrue gradually over a period of 40 years. The
   cost of alternative schemes of water supply (supply by tank
   trucks to distribution points) without the SSP was considered to be substantially higher. According to the Bank, the
   phasing of M&I water was projected according to tentative
   plans made by the Gujarat Water Supply and Sewerage Board.
   Withdrawals would commence almost simultaneously with
   the first irrigation and would not reach their final levels until 30 years later (WB 1985c: 56).
5. *Other Benefits:* Fish production in the reservoir would rise,
   and reforestation in the catchment area and parts of the command area would offset the economic costs of submergence
   and of changes in the downstream impact. The availability of
   water in the reservoir and in the command area will generally improve the health conditions of the people. Frequent
   floods in the downstream region will be controlled by the dam,
   and the SSP's early flood warning system (part of the hydro-
   meteorological network) will contribute towards this goal.

Estimated Project Costs

The TECS appraisal defined 'costs' as including 'all project related
costs such as costs of submergence, rehabilitation, protective
plantation, command area development, drainage and conjunctive use of groundwater, and also repair and maintenance of the
project facilities' (TECS 1983: Preface). It estimated a total cost
of Rs 48,870 million over a period of 23 years. The item-wise
and commodity-wise break-up are given in Table 4.5.

Table 4.6 shows the Bank's cost estimates in its appraisals of
the power and irrigation component of the SSP. The World Bank

Table 4.5 TECS Estimates of Project Costs and Expenditure in 1983 (1981–82 prices)

Item	Rs Million	Percentage	Commodity	Rs Million	Percentage
Dam	5650	11.6	Cement	4371	10.0
Rockfill dykes	560	1.1	Steel	2391	5.0
RBPH	2770	5.7	Petroleum products	2123	4.0
CHPH	970	2.0	Other materials	3185	6.5
Main Canal	12060	24.7	Power	456	1.0
Branch Canals	6060	12.4	Mechanical equipment	7349	15.0
Distribution System	10290	21.1	Electrical equipment	2298	4.5
Command Area Development	6040	12.3	Transport equipment	1541	3.0
En route Storage	430	0.9	Unskilled labour	9167	18.5
Drainage	3050	6.2	Skilled Labour	13263	27.0
Costs in Rajasthan	990	2.0	Others	1758	3.5
Total	48870	100.0	Total	47876	100.0

Source: TECS (1983: 73–77).

**Table 4.6 Estimates of Costs in 1985 by the World Bank
(1984 prices in rupees million)[14]**

Components	Local	Foreign	Total
Main Dam	3790	1679	5469
Rockfill dams and link channels	92	17	109
RBPH	2726	2426	5153
Garudeshwar Weir	352	149	501
CHPH	884	176	1060
Transmission System	142	85	228
Vadgam Saddle Dam and bypass tunnel	196	61	257
Hydro-meteorologcial network	142	33	175
Training and technical assistance (power)	6	23	29
Land Acquisition and rehabilitation	1681	–	1681
Main Canal	–	–	12175
Branch Canals	–	–	9401
Distribution and Drainage Systems	–	–	20976
Training and technical assistance (irrigation)	–	–	45
Command Area development	–	–	5380
Total	–	–	62639

Source: World Bank (1985c: 30; 1995c: Table 6).

estimated a total base cost[15] of Rs 62,639 million (1984 prices) for the SSP. It is worth noting that the appraisal set aside an estimate of Rs 1,680 million as land acquisition and resettlement costs. Land acquisition in the reservoir area and for infrastructure in the SSP command was priced using the estimated economic value of annual production foregone, based on projections of agricultural net returns per hectare in the future sans project. The cost of resettling the displaced was included insofar as it did not represent transfer payments. Other recurrent costs included value of foregone forest production in the reservoir area due to submergence; operation and maintenance costs of the SSP's irrigation system and the dam and power facilities; and the cost of pumping groundwater and lifting water for the SSP's branch canals at Saurashtra and Kutch (WB 1985c: 54, WB 1985f: 162).

In the Bank's estimate, about 2.5% of the total cost was set aside for resettlement and rehabilitation. During the appraisal process the Bank's attention was drawn to displacement risks in the SSP and to the inadequacies in the compensation package of ARCH-Vahini—an NGO working on issues of health and environment in Mangrol, Gujarat.[16] As early as 1983, the NGO had written to the World Bank about the plight of adivasis (ethnic minority groups) affected by the project. In the same year the Bank recruited Thayer Scudder, an international expert on displacement and resettlement, for a mission on displacement and resettlement in the SSP. Scudder noted that India's past record on resettlement did not meet Bank standards. The country lacked a national resettlement policy. Its legal instruments for acquiring property were inappropriate and the means to implement resettlement programmes were inadequate. He also emphasised that the Narmada Water Disputes Tribunal Award did not apply to the majority of those displaced, i.e. the landless (see Scudder 1989). Scudder's 1983 report was the first critical assessment of displacement and resettlement in the SSP.

The World Bank, having appraised the problems of displacement, negotiated a resettlement package with the three riparian states, which further improved the provisions in the NWDT award of 1979. Although the Gujarat government had in 1983 devised what was called the 'Sardar Sarovar Development Plan', this had made no reference to the issue of displacement, despite the fact that it had commissioned a socio-economic study of the 19 villages to be submerged in Gujarat.[17] The Bank's Staff Appraisal Reports incorporated the findings of this study[18] and noted the following:

> While the Tribunal's mandate covered many aspects of the interstate aspects of the resettlement effort, it was not explicit on the compensation to be given to oustees in Gujarat itself, nor on how landless people were to be compensated for displacement and consequential loss of livelihood. It was required therefore that the Tribunal compensation package be expanded. Of particular concern was the need to ensure that the landless, most of whom are tribal peoples ... had their standards of living protected. In addition, planning and financing of the resettlement effort had to be integrated into the project and its timing made consonant with rising water levels resulting in dam construction. Also arrangements for the monitoring and evaluation of the plan had to be established. During project appraisal, an overall program for resettlement for oustees was drawn up and time bound schedules for resettlement of all affected villages and the costs thereof assessed (WB 1985a: 20).

It was this position that led the Bank to revise NWDT estimates of land submergence and the number of affected people,
and to negotiate a resettlement package in the project. The main
objectives of the resettlement plan were (*a*) that oustees regain
or improve their livelihood promptly after displacement through
the provision of appropriate compensation and adequate social
and physical infrastructure; (*b*) that they be located as village
units, village sections (hamlets) or family units depending on
their preferences; and (*c*) that they be fully integrated into the
community in which they settle. The resettlement plan was to
also ensure adequate participation of the oustees. Each landed
oustee was entitled to irrigable land in the state where s/he chose
to resettle, equal to the size s/he owned prior to displacement and
subject to ceiling laws of the state. Oustees owning less than 2
ha of land were entitled to a minimum of 2 ha of irrigable land
acceptable to them. Landless oustees were to be rehabilitated
in agricultural or non-agricultural sectors and be eligible for
stable means of livelihood in accordance with the overall objectives of the resettlement plan.

As the appraisals show, however, besides drawing up a rehabilitation plan, the Bank did not intend to financially support
the SSP resettlement costs either partly or substantially. The
outlay for resettlement (see Table 4.6) was to be financed do-
mestically by the different governments.

The NWDT had not considered any environmental effects of
the SSP, neither did environmental costs figure explicitly in the
TECS appraisal or in the Bank's appraisal of 1985. The M.S. University at Baroda conducted the first environmental study for
the government of Gujarat in 1983, when the environment impact assessment of major projects and environmental clearance
from the central government had become mandatory. In the same
year, the department of Environment and Forests (DoEF), Government of India, refused environmental clearance to the SSP,
noting serious deficiencies in the University's assessments.

The Bank's environmental appraisal of 1985 mentioned both
negative and positive effects of environmental 'changes' in the
SSP. For instance, it said that downstream effects would be adverse (e.g. degrading the river streambed) as well as beneficial
(e.g. controlling low-intensity flood damage). Similarly, while
the project affected fisheries, forests and wildlife habitat, 'the

creation of a more positive eco-cycle on the periphery of the reservoir may be expected in terms of improved soil moisture for nearby forests and other bio-mass and improved wildlife habitat' (1985a: 21).

The appraisal also mentioned that in 1985 a comprehensive environmental protection programme covering fish and fisheries, forests and wildlife, and public health, had been *outlined* and that the concerned states and the Bank had *agreed* with its objectives. For the environmental impact of the irrigation component of the project, a two-year study was to assess the effects on issues such as salinity, waterlogging and water-borne diseases in the SSP command area (WB 1985b: 20). This appraisal and subsequent loan agreements included 16 covenants dealing with resettlement, three on health and three on environment.

In January 1986, the Bank's board approved the Sardar Sarovar Dam and Power Project for an IBRD loan of US$ 200 million and an IDA credit of US$ 100 million (SDR 99.7 million equivalent) and the Sardar Sarovar Water Delivery and Drainage Project for an IDA credit of US$ 150 million (SDR 149.5 million equivalent).

'Benefits Outweigh Costs': The Received Wisdom

All the economic appraisals indicated that the SSP was a worthwhile investment. In cost-benefit analysis the worthiness of public investment is measured in terms of the benefit-cost ratio (BCR) or the internal rate of return (IRR). The BCR is a measure of return to investment for every unit of capital invested over the life of the project. In other words, it calculates the inter-temporal stream of benefits and costs using different market rates. It estimates the present value of benefits to the initial investment. The normal decision rules are to accept the project if the BCR is more than 1 and to reject it if the BCR is less than 1.

The IRR refers to the value of the project to the national economy. Porter et al. (1991: 107) describe lucidly the significance of the term to economic appraisals.

> The IRR measures the worth of a project by finding the discount rate that makes the net present value (NPV) of the net benefits equal to zero. To discount is to recognise that the value of benefits will be worth less in the future than they are now.... There is a discount rate which will result

in an NPV of zero over the life of the project, that is when all the benefits minus all the costs discounted over the life of the project are equal to zero. By iteratively trying different discount rates until the NPV is near to zero, we can ascertain the IRR of the project. The IRR is therefore equal to the maximum interest that a project can pay for the resources if the project is to recover the investment in it...

In other words the IRR is a measure of the return to the capital invested in a project. Normally, a project with returns less than the opportunity cost of capital is rejected. 'Opportunity cost refers to the benefits foregone by using capital for one purpose instead of the next best alternative use' (ibid.). Thus, an acceptable IRR should normally be (more than) the interest rate at which finance is available. This, however, is not a strict rule for public investments where other factors such as social objectives come into play. For instance, a public investment in water resources with a lower IRR may be preferable for a drought-affected region.

The Gujarat government's 1983 appraisal of the SSP shows the estimated BCR (economic prices) at 12% discount rate to be 1.94 and at 10% discount rate to be 2.36 (1983: 549). TECS appraisal for the NPG estimated the BCR at 1.84 (economic prices). At market prices (1981–82) the BCR was 1.39 (TECS 1983: 79). Table 4.7 shows TECS estimates of the BCR at economic prices.

Table 4.7 TECS Benefit-Cost Ratio

Components	Benefits— Present Value (in Rs million)	Costs— Present Value (in Rs million)	Benefit— Cost Ratio
Power	7540	2483	3.04
Water Use	23094	14204	1.63
SSP	30633	16687	1.84

Source: TECS (1983: 80).
Note: Figures on water use up to and beyond Mahi have been integrated.

This appraisal also estimated the IRR of the project to be 18.3% at 12% discount rate. It assumed a project horizon of 51 years (up to year 2031–32).

In 1985, the Bank estimated the IRR for the SSP at 13% covering 50 years of the project's life and at 12% discount rate (WB

1985c: 56). Although in quantitative terms the Bank's assessment differed from that of the Gujarat government's—an issue
to which we shall return later—in both estimations the figures
were above the 'opportunity cost of capital', i.e. the economic
yardstick of investment worth. In this respect the goodness of
the SSP had been demonstrated.

In addition to this, the Bank carried out a sensitivity analysis
taking into account some of the major uncertainties and risks
in the project. Sensitivity analysis (also called risk analysis) is a
decision tool that helps decision-makers to consider underlying
variables and their inter-dependencies, and the uncertainties
characterising the project. It shows how robust or vulnerable a
project is to changes in the variables, and also shows if the net
present value (NPV) or the IRR is highly sensitive to changes in
some variables.

The sensitivity analysis in the SSP was conducted given a
number of assumptions made in the project appraisal about yields,
prices, physical parameters related to the implementation period
and project costs, and the possible delay in construction of the
major upstream regulation facility. In the Bank's view, the
timeframe and benefit build-up in the SSP were unusually long.
Returns to investment would therefore depend on the
interaction of many variables.[19] Hence, the risk analysis was
carried out 'in order to furnish the decision-maker with information on the variability (distribution) of the economic rate of
return especially the probability at which the ERR would fall
below or above the single point "best estimate"' (WB 1985f: 165).

In the risk analysis—defined as the potential failure of the
SSP to achieve certain minimum economic criteria, customarily
expressed as economic internal rate of return (ERR)—the Bank
claimed to incorporate the following factors:

(a) cost of the dam, powerhouses, main canal, transmission sys-
     tem, branch distribution and drainage system, command
     area and groundwater development;
(b) implementation period for the above;
(c) errors in estimation of economic conversion factors;
(d) deviation of agricultural gross returns (yields), cost of culti-
     vation, water demand and cropping intensities from base
     assumptions;

(e) deviation of economic prices of agricultural outputs and inputs and of the economic value of power from base assumptions;
(f) change of build-up rate of agricultural benefits from assumed rates;
(g) loss of agricultural production due to drainage problems;
(h) deviation of river runoff and of actual water availability to Gujarat and Rajasthan from assumed volumes and of system efficiencies from those assumed;
(i) delays in construction of the NSP upstream in MP; and
(j) variations in the share of Gujarat's water allocated to M&I uses.

The risk analysis[20] yielded a random sample of IRRs; the minimum ERR observed was 7.4% and the maximum 18.9%. It was noted that the band of IRRs had a spread of only about 5–6 percentage points around the mean and that IRR of less than 10% would occur only with a chance of one in six. Table 4.8 shows the probability of IRRs falling within certain intervals:

Table 4.8 Sample Distribution of IRR (in percentage)

Range of ERR	Probability	Cumulative Probability
Less than 8	2.2	2.2
8 to 10	13.4	15.6
10 to 12	34.4	50.0
12 to 14	34.4	84.4
14 to 16	13.4	97.8
Greater than 16	2.2	100

Source: World Bank (1985c: 62).

From this it was deduced that the risks of the SSP appeared low as its multipurpose character spread the risks across different functions; in particular, the power function was inherently less risky than the irrigation function. In the Bank's assessment, the SSP was expected 'to yield satisfactory returns to investment with a reasonably low overall risk, considering its size and time frame' (WB 1985c:62).

Appraisals of the SSP: A Critical Perspective

The three appraisals of the SSP regarded it as a worthwhile investment, its benefits *far outweighing* its costs. This opinion was arrived at using scientific and objective tools. The claim is that such tools do not entertain political criteria or considerations in demonstrating the goodness (or otherwise) of a project. The appraisal mechanism followed a simple set of procedures. On the cost side inputs refer to the physical structure of the project—cement, steel, electricity, transmission of power, delivery of water, labour and so on. Costs of agricultural inputs (costs of cultivation), i.e. fertilisers, seeds, water charges, electricity charges, wages, etc. incurred by farmers in irrigated agriculture are added. Costs of submergence (land and output therein going out of production), rehabilitation, protective plantation, command area development, repairs, and operation and maintenance are included. On the benefit side, crop yields from irrigated agriculture are estimated. Hydropower benefits are estimated using the least-cost approach (compared to gas-based and thermal alternatives) and the consumer's willingness to pay. Benefits of M&I water supply are evaluated by estimating the cost savings of supplying the identified demand from Narmada water rather than from other sources. [21]

Nevertheless, questions arise both with regard to the specific application of tools in the SSP—the manner in which the appraisals were conducted—and with regard to the general claim that such appraisal procedures are politics-free and value-free exercises. One is inclined to begin with a very fundamental question. When there are several appraisals, which should one believe and which not? How is it that, despite significant differences in values attached to the (input–output) components of the project between each appraisal, their IRR remains above the opportunity cost of capital?

In 1990, the Bank carried out a mid-term economic appraisal of the SSP 'based on adjustments to the original economic analysis to account for changed quantities, prices and phasing since 1984' (WB 1991b). It considered the dam construction period to be 12 years compared to 10 years as was assumed at the first appraisal in 1985. The reappraisal process identified and accounted for new costs and benefits. [22] On the cost side it increased resettlement costs by 300% from the original estimate; a 20% flat rate of the total resettlement cost was included for compensation for canal-affected people. Environmental costs were also added—public health costs, wildlife costs, costs of moving religious and cultural monuments, etc. Despite this increase, it was noted that 'resettlement...represented 3% of the total project cost and environment [even] less' (WB 1990: 6).

The reappraisal also changed its calculations on the benefit side. Agricultural yields in less suitable soil in the command area were adjusted downwards by 25%. Simultaneously, the benefits of more high-value crops and higher cropping intensity were added. Estimates on the basis of which power benefits had been initially calculated (Rs/Kwh) were upgraded. 'Environmental benefits' from fuelwood, fisheries, public health, flood control and wildlife were added. While forest loss was calculated at Rs 2,190 per hectare per year, the benefits of fuelwood production were estimated at Rs 4,132 per hectare per year. Despite numerous alterations in the assumptions and estimation, the IRR remained at 12.10%—above the opportunity cost of capital.[23]

Just two years prior to the reappraisal, the Gujarat government conducted its own reappraisal, recommending that the project implementation schedule be reduced from 17–22 years to 10 years. According to the report, 'in the 17–22 year schedule, the benefits are delayed considerably and cost escalation is higher. The assessment established that the rescheduled project is economically viable but if delayed substantially would become economically unacceptable' (GoG 1989). Table 4.9 indicates the comparative IRR of this estimate.

Table 4.9 Comparative Costs and Benefits Discounted at 12% in 1989 (in Rs million)

Implementation Schedule	Present Value of Costs	Power Bene-fits (PV)	Irrigation Bene-fits (PV)	M&I Bene-fits (PV)	Total Bene-fits	Net Bene-fits	Internal Rate of Return
10 years	48570	13980	32970	7530	54480	5910	13.12
17–22 years	44530	13980	21190	5510	40680	–3850	11.13

Source: GoG (1989).
Note: Base Year—1986–87.

Which of these appraisals should we believe? Or does it matter? Porter et al. (1991: 125) express the view that 'the power of techniques like cost-benefit analysis lies not in their actual results but in the promises that these results are available through their application. The essence of the promise of these results lies in the legitimacy of the ritual.' Should we see SSP appraisals as rituals seeking legitimacy for what was already intended? In other words, are the appraisals merely rubber-stamping what has already been decided? In that case, should the appraisals, instead of being seen as neutral and objective decision-making tools, be seen rather as legitimising the interests and values of planners, donors and developers? Do they then not justify an already trodden path of development?

The answer to all these questions appears to be an emphatic affirmative when SSP appraisals are considered. But we have already seen that the decision to implement the SSP was determined by factors other than the criterion of internal rate of return. In fact, comprehensive cost-benefit calculations were made only *after* it had been decided to go ahead with the project.[24] The appraisals thus could not have judged whether the SSP was the best option among alternatives. The Bank's 1985 appraisal reports attempt to justify this oddity.

> [In] the binding rules of the society as formulated by the NWDT... consumptive water uses have priority over power-generation. It is supported by the notion that water cannot be replaced by any other factor in agriculture and human consumption whereas power can be produced from alternative resources.... No project is conceivable without the irrigation and M&I water supply components. Given the government of Gujarat's decision that the main objective is providing irrigation and M&I water supply, it is unrealistic to conceive an alternative project satisfying this objective as required by the classical method of appraising multi-purpose projects because any alternative means of providing such water is prohibitively expensive (WB 1985c: 56–57).

The fact that appraisals were conducted after the decision had been taken to construct the project may be seen as reinforcing the use of methods as ritual and legitimising tools. One could say that they simply justified decisions made on other grounds by achieving a satisfactory rate of return. Renard and Berlage (1992: 50) argue how cost-benefit techniques are easily amenable to the motivation and manipulative power of actors involved in decision-making:

A cost-benefit exercise can be easily manipulated. There are so many data to be guessed or estimated, on the basis of so many assumptions...that this by itself is often sufficient to...influence the outcome one way or the other. [One] can choose between different textbooks which...can be misused in their own peculiar ways. And there are enough unsettled issues...to...further opportunities of steering the evaluation in a desired direction...[The] CBA [is] often applied cosmetically to justify projects already selected on other grounds. If they are compared then it is only to those alternatives which they can beat...

Economists respond to such criticisms by suggesting that, theoretically at least, it is possible to reject a project on the basis of proper economic appraisal even if political decision-makers favour going ahead. How far can we proceed with this proposition given the appraisal experiences in the case of the SSP?

These appraisals have drawn criticism on a number of counts. The study conducted by Paranjpye (1990) shows how TECS appraisals relied on a database of 480 farmers, statistically insignificant considering that an estimated 340,000 households were engaged in agriculture in the command area. Although the command area is extremely diverse, farm-level data for each of the 13 regions were not collected in any systematic manner. In fact, district-level data were used as indicators. Paranjpye also notes that barring wheat, district-level data showing irrigated and dry-land yields for other major crops were not available. TECS appraisals simply assumed that irrigated yield would be twice the dry-land yield. On the cost side, Paranjpye writes that costs of rehabilitation, compensatory afforestation, command area treatment and downstream effects were grossly underestimated. Moreover, benefit distribution and equity issues were totally ignored.

The 1985 appraisal of the Bank, according to Paranjpye, used the same set of farm-level data and therefore committed the same errors as TECS. These appraisals also ignored and underestimated major costs in the project, such as submergence and benefits foregone due to the network of canals, and the loss of historical and cultural monuments. Paranjpye also holds that power benefits estimated in the SSP were significantly higher than the average tariff realised by state electricity boards, which makes the IRR artificially attractive.

Several other issues could be added to Paranjpye's critique. How do we value the suffering of displaced people? How do we put a value to environmental despoliation or the loss of a river ecosystem? Did the appraisals include losses of common property resources? Identifying some of these problem areas in the cost-benefit estimates is central to Chapter 6 which undertakes a detailed and critical assessment of the assumptions, uncertainties and risks in the SSP. Here we are concerned with the question of political economy. Insofar as cost-benefit appraisals advance a set of truth claims regarding the project, what interests and values can possibly underlie such claims?

Normatively speaking, the outcome of a cost-benefit analysis is essentially to prescribe which course of action amongst a given set is preferred. A number of value judgements need to be made. 'To make the best use of resources is to imply some value judgement about what the desirable end of economic activity is' (Pearce & Nash 1981: 9). Since it is difficult to find a set of value judgements that will have universal appeal, the CBA operates on the assumption that the preferences of decision-makers reflect broad societal values. State objectives become a proxy for societal welfare. In assessing the merits of various projects, the CBA will thus take into account the state's objectives, what it wants to do with resources, and how it wants to distribute them. The analysis ought to echo state preferences in resource development.

The extent to which governments/states represent societal welfare is highly debatable. For instance, from a Marxist political-economy perspective, states represent the interests of a powerful minority in a society at the expense of others (Barnett 1981; Stewart 1975). Others consider them to be benevolent executors of people's will (Helmers 1979). While these two positions may be deemed the extreme ends of a spectrum, the issue brings to the fore the role of power and politics, which techniques such as cost-benefit analysis cannot address or explain.

On the one hand SSP appraisals reflect state preferences on resource development. The SSP is appraised for the state's interests and objectives of industrial growth, agricultural modernisation, income generation. At the same time the project becomes the source of competing interests and conflicting values as political tensions and power struggles build around the CBAs. It comes under pressure to accommodate the interests of actors, other than decision-makers, to be included in the benefit stream, to be omitted from the cost stream or to be recognised in the cost stream with claims to compensation. The appraisals seem to negotiate these pressures by periodically accommodating some interests and ignoring or suppressing others in order to be able to produce the desired outcome. This limitation not only undermines the usefulness of the appraisals, especially when the interests of vulnerable and subordinate groups are at stake, but also exposes the so-called neutrality of the appraisal process. Little wonder, then, that the SSP appraisals have been criticised for hiding costs and exaggerating benefits. They have also become the focal point for political conflict, protests and collective action.

Whose Benefits and Whose Costs: The Distribution of Stakes in the SSP

Given the claims and counter-claims around the appraisals, it is important to map the likely losers and gainers in the SSP indicating the opportunities for and risks to an identifiable group of people, the mapping takes into account major issues of contention such as likely impact, uncertainties and delays in the implementation period, and the likelihood of adequate and just compensation.

Potential Beneficiaries

1. *The Industrial Sector:* Industries are likely to be the major beneficiaries of the SSP. They expect electricity and water from the project, profit from the supply of materials for project construction, and benefit from backward linkages with agriculture (particularly processing and agro-industries). At a macro level, the Indian industrial sector is the largest consumer of electricity (of about 70% followed by agriculture of about 12%). Substantial power benefits from the SSP will accrue to the sector. The World Bank estimates the percentage distribution of power benefits to industrial consumers at 55% (CWB 1991b). Water supply as planned in the project will also directly benefit the industrial sector (0.2 MAF). In anticipation of power and water benefits, industry has responded

enthusiastically to the Gujarat government's initiative of building what is called the 'golden corridor' on the Ahmedabad–Delhi route; by 1995 the corridor had attracted about Rs 75,000 crore worth of investments (Desai 1995, cited in Mehta 1997). The Jhagadia industrial estate is being developed near the dam in Gujarat. Industrial estates in Kutch and Saurashtra also expect power and water benefits from the project. The location of these estates has been a major consideration in finalising the alignment of the two major branch canals of the project.

Agro-processing units expect to benefit from the SSP. Anticipating large-scale sugarcane cultivation in the command (following the Ukai–Kakrapar experience in south Gujarat), sugar cooperatives and factories have come up in Ahmedabad and Baroda. As a result of irrigated agriculture, a dramatic rise in demand for agricultural inputs such as fertilisers and pesticides is also likely to generate benefits for the industrial sector. The supply of inputs, e.g. cement and steel, transport, mechanical and electric equipment during the construction phase of the SSP would benefit domestic industry whereas inputs pertaining to heavy machinery, power turbines and even cement will benefit global transnational industries.

2. *The Farming Community:* The benefits of irrigation, flood control and electric power will accrue to farmers in the command area, albeit unevenly. About 3,400 villages are likely to reap irrigation benefits assuming that the command areas develop as planned. However, farmers in Broach, Baroda and Ahmedabad districts, at the head end of the command, will be first to benefit from the project. Farmers' organisations and cooperatives in these areas, mostly controlled by rich Patidars, have invested heavily in security bonds and debentures issued by the Sardar Sarovar Narmada Nigam Ltd. (SSNNL). Agricultural production at the head end of the command (in Baroda and Ahmedabad districts) is already capital- and energy-intensive. Farmers in this region mostly rely on private groundwater, so the provision of assured canal water will be a major benefit to them. Farmers in the Broach district will also benefit from the project's flood control provisions.

Beyond the river Mahi, farmers in the districts of Mehsana, Sabarkantha and Banaskantha anticipate water from the project. However, irrigation benefits to these areas will be subject to the pace of development of the irrigation infrastructure. The same is true for farmers in Saurashtra and Kutch, where the branch canals have been designed to navigate low-lying areas including the famous Saurashtra depression. The supply infrastructure, requiring lifts and falls, will be complex, time-consuming and expensive. It is worth noting here that, despite delays if not uncertainty in the supply of water to Kutch and Saurashtra, support for the SSP in these areas has been steadfast. As Mehta (1997: 8) notes, 'in Kutch, it is a taboo to talk against the project or even to suggest that SSP is not the panacea that it is made out to be.'

Farmers in the Saurashtra region suffer from perennial drought conditions. The more affluent among them often lease agricultural land in the water-endowed regions of Gujarat. Known for their 'remarkable farming skills' (personal communication, Sanat Mehta, Member of Parliament and ex-Chairman SSNNL), their seasonal movement to the head end of the project is likely to become permanent, at the cost of small and marginal farmers. Experiences in command heads of other projects in the country show higher land concentration and the crowding-out of small and marginal farmers by rich farmers (see Singh 1990; 1997).

The cropping pattern assumed in the appraisals shows a marginal increase in sugarcane production.[25] Project authorities claim that the cultivation of water-intensive crops like sugarcane will be discouraged in the command through the controlled supply of water, agricultural extension, and the mediation of local NGOs. Considering the political clout that goes with the production and processing of sugarcane, it is unlikely that authorities will be able to convince farmers of their plans. As private groundwater develops in the command, sugarcane will in all probability become a dominant crop, at least at the head end.

The farming community also anticipates a reliable electricity supply from the SSP. In Gujarat, farmers rely mostly on diesel pumps which have high operational costs. The project makes it likely that farmers will switch to electric pumps which are cheaper; moreover, electricity to rural areas is heavily subsidised.

3. *Wage Labour:* The demand for wage employment will rise with irrigated farming in forward and backward linkages in sectors such as transportation, trade, storage and services.[26] During project implementation—according to the TECS estimate—about 18% of the project's financial cost will be for wages, with unskilled labour amounting to about 3.7 million person years. The wage labour class, mostly comprising Dalit (oppressed caste) and adivasi communities, will benefit from these employment opportunities. Whether there will be any significant positive impact on wages remains to be seen (see Banaji 1995; Breman 1985).[27] Likewise, such 'benefits' have to be weighed against reports of repressive labour regimes (e.g. denial of minimum wages and of the right to collective bargaining, etc.) (see Srinivasan et al. 1989).

4. *Professional and Educated Class:* If TECS estimates are any indication, skilled labour constitutes 27% of project costs, indicating a rise in demand for blue collar and white collar jobs[28] for project implementation, maintenance and consequent to its multiplier effects. SSNNL, the corporation in charge of construction and implementation of the project, has become one of the largest bureaucracies in Gujarat. In addition, the project headquarters in Kevadia, the numerous project canal divisions in the state, the Narmada offices in Baroda, Ahmedabad and other cities in Gujarat, generate enormous demand for a skilled labour force.[29]

The SSP also generates employment opportunities for skilled workers in the NGO sector which is closely involved in the SSP, providing critical inputs in resettlement, in water management, and in the formation of water-users' associations in the command area. NGOs also contribute to social and environmental impact assessments and action plans.

5. *Rent-seeking Class:* The benefits from the SSP accrue to 'rent-seekers' in the form of subsidies, consultancy and contracts. By stretching the definition of rent-seeking, kickbacks and votes could also be included in this category. Beneficiaries are contractors, engineers, bureaucrats, politicians, international consultants and experts in financial and donor agencies.

In assessing water resource investments, studies on irrigation projects mention how the rent-seeking chain spreads from the World Bank to the irrigation bureaucracy, down to the canal operator at the village level (see Repetto 1986; Wade

1975, 1976). Political economy scholarship views this category of people as major project beneficiaries who wield tremendous political clout, power and patronage (see Singh 1997).
6. *Other Beneficiaries:* Doubts over drinking water benefits notwithstanding, if and when they materialise, people of selected cities and villages in Gujarat would benefit from improved drinking water supply. Women in these areas may be considered immediate beneficiaries as they expend labour time in ensuring household water supply. Drinking water supply, however, will be the last benefit to be derived due to the massive scale of investments required, over and above those provisioned, to put in place the necessary water supply infrastructure. In Kutch and Saurashtra regions, where drinking water is scarce and where all villages have been brought under the purview for drinking water supply, benefits cannot be expected until 20 years after completion of the SSP.

Since the SSP reservoir would be used for such purposes as fishing and tourism, these activities are expected to generate some benefit for the local population. The reservoir extends over an area of 37,000 ha with a linear stretch of 214 km. Project authorities claim that fisheries development in this large water body will yield 80 kg/hr/yr[30] (Alagh et al. 1995: 270), largely benefiting households in the submergence zone as well as some 15% of the affected fishing communities downstream. However, local fishing communities often do not enjoy these benefits as fishing rights in the project reservoir are auctioned to rich private contractors. The Bargi project on the Narmada in MP is a good example: the affected local community has staked its concerted claim to cooperative fishing rights in the reservoir.

Bearers of Project Costs

1. *Farmers in the Reservoir Zone:* The Sardar Sarovar reservoir will submerge (partially and fully) about 245 villages. These villages can be divided into three types: mainstream caste villages in the Nimad plains of MP (about 140), adivasi forest villages in the upper hills spreading across the three states of MP, Gujarat and Maharashtra (about 85), and intermediary villages (about 15) lying between the Nimad plains and the

adivasi villages. In 1985 it was estimated that 41,000 ha were to be submerged,[31] which meant the eviction of 67,000 people[32] in the three states: 45,000 in MP, 11,700 in Maharashtra and 10,500 in Gujarat. In 1993, revised estimates doubled the number of affected people to about 127,000, i.e. 89,700 in MP, 19,000 in Maharashtra and 18,000 in Gujarat.[33]

The impact of the reservoir will differ in time and across regions, classes, ethnic groups, gender, age groups, villages and households. A more detailed analysis of the differential impact of the Sardar Sarovar, taking some of these aspects into consideration, is the subject of Chapter 8. Suffice it to mention that the Nimad valley in MP, bearing the brunt of submergence, has fertile land, assured irrigation and prosperous farmers. The latter would be the worst losers under the SSP. Since the land compensation package is subject to ceiling laws of the three states, families with large holdings face the risk of marginalisation.[34] In the intermediary area, although land is far less fertile, some farmers have access to irrigation. The displacement impact will be similar to that of Nimadi farmers, although the average holding size is smaller and land productivity less compared to the situation in Nimad.

In the adivasi villages, agriculture is rainfed and is carried out on hilly slopes. Few households own land, the majority depending on 'forest land' for cultivation. Forest land, termed 'encroached' land in official discourse, is crucial to the survival portfolio of virtually every peasant household in the adivasi region. These households also depend on other forest products. The risks to adivasi peasants have been the subject of debate in the SSP and will be addressed later in this book. Here we need only draw attention to the fact that the risks to adivasi peasants are somewhat cushioned through compensation policies of project authorities, thanks to sustained political pressure. However, the compensation package does not offset cultural losses, social disarticulation, and common material and social resources.

2. *Landless Labourers and Marginal Farmers in the Reservoir Zone:* More than 50% of project-affected people in the reservoir zone are landless agricultural labourers and marginal farmers. Being mostly adivasis and Dalits, they will lose their jobs and the security of the village moral economy. Together

with marginal farmers, this group depends heavily on village commons and similar resources for their subsistence. Past experiences with displacements show that the landless are always left out of the purview of compensation as compensation policies by definition are land-centred. The risks of impoverishment among this category of people are high. In the SSP, however, revisions to compensation policies and packages in the post-appraisal years seem to cushion risks to this group. If compensation is properly and justly administered, some households may even stand to benefit because under the Gujarat government's policy, an affected landless agricultural labourer is entitled to 2 ha of land as compensation in the project command, as are his/her adult sons. More on this in Chapter 8.

3. *Women in the Reservoir Zone*: In one of the early research studies on affected women in the submergence zone of the SSP, Mehta (1992: 115) observes the following:

> the problems attached to displacement that directly affected women were often not perceived by the women as women's problems. It was difficult for the women to identify women-specific problems as opposed to those [which] concerned the community at large....The interests of the family or larger group outweighed those of the women.

This observation on women's interests being subservient to larger household interests should be seen in conjunction with the fact that women in the reservoir area do not constitute a distinct interest group. As might be expected, their identity is fragmented across class, region, caste and age group. The political implications of women's inability to mobilise themselves and promote their specific interests in displacement, or the failure of mediating NGOs and action groups to politicise the specific vulnerability of women, are issues of further research. Women, however, will certainly lose access to common resources (fuelwood and water) and social elements (neighbourhood and kinship networks), essential for household productive and reproductive functions. Given that women in the affected areas are involved mainly in reproductive activities, the relative psychological stress of displacement and uncertain future is likely to affect them more than men (see also Parasuraman 1993).

The 1985 appraisal report of the World Bank did not address women's specific needs. Ten years later (during which period there was an apparent increase in the Bank's sensitivity to women's issues),[35] the Bank's 1995 project completion reports (PCRs) on the SSP failed even to mention women as a specific category of losers. As the PCRs also include 'lessons learnt' from the SSP, one can conclude that the Bank is either ignorant or simply lackadaisical about the impact of displacement on women. It should therefore be stated once again that compensation policies and resettlement plans almost never address the specific problems of women. In pursuing compensation, men attach top priority to land quality, quantity and also jobs. Access to common property resources, social networks, provision of drinking water, etc. mean little to them. Compensation— whether cash, land or jobs—almost always goes to men as husbands, fathers or sons. It is worth noting, however, that in the SSP, the compensation policy of the Maharashtra government directly entitles women—specifically, adult unmarried daughters—to compensation.

4. *Other Categories of People in the Reservoir Zone:* Other likely losers include landless people engaged in non-farm occupations, particularly in the Nimad valley. Artisans, including fisherfolk, various service providers such as barbers, washerpeople, carpenters, tailors, leather workers and those in petty trade and shopkeeping, are all affected. The loss to them is not in terms of productive assets, which theoretically can be removed lock, stock and barrel, but of established clientele in the village economy.

Children and youth with access to education in the reservoir area also form a potential category of losers, but their loss is likely to be offset by ensuring access to education under the resettlement policies. In fact, some youth in remote affected villages foresee benefits from displacement, e.g. directly through land entitlement (adult sons, unmarried adult daughters) or indirectly through access to more 'modern' amenities in and around the resettlement sites.

5. *Canal Affected People:* The 50,000 km canal network of the SSP will affect a large group of farmers in Gujarat. According

to project authorities, the first 144 km stretch of the main canal required the acquisition of 3,520 ha of land (Patel 1992). A 1991 study conducted by the Indian Institute of Management, Ahmedabad, covering 300 km of the main canal, listed 165 villages affecting 11,000 farmers and needing the acquisition of 6,000 ha of land. It is estimated that between 60,000 and 85,000 ha of private land will ultimately be required for the entire canal network. According to SSNNL estimates, 170,000 farmers will be affected, of whom 145,000 will lose less than 25% of their holdings. About 2,000 farmers will lose 75% of their holdings, while 750 farmers lose everything (Alagh et al. 1995: 305). Following from this, project authorities assume that since a large majority of farmers will be only partially affected, there will be no large-scale displacement *per se*, but only shifting and realignment of homesteads and land in the vicinity. The IIM study, however, estimates that about 10,000 farmers will lose all their land.

In principle, the command area and the reservoir area are not treated equally as the provision of irrigation in the former is expected to offset the loss of land due to the canals. Concomitantly, the compensation package for canal-affected people for land acquired is cash-based and undertaken under the Land Acquisition Act, 1894. Canal-affected villages and their supportive action groups, however, have demanded similar entitlements to their reservoir counterparts. This has culminated in the improved provision of compensation for those losing more than 75% of their holdings.

6. *Downstream People:* From the dam site in Kevadia to the Arabian Sea, the Narmada flows for about 175 km (about 10% of its catchment). As in the canal systems, the impact of the dam on the population downstream will be spread over a period of time. Water availability downstream will depend on the development of the command area and the canal networks. When the command area is fully developed, the riverbed downstream is likely to remain dry, barring four months' monsoon. This prompted one official to suggest that it will be suitable 'to play cricket in it' (cited in Soni 1993:7). Apart from the problem of dry bed, there is that of excess water flows that will confront downstream users. Drawdowns of reservoir release water downstream will create artificial flooding. With careful

planning, however, this problem can be converted into some benefits for downstream farmers. [36]

Under the SSP, at least 5,000 families of the fishing community (an estimated 25,000 people using a multiplier of 5) and about 300 families (1,500 people) engaged in boat transport will be directly deprived of their means of livelihood. Both industry and agriculture downstream will also be affected because of lost access to water. Furthermore, about 30% of downstream villages that depend on the river for their daily supply of water will be affected.

The downstream impact of the SSP on the esturine and riverine fisheries can be gauged from the following. The esturine yield is estimated at 250 kg/ha/yr, whereas riverine yield is 530 kg/ha/yr. Fish include mostly the high economic value varieties such as *hilsa*,[37] giant prawns and mullet. Moreover, the fishing regimes in the Narmada estuary supply seed stocks to freshwater inland fisheries. A study conducted by the Central Inland Capture Fisheries Research Institute (cited in Soni 1993) states that, on average, the annual income of a fishing family (from fishing) is Rs 10,500. Thus, the downstream losses for those dependent on fisheries include reduced yield and income.

7. *Other Losers:* Apart from villages affected by the construction of project headquarters and staff colonies for the SSP in the early 1960s, as already discussed, land may also be required and acquired for purposes such as catchment area treatment and compensatory afforestation as part of environmental mitigative measures. In the SSP, the reservoir catchment area earmarked for priority treatment is approximately 155,000 ha. Forest lands, government wastelands and even village commons could be used for such purposes, thereby affecting people dependent on these lands.

Three conclusions can be drawn from the foregoing. First, there are more interests and stakes and consequently claims and preferences than are acknowledged in the project appraisals. Some interests filter through to the cost-benefit net through collective action and political representation. Hence, after three years of sustained political protest, the 1990 World Bank reappraisal increased the resettlement costs by 300% and included a specific sum as compensation for canal-affected people. This only strengthens the argument that the appraisal process is political in nature and is susceptible to pressure from those who enjoy political power and clout vis-à-vis the decision-makers of the day. As will be seen later, this holds true for project evaluations as well.

Second, the distribution of benefits in the SSP favours developed and better-off constituencies in society. Although water and power benefits seem to have an even spread in terms of prospective beneficiaries, it is very likely that they will accrue first to those who are already better off. The likely losers form a heterogeneous set of people, including some with considerable material and political resources and some without any. Promises of, and interests in, compensation make it necessary to differentiate between those who lose, those who can be compensated, and those who stand to gain.

Third, and closely linked to the second, are the incremental gains in government policies on resettlement. They cushion some displacement risks, but when compared to other projects in the three riparian states, the entitlements of losers under the SSP are definitely advanced.[38]

Environmental Appraisal in the SSP: Political Interests

The Department of Environment and Forests (DEF), Government of India, had refused environmental clearance to the project, expressing concern over forest loss and dissatisfaction with the environment impact assessments. It also refused permission to clear forest land for resettlement purposes under the 1980 National Forest (Conservation) Act. In addition to the DEF, international NGOs and some national environmental groups voiced their concern regarding the environmental impact, particularly forest and wildlife loss, on the SSP. However, grassroots action groups in Gujarat and Maharashtra were then mobilising adversely-affected people for land-based compensation. They demanded the release of forest land for resettlement, particularly for the adivasi villages. As a result DEF came in for criticism on its rigid stand on conservation. In 1985, when the DEF was upgraded to the Central Ministry of Environment and Forests

(MoEF),[39] it came under pressure from grassroots groups, the Central Ministry of Water Resources, state governments and the World Bank to release forest land. Meanwhile, project authorities went ahead to complete the appraisal process and sign the loan agreements with the World Bank without seeking the mandatory environmental clearance from the MoEF.[40]

In 1986, the Government of India passed the Environmental (Protection) Act. The SSP (together with the NSP in Madhya Pradesh) now required both forest clearance and environmental clearance from the MoEF. The latter was concerned, however, about the incomplete studies and assessments and work plans on rehabilitation, compensatory afforestation, catchment area treatment, command area development and wildlife. It also stressed the need for 'a review of the storage planning of SSP and NSP' with a more accurate study of Narmada water yield data than had been agreed to by the states and assumed by the NWDT. The MoEF was also not convinced that the hydrological estimate, i.e. annual utilisable water flow in the Narmada, which had been sorted out through political agreement between the chief ministers of the states, was accepted by the NWDT.

In June 1987, the MoEF accorded *conditional* environmental clearance to the SSP and NSP. Its office memorandum mentioned that 'field surveys are yet to be completed' on the rehabilitation master plan, catchment area treatment schemes, compensatory afforestation plan, command area development, survey of flora and fauna, seismicity and health-related aspects. It put forth the conditions that 'the NCA will ensure that environmental safeguard measures are planned and implemented *pari passu* with progress of work on projects' and 'the detailed surveys/studies assured will be carried out as per the schedule proposed and details made available to the department.'[41]

A few months later, the MoEF sanctioned the diversion of about 13.5 thousand hectares of forest land[42] for the SSP but again put stiff conditions and deadlines for the state governments. It demanded full details of non-forest land identified for compensatory afforestation, including plot numbers and names of villages. Additionally, it asked for full details of non-forest land available for the rehabilitation of all oustees. Plans for the treatment of catchment areas, doubling forest cover through compensatory afforestation in degraded forest lands in the project impact region, and tree planting on either side of the canals and roads on the foreshore of the reservoir and other wastelands were further conditions set by the MoEF.[43]

Why did the MoEF tone down its earlier stand on the projects and issue 'conditional clearance'? For a large part, it gave in to pressure from the Bank and project authorities. Between 1985 and 1987 the Bank, on its own admission, exerted 'much pressure on GoI for releasing the environmental clearance so that construction work could start' (WB 1995c: 3). By this time over Rs 3,500 million had already been spent on the SSP. Tender contracts totalling Rs 6,000 million had been called and opened and were awaiting acceptance, for the dam, powerhouse and canals. A bilateral credit had also been signed with Japan. According to official estimates, the country was losing Rs 50 million a day on account of foregone benefits from the project. However, these economic factors should be seen together with political developments in the country if a clearer picture is to be obtained of why environmental clearances were accorded to the SSP despite serious reservations of the MoEF.

According to the MoEF, the necessity for environmental and forest clearance was felt at the highest political level, i.e. the Office of the Prime Minister (MoEF submission to FMG:1993). In 1987, the Prime Minister and his Congress party were facing trying times. The former was believed to be involved in the infamous Bofors scandal[44] and discontent simmered within the party.[45]

In Gujarat, the politically winning strategy of the Congress— the famous Khastriya, Harijan, Adivasi and Muslim alliance (popularly known as KHAM)—was beginning to wear thin. In the mid-1980s, a peasant front, the Bharatiya Kisan Sangh (BKS), had made strong inroads into drought-affected north Gujarat, demanding lower electricity charges and higher prices (Omvedt 1993: 262). Supported by the Bharatiya Janata Party, the BKS had the backing of about 500,000 farmers (Banaji 1995: 232). In south Gujarat, the Khedut Samaj had formed an alliance with the Bharatiya Kisan Sangh in the north as well as with the Shethari Sangathana in Maharashtra. Patidars dominated both peasant movements in Gujarat.

In March 1987, Gujarat witnessed violent farmers' unrest. The agitators demanded removal of the ceiling on agricultural and urban land, lifting of the ban on farmers buying land more

than 8 km away from their homes, exemption from the stipula-
tions of the Minimum Wages Act (including withdrawal of cases
against employers who did not comply with the Act) and, above
all, implementation of the SSP on a war footing (Banaji 1995: 239).

Against the background of these developments, the Prime
Minister's decision to clear the SSP and NSP made political
sense, both to improve a tarnished image and to elicit political
support. In May 1988, responsibility for project implementation
was transferred from the Narmada Development Department
(a Gujarat government agency) to the Sardar Sarovar Narmada
Nigam Limited (SSNNL), a parastatal corporation organised
along functional lines. A few months later, in October 1988, the
Planning Commission gave the mandatory investment clearance
to the SSP.[46] By 1987–88 full construction work on the project
had commenced.

Notes

1. In terms of volume of concrete (gravity dams) the Navagam dam is
    second only to the Grand Coulee in the USA. Its spillway discharge
    capacity of 3 million cusecs places it after Gazhonba dam in China
    and Tucuri dam in Brazil. As far as its reservoir storage is con-
    cerned the Sardar Sarovar is the fourth largest in India after the
    Nagarjuna Sagar (Andhra Pradesh), Rihand (Uttar Pradesh) and
    Bhakra (Himachal Pradesh).
2. In anticipation of the NWDT Award, the World Bank's reconnais-
    sance mission in 1978 had evinced interest in the Narmada devel-
    opment projects. By 1980, it had taken an active interest in funding
    the projects. To fulfil the Bank's requirements for loans and aid,
    the Narmada Planning Group—a policy-making advisory body in
    the SSP (hereafter NPG)—commissioned studies on project features,
    environmental and social impact in the command and submer-
    gence areas, and project benefits and costs. The United Nations
    Development Program (UNDP) and the Bank provided the NPG
    with financial assistance for project preparation. The project con-
    cept and design closely followed the NWDT award. The project
    development plan was prepared in 1983 by the Gujarat govern-
    ment. The Narmada Development Department, Government of
    Gujarat, in charge of project implementation prepared detailed
    designs and cost estimates for the major project structures. The
    Bank's own appraisal took from March 1983 to August 1984. Credit

negotiations took place in November 1984 and the projects were formally presented to the Bank Board in March 1985.

3. Each village service area (VSA) will be divided into *chaks* (30–60 ha) and each *chak* into sub-*chaks* (8 ha). A minor canal will carry water from the VSA to the head of *chaks* and a sub-minor from *chak* to the head of sub-*chak*. Field channels will carry water from sub-*chaks* to individual fields (Alagh et al. 1995).

4. Covering the states of Gujarat, Goa, MP, Maharashtra, and the Union Territories of Daman and Diu, and Dadra and Nagar Haveli.

5. Distribution weighting can be attached to costs and benefits but how and what is open to contest, methodologically and in value terms.

6. Cropping intensity is estimated in terms of gross cropped area as a percentage of net cultivable area. The net cultivable area in the command (cultivable command area) is 2.12 m. ha.

7. The TECS report estimated the increase to be 128% (TECS 1983:51) whereas the WB appraisal put it at 135–140% (WB 1985f: 176–78).

8. Irrigation intensity is the ratio of irrigated to irrigable area. Some other definitions see it as gross irrigated area as a percentage of gross cropped area.

9. In the economic appraisal undertaken by TECS (1983), sugarcane is listed as a major crop whose annual incremental production was estimated to be about half a million tonnes (+250%). The TECS report did not consider production increases for vegetables, fruit and fodder. In the economic evaluation of the Gujarat government (GoG 1983), sugarcane does not feature as a major crop, whereas the Bank (1985c) estimated that its incremental production would be insignificant.

10. While there is no major difference between the GoG, TECS and the World Bank in estimates of the incremental production of food crops, the estimates on percentage increase in cash crops are different—Bank data show a 95% increase whereas computation of TECS data yields 185% increase. This is despite the fact that TECS data do not include vegetables, fruit and fodder which, according to Bank estimates, will have the maximum incremental yield.

11. TECS data show an increase of 500% in the *net* value of agricultural output (net value = gross value–cost of cultivation) with the project compared to the situation without the project (TECS 1983: 53).

12. This is the Bank's appraisal figure used in subsequent government reports (see NCA 1990). The TECS appraisal was based on total installed capacity of 1,000 MW.

13. The total installed capacity (1983 figures) in the western grid region is about 10,045 MW, consisting of 78% thermal, 18% hydro and 4% nuclear. About 51% of total regional capacity is located in Maharashtra, 28% in Gujarat and 21% in MP. In 1981–82, industrial electricity

consumption represented 62% of the region's total, agricultural
uses 13% and the remaining 25% for domestic and commercial
(domestic and commercial light, railway traction, public water
works, lighting etc.) consumption (WB 1985f).

14. There appears to have been a major change in appraisals in World
Bank reports on the irrigation component of the project. In the
IDA report and recommendation to the Executive Directors, the
Bank (6 February 1985) estimated the costs of the irrigation component to be much lower. The figures in that report for irrigation
items, in Rs Million, are as follows: main canal 3,311; branch canals 792; distribution and drainage systems 388; and technical
assistance 14, largely because at that time the Bank was only going
to help finance 144 km of the main canal and the distribution network.
The report also made no mention of command area development.
However, the figures were later revised, as shown in Table 4.6.

15. This excludes physical and price contingencies to the tune of Rs
73,760 million. The total project cost including physical and price
contingencies was Rs 136,407 million.

16. ARCH-Vahini consists of erstwhile members of Action Research
for Community Health and the Sangharsh Vahini (volunteers for
struggle), a youth group formed in Bihar and Gujarat during the
movement against the declaration of Emergency in the 1970s, also
popularly called the 'JP movement' after its leader, Jay Prakash
Narain.

17. By 1983, the study—conducted through the Centre for Social Studies in Surat—comprised 17 village reports. The same year the main
findings were presented to the Bank noting (a) that an adivasi
village was not a single homogenous unit and that its ethnic and
secular differences meant that en masse rehabilitation of oustees
was not desirable; and (b) that such villages, while being separate
and distinct, change with progress in wider society. They must,
however, have autonomy to decide the pace of such change (see
Joshi 1987).

18. In Maharashtra the Tribal Research Institute had conducted socio-
economic studies of adivasi people displaced in public works
projects. However, there was no evidence of this data being incorporated into either the Bank's pre-appraisal works on resettlement
or the GoM's resettlement plan at the time of project appraisal in
1985. As far as MP was concerned, it had not completed any socio-
economic study by the time of the Bank's appraisal (WB 1995b: 3).

19. Some of these variables are costs of individual components in real
terms, resource availability, agricultural yields and inputs, volume of energy produced, economic values attached to benefits,
implementation speed (delays in construction), irrigation system

efficiencies, water availability and requirements, and speed of adoption in build-up of agricultural benefits.

20. According to the appraisal reports of the World Bank, the simulation model was run 500 times, each time with a different set of parameters and by calculating the ERR for each run.

21. The Bank also argued that since M&I water allocation was relatively small (1,310 million cubic metres), any errors in estimating its value would not significantly affect the overall result of the economic analysis (WB 1985f: 158).

22. It should be noted that the popular movement against the dam had by then intensified.

23. Economic appraisal was also undertaken as part of the Project Completion Report (1995). While this appraisal is said to have been completed in April 1993 (WB 1995c: 2), it is actually a reproduction of the mid-term reappraisal conducted in 1990.

24. Ideally the economic appraisal should indicate that between mutually exclusive projects, one should choose the alternative which shows the most favourable IRR or CBR. However, this principle need not be a hard-and-fast rule. It is quite legitimate to estimate whether a project (selected on other grounds) shows a favourable benefit-cost ratio or internal rate of return (or positive net present value). The point is that both conditions are susceptible to manipulation. Favourable cost-benefit appraisal can be applied cosmetically to justify projects already selected on other grounds. As far as comparison of alternatives is concerned, an appraisal can compare the (already selected) project with only those alternatives that it can beat on economic grounds.

25. Agrarian studies show that sugarcane production in India has a direct relationship with political clout and power. Breman (1985) refers to the displacement of cotton by sugarcane after the coming of the Ukai project in south Gujarat. See Attwood (1992) for a discussion on the political economy of sugarcane in western India.

26. Studies elsewhere point to this benefit. For example, in the Bhima Command Area Development project, average days of employment for male farm labourers increased from 180 per annum to 250 after irrigation (Jazairy et al. 1992:134). In the SSP, job opportunities are also created in programmes of environmental protection, such as catchment area treatment and afforestation.

27. Also see Gorter (1989). Following Breman (1985), he shows that growing demand for labour in the sugar factories, sugarcane and paddy fields in south Gujarat in the Ukai Command has led farmers to recruit labour from other parts of south Gujarat and Khandesh in Maharashtra, an indication of changing labour relations with perennial canal irrigation.

28. It is difficult to establish the extent to which these benefits accrue to Gujaratis; one can, however, expect them to be the major beneficiaries. Furthermore, between blue collar and white collar jobs one can assume a 70:30 ratio considering the pyramid structure of public sector organisations in India.

29. Two points need to be made. First, the listing does not include the Narmada Valley Development Authority in MP, which, apart from overseeing the MP projects, coordinates the resettlement component in the SSP with the Sardar Sarovar Poonarvasat Agency in Gujarat. Second, the creation of these bureaucracies does not automatically imply additional recruitment of skilled personnel. A sizeable section of personnel are under deputation from other projects and departments.

30. This seems to be a gross exaggeration. Soni (1993:13) cites the study conducted by the Central Inland Capture Fisheries Research Institute which states that 'the average yield from Indian reservoirs at the present rate hardly exceeds 15 kg/ha/yr.'

31. The World Bank's 1985 reports are inconsistent as far as the total submergence area is concerned. In the Staff Appraisal Report the figure mentioned is 37,000 ha—18% of which is forest land, 33% cultivable land and nearly 50% is waste or of low utilisation potential (WB 1985c: 28).

32. This figure was subsequently revised to 74,230 affected people (11,850 families) as per the NCA's supplementary report of 1984.

33. Officials attribute the doubling of the figure to population growth between the 1981 and 1991 censuses.

34. *Benami* holdings in the area are common. To avoid land ceiling laws a family may hold land under fictitious names. In the Nimad plains big landholders own land that is 'legally' in the names of their 'family agricultural workers'.

35. See Srinivasan (1997) for a critique of the World Bank's policy of mainstreaming gender.

36. Controlled release of water downstream (preferably all year round) is considered the best solution to minimise the downstream impact. Adams' (1993) study of the Bakolori dam in Nigeria comprehensively documents the problems faced by downstream users, particularly farmers.

37. *Hilsa* is an anadromous migratory variety like salmon, migrating upriver from the sea to breed.

38. However, compared to the policies of Gujarat and Maharashtra, the resettlement policies of MP are inadequate. See Appendix 2 for a comparative picture.

39. The upgrading of a five-year-old department to a fullfledged ministry underscores the importance that environmental issues had come

to bear on the national psyche and agenda.

40. Under the 1985 agreement between the World Bank and Government of India, the latter was to 'take all actions necessary to release forest lands reserved by the Forest (Conservation) Act within the boundaries of the states of Gujarat, MP and Maharashtra if required for the purpose of implementing the Sardar Sarovar Dam and Power project *including the resettlement and rehabilitation programs and plans*' (WB 1985a: 21 emphasis added).

41. Ministry of Environment and Forests, Government of India, Office Memorandum No.3-87/80-IA of 24 June 1987.

42. The break-up in the three states: Gujarat 4,165 ha, MP 2,731 ha and Maharashtra 6,488 ha.

43. MoEF, GoI Letter No. 8-372/83-FC of 8 September 1987.

44. In early May 1987, Swedish radio announced the discovery of a scandal involving payments of millions of dollars in kickbacks by their Bofors company for a weapons contract to unnamed Indian middlemen who were influential in Congress party circles.

45. Between April and July 1987 first came the resignation of V.P. Singh (later to become Prime Minister) as Defence Minister and then the expulsion of four senior leaders from the Congress, including Singh.

46. The clearance was conditional. The Commission sought the compliance of states with the conditions and requirements of the MoEF, asked for priority funding for the project from the states during the Eighth Five-Year Plan period, and asked Gujarat state to draw up detailed implementation schedules for completion of micro-level canal networks in the command area.

5

Collective Action and the Development Discourse: Against the SSP, Against Development

Introduction

This chapter sets the backdrop for a fuller analysis of the significance of the Narmada movement, a social movement which links local actions against the SSP with the 'discourse' of development at a macro level.

The questions addressed here are: What issues fuelled the Narmada movement? How does it represent displacement and despoliation in the SSP? What understanding of development does it construct and convey through its politics and practices? The period covered runs from 1986 to 1996, the period termed the construction phase, with the qualification that, while construction work on the dam and canals indeed gathered momentum during this period, work on the dam was suspended in early 1995 by order of the Supreme Court of India and was only partially resumed in 1998.

The primary data are drawn mainly from campaign documents, newsletters, correspondence, and interviews and informal discussions with leaders and activists involved in the campaign.

Collective Actions around Displacement and Despoliation: Historical and Contemporary Evidence

Collective actions over dam-induced displacement in India have a rich history. Local people have periodically resorted to protest actions, both during colonial rule and in the post-Independence era, although a sharp decline in such protests despite a steady increase in displacement was discernible in the era of development optimism of the 1950s and 1960s. From the 1970s onwards, there has been a renewal of political action around large dams, not only regarding displacement but increasingly over environmental despoliation.

The Mulshi Satyagraha

One of the early protests was the Mulshi Satyagraha,[1] a movement organised between 1921 and 1924 against the proposed Nila-Mula hydroelectric project near Poona (Pune) in Maharashtra. This early protest action has been of some interest to contemporary scholars of the Indian environmental movement (see Gadgil & Guha 1995: 68–70). The movement can be summarised here based on an excellent doctoral thesis by Livi Rodrigues (1984) on peasant movements in western Maharashtra in which a chapter is devoted to the Mulshi Satyagraha (pp. 180–236).

The Mulshi works proposed by the Tata Power Company (TPC) and approved by the colonial government of Bombay (Bombay Presidency) in 1919 involved a dam at the confluence of rivers Nila and Mula near the village of Mulshi, and support structures for a powerhouse mainly to supply electricity to the textile industries in Bombay. The project affected around 4,000 ha of land in over 50 villages and threatened to displace 10,000 people in the area. The colonial government and TPC justified the construction on grounds that it would replace fossil fuel power that was expensive for the mill industry, create irrigation potential, and generate employment opportunities in the cotton mills.

The Bombay Presidency gave full support and concessions to the TPC and set out to acquire land for it.[2] However, the majority of affected people in the Mulshi Peta area refused to move from their lands. They were mostly small-holding Malwa peasants (with average holdings of 1 ha): owner-cultivators, tenants and sharecroppers. A small class of rich peasants and one of money-lending landowners (Sahukars) were among others affected by the project.

The pressure for political action came from the Malwa peasantry. In the initial years (1919–1921) the strategy adopted was one of legal appeal and individual resistance. The publicity given by the nationalist Marathi press soon drew Congress activists to the affected villages. To them the mediation meant a mass base for the Congress movement; to the affected people the mediation meant outside support and the opportunity to organise mass action. Protest action through satyagraha soon became imminent as a militant mood developed at the grassroots, especially among the Malwas. Refusing to part with their land or to accept compensation, the Malwa peasantry rallied under the slogan *jamin ya jaan!* (Land or Life!).

The affected money-lending class and the absentee landlords wanted to defer the satyagraha and were inclined to use it as a bargaining chip for maximum compensation. Rodrigues (1984: 198) notes the specific interests and dilemmas of this class:

> If the satyagraha was started...and the government suppressed it fixing a nominal compensation for the land acquired, then this would be a loss to the sahukars. On the other hand, if the satyagraha were successful, if the project was halted, if the Malwas retained their land, their hand would be strengthened and it would be difficult for the sahukars to continue at Mulshi.

Their preference for 'constitutional agitation' rather than 'passive resistance' of the satyagraha, however, was countered by pressure from the peasantry. The first phase of the satyagraha commenced in April 1921.[3] The dam site was occupied by volunteers and work successfully stalled for 15 days. The TPC agreed to suspend the work for six months and the satyagraha was temporarily called off.

This first phase was clearly a united front. Different factions (radical and constitutional; Tilakites and Gandhians) of the Maharashtra Congress and different class and caste groups in the Mulshi valley had joined ranks in the struggle. Support groups had also been formed to mobilise public opinion in the urban centres, particularly Poona. However, the landlords entered into negotiations over compensation with the TPC and extracted an acceptable package. This deal was supported by a faction of the Congress and the colonial government soon announced that work in the project would continue:

> [The] government will spare lands for the ryots who come forward and express their willingness to settle elsewhere. To those who prefer to take cash a very liberal compensation will be paid...long before the people leave their homes.

The agreement however did not deter the peasantry from going ahead with the second phase of the satyagraha a year later in April 1922, this time under the leadership of 'Senapati' Bapat,[4] but work on the project could not be stopped. The government came down heavily on the volunteers; many were arrested and imprisoned for one to six months. The numbers of those accepting compensation rose steadily.[5] At a certain stage of the agitation, Bapat mobilised outside support and expressed the opinion that 'the Mulshi issue should be fought entirely by young volunteers from outside the Peta, the Malwas themselves merely standing by' (ibid.: 215). This observation by Bapat indicated the felt need for a wider support base and actors for peasant-related actions to succeed. Ideologically, therefore, the Mulshi agitation was linked to the broader struggle against colonialism and imperialism. However, Bapat and the local leaders seem to have had little success in their endeavour to broaden the struggle base and the language of protest.

When the third campaign began in September 1922, the volunteers, including Bapat, were quickly arrested. Incidents of stone-throwing and violence led Gandhi's followers to delink themselves from the agitation and in fact to condemn it as being the 'opposite of satyagraha' (ibid.: 223). The campaign then entered a period of crisis; leaders were put behind bars and the bulk of the Malwa peasantry seem to have accepted compensation. At a local leader's request, Gandhi observed:

> It appears to me that the movement has got to be dropped for two reasons, or rather three: (a) I understand that the

vast majority of farmers have accepted compensation and that the few who have not, cannot perhaps even be traced; (b) the dam is nearly half finished and its progress cannot be permanently stopped. There seems to be no ideal behind the movement; (c) the leader of the movement [referring to Senapati Bapat] is not a believer out and out in non-violence. This defect is fatal to success (ibid.: 218).

Following from this advice, the Mulshi Satyagraha Committee was dissolved and the agitation wound up. Bapat tried to revive some spirit thereafter but was soon imprisoned for seven years for acts of violence. The first organised struggle against dam-related displacement in India (in the 20th century) thus came to an end. However, it set the shape of future struggles involving peasants' land rights and industrial development. Senapati Bapat rightly predicted that 'the problem of rehabilitation of the farmers would in due course, arise everywhere in India, once similar schemes were accepted and implemented in various parts of the country' (ibid.: 221–22).

Protests over the Hirakud Dam

In 1945, the government of Orissa proposed the Hirakud multi-purpose project as a panacea for the repeated flooding of the river Mahanadi and the devastating effects that this had on the densely-populated coastal delta.[6] Although the Tikarpara site about 200 km downstream 'promised the cheapest storage and greatest head of water for hydro-electricity' (Hart 1956:120), it was opposed by native rulers of the feudatory states—the *gadjats*[7] in the Athamalik region fearing large-scale submergence. The site was therefore shifted to Hirakud upstream near the town of Sambalpur in western Orissa, and the foundation stone of the project was laid in March 1946.

The main purpose of the Hirakud dam was flood control in the coastal delta (250 km downstream). Benefits from irrigation (in western Orissa tracts) and power generation were also envisaged. However, the project involved the submergence of about 250 villages, 113 fully and the rest partially, affecting 22,150 families, mostly in western Orissa. About 100,000 people would be affected. Whereas initially the total cultivable land to be submerged was estimated at 28,000 ha, the subsequent increase in dam height by two metres (to 189 metres) led to the additional submergence of over 12,000 ha of cultivable land. As a flood-control project for coastal Orissa, the Hirakud project therefore generated interest conflicts between coastal and western Orissa, situated respectively downstream and upstream of the project.

The anti-dam agitation was led by local political leaders (including members of the Congress party who resigned in order to join the agitation) and some retired administrators from western Orissa. As noted by Pattanaik et al. (1987) and Baboo (1991), the language of protest evoked feelings of regional discrimination. If coastal Orissa needed to be saved from floods, why should people in western Orissa be flooded, the leaders asked. During the course of the agitation the problem of displacement soon became articulated in regional–cultural terms, which formed the basis for the political demand for a separate state of Sambalpur, seceded from Orissa (Pattanaik et al. 1987: 52).

The agitation—involving strikes, protest meetings and processions—was backed by powerful *gauntias* (feudal zamindars) who controlled the bulk of the land in the submergence area.[8] Eminent Congress leaders at the national and state levels, however, projected the Hirakud dam as a developmental project for Orissa, rather than a flood-control project for coastal Orissa. Gandhi disapproved of the opposition to the project by local Congress leaders in western Orissa, which led local Congress leaders in Sambalpur to disassociate themselves from the agitation. Congress leaders also targeted the *gauntias* and rajas of gadjats for engineering the agitation for their own vested interests and for appropriating most of the compensation money. In 1947, the Orissa Legislative Assembly approved a resolution in support of the Hirakud dam. Nehru inaugurated the construction work in 1948, calling it a project that would modernise Orissa through irrigation and electricity generation. The anti-Hirakud dam agitation died soon thereafter.[9]

Protests over Other Projects: An Overview

The Mulshi Satyagraha and the Hirakud agitation were the precursors to anti-dam protests in the post-Independence era. Large projects of the 1950s such as the Bhakra in Punjab (1948–63),

Tungabhadra (1945–58) and Nagarjunasagar (1956–74) in Andhra Pradesh, and Rihand (1952–63) in Himachal Pradesh, however, met with little sustained opposition (Gadgil & Guha 1995: 71). Between them, the four projects displaced an estimated 140,000 people (IIPA 1988, cited in Singh 1997). The building of hydro-projects intensified in the 1960s and 1970s for a number of reasons, including food shortages and dependency on food aid, and oil price shocks. As dam construction activities increased so did political protest actions around them. Noted flashpoints included the Subarnarekha (Areeparampil 1987) and Koel Karo projects (Bharati 1991) in Bihar, the Bhopalpatnam and Inchampalli projects spread over MP, Maharashtra and Andhra Pradesh (Cholchester 1986), the Tehri project (Roy 1987) in UP and the Srisailam project (FFC 1986) in Andhra Pradesh.

The protests mainly concerned involuntary displacement, forceful eviction and inadequate compensation. Increasingly, however, environmental concerns regarding forest submergence, wildlife loss, waterlogging, salination, siltation, health hazards, reservoir-induced seismicity in these projects, filtered through to the language of protest (see Dhawan 1990; Goldsmith & Hildyard 1984, 1986), albeit in varying degrees of intensity in different protest movements.

There have been a few successful stories of protests that re-sulted in project modification or suspension. The opposition to the Silent Valley hydroelectric project in Kerala in the 1970s is a case in point. The project involved no displacement of human population but was to submerge vast tracts of pristine rain forests in south India. The opposition was led by city-based conserva-tionists on the one hand and the Kerala Sastra Sahitya Parishad, a forum on popular science, on the other. Although the govern-ment of Kerala was keen to go ahead with the project, it was scrapped in the early 1980s at the behest of Prime Minister Indira Gandhi. In the 1970s, the proposed Bedthi dam in Karnataka had to be abandoned after sustained opposition from rich spice farmers who successfully mobilised political and public support against it (Sharma & Sharma 1981).

Another environmental campaign in 1977 on the Tawa dam in the Narmada valley in MP involved not those displaced but the supposed beneficiaries of irrigation. In what was known as the Miti Bachao Abhiyan (Save the Soil Campaign), farmers in the command area of the project protested over serious waterlogging caused by excess seepage from low-lying canals. The dam actually led to a *reduction* in farm yields *after* irrigation (Mishra 1986) and the agitation forced the government to sink tube wells in the command area in order to flush out excess water.

In the 1950s and 1960s, the developing world had witnessed large river projects, such as the Kariba in Zambia-Zimbawe, Akosombo-Volta in Ghana and the Aswan High Dam in Egypt. As hydropower investments increased in the 1970s following the oil price shock, protest actions over dams around the developing world seem to have followed a similar trajectory as in India. Among major hydroprojects, the Nam Choan dam in Thailand (Cox 1987; Hirsh 1987), the Bakun project in Malaysia (Blackwelder & Carlson 1986; SCSS 1987), the Chico projects in the Philippines (Drucker 1985; Fiagoy 1987; Hilhorst 1997), the Itaipu, Machadinho and Ita projects in Brazil (McDonald 1993; Navarro 1994), the Riam Kanan, Kotopanjang and Ci Tanduy projects in Indonesia (Aditjondro & Kowaleski 1994) and the Kaptai project in Bangladesh (Oliver-Smith 1991), have all drawn scholarly attention to the protest actions that they generated.

Apart from the Nam Choan project where the adverse impact on wildlife brought conservationists to the forefront of opposition (like the Silent Valley in India), other protest actions have been linked to problems of displacement and inadequate compensation. Clearly, then, displacement appears to be a necessary if not sufficient condition for collective action. The reasons for such protests and resistance are not difficult to imagine. In the 1990s, studies on development-induced displacement[10] generated conclusive evidence about the adverse impact on affected communities in particular. Displacement is known to cause disruptions of production systems and kinship groups, the loss of assets and jobs, the disruption of local labour markets and ties between producers and consumers, the dismantling of social and food security, credit and labour exchange networks, and the deterioration of public health among displaced communities (WB 1994a). It unleashes a process of economic impoverishment and socio-political disempowerment whereby communities lose control over their material environment and cultural identity.

In India, estimates indicate that over 20 million people have been displaced in development projects (Fernandes et al. 1989).[11]

Roughly 65% of development-induced displacement is caused by large river valley dam projects. The displacement toll of the 300 large dams that are constructed every year world-wide is estimated at 4 million people. South and East Asia account for 80% of this displacement.

Displacement resulting from development activities is often justified as a cost borne by some people for the greater public good. Clear evidence to this effect is the justification of the SSP. However, the mapping of affected people in the SSP shows that offsetting losers' losses may be inadequate. There is also conclusive evidence from other projects that project appraisals ignore displacement costs; compensation packages are extremely inadequate; resettlement policies are absent at worst and ad-hoc at best; and resettlement sites lack basic amenities. Project authorities tend to view displacement and resettlement as project bottlenecks to be removed rather than as social engineering challenges that need to be addressed. Promises of compensation and resettlement made to affected people before displacement remain unfulfilled. In practice, displaced communities experience acute marginalisation.

It is this threat of marginalisation that causes strong resentment among affected communities who then express themselves through protest actions and/or resistance movements. Whereas the communities themselves are often involved in spontaneous and organised political actions, in recent years the involvement of NGOs and of social and environmental action groups has contributed towards the increased politicisation of these issues and more organised forms of protest. This is not to argue that displacement automatically leads to political action, or that all those who face displacement engage in protest and resistance actions. As Oliver-Smith (1991) points out, factors such as patterns of internal differentiation within communities, multifaceted relationships with the immediate environment and with the state, availability of local and non-local allies, and the quality of compensation and resettlement, are crucial to an understanding of the political actions of affected people. Despite these variables and their operations in different contexts, it can currently be said that displacement brings project authorities into conflict not only with directly affected local people but with a host of actors spanning the globe. In the 1980s and 1990s, the displacement problems of local people were brought to the national and global agendas by actors committed to different degrees to the causes of environment and rights. Large dams today arouse radical environmental response from a wide range of action groups and coalitions. In the 1990s, to an already long list of controversial dams were added the Maheswar and Allamati dams in India, the Arun-III project in Nepal, the Three Gorges dam in China, the Ralco dam on the Bio-Bio river in Chile, the Manatali dam in Senegal and the Hidrovia projects in Paraguay.

Collective Actions as Environmental Movements

Theorisation on such forms of collective action over displacement and despoliation have moved alongside their intensification around river valley projects. In the South, in the 1980s, scholars linked these actions to the broader corpus of nature-related conflicts over water, forests and land. Some have viewed them as resistance actions aimed at economic activities that destroy the environment and impoverish local communities (Gadgil & Guha 1995: 2). Others see in these actions an environmental awareness and abilities to redefine the concepts of development and economic values, of technological efficiency and scientific rationality (Kothari 1995: 4, 25; Shiva 1991: 24). Scholars are unanimous, however, that displacement, suffering and despoliation caused by development that is narrowly conceived and based only on short-term commercial criteria of control and exploitation of natural resources, and which almost exclusively serves the needs and interests of the rich minority, constitute the material basis of collective action.

Major development projects thus become sites of 'environmental' conflict. Large river valley projects, industrial mining, mechanised fishing and commercial forestry are therefore major action fields of movement. As these projects entail large-scale resource transfers, displacement of people and disruption of their land and livelihood, and environmental despoliation, they sow the seeds of conflict (Gadgil & Guha 1995; Sethi 1993a; Shiva 1991; Shiva & Bandhopadhyay 1989). Environmental movements can therefore be seen both as mediation in, and reflection of, these conflicts.

Guha (1988, 1989) has argued that environmental movements in the South are distinct from those in the North. To him, the essential struggle in the developing world over environmental resources is not how they should be used but who should use and benefit from them. That is to say, environmentalism in the South does not connote the quest for quality life but for subsistence. This 'environmentalism of the poor' leads him to characterise environmental movements of the South as essentially 'peasant movements'. In his analysis of the famous Chipko movement[12] over commercial forestry in the Himalayan foothills, Guha (1989) distinguishes its 'public profile' as a celebrated environmental movement in the world, from its 'private face' as a peasant mobilisation (ibid.: 178). He locates Chipko's environmental face as the contemporary manifestation of a century-old tradition of peasant struggle over forest rights.

Guha (1988) also notes three distinct ideological and strategic orientations of contemporary environmental action groups in the Chipko movement. The first, 'Gandhian deep-ecology', appeals to the superiority of religious and ethical traditions of the pre-modern world. The second, 'appropriate technology', is less strident in its opposition to industrialisation and strives towards constructing and diffusing technologies which are resource-conserving, labour-intensive and socially liberating. The third strand, 'ecological Marxism', integrates the struggles over environmental resources with those for radical social transformation, including modes of production and distribution.

In their very different ways the three wings of Chipko have questioned the normative consensus among Indian intellectuals and political elites on the feasibility of rapid industrialisation and technological modernisation. Of course the environment debate is, world-wide, as yet in its very early stages. The linkages between technology and ecology and politics and culture will undoubtedly undergo significant changes in the years ahead. In the Indian context the Chipko movement and its legacy have helped define these issues with particular clarity and sharpness. It is likely that the continuing evolution of Chipko and of its three contending subcultures will help define the outcomes as well (Guha 1989: 184).

Sethi (1993a) offers a somewhat similar classification of environmental struggles. In his view, they denote a redefinition of usufruct and control rights over the resource in question (close to 'ecological Marxism'), an environmental response seeking correctives through legal and policy modifications (approximating 'appropriate technology'), or, more radically, an ecological reaction, rejecting the dominant development paradigm and seeking fundamentally to alter existing conceptions on and modes of resource use (similar to 'deep-ecology Gandhians').

What implications can be drawn from this for an analysis of the Narmada movement? First, displacement and deprivation of livelihood ensuing from development projects condition political actions around such issues. However, collective actions over development projects often involve actors other than the affected people themselves; external actors tend to link local issues to wider political goals and visions, for example, against colonialism in Mulshi or regional self-determination in the Hirakud. The 'locality' of protest therefore is a flexible one as it depends on the levels at which external actors operate. Second, neither affected people nor the external actors who ally with them are homogenous categories: there is a considerable heterogeneity of interests and resources, strategic choices and visions in these conflicts. Collective actions denote the active processes of human agency involving the social construction of identities and meanings on the one hand, and the political mobilisation of organisational resources and networks on the other. In fact, Guha's assessment of collective actions in the Chipko accords centrality to political practices in which identities are formed and meanings produced.

The Emergence of the Narmada Movement

Environment, Displacement and the Demand for Fair Rehabilitation

The success in the late 1970s of the Chipko movement and the Silent Valley project in Kerala inspired a group of environmentalists to initiate research and documentation on the social and environmental impact of large dams. In the early 1980s this began to take the form of a campaign against large dams.[13] In July and August 1983, concerned over the impact of the SSP, a Delhi-based environmental group called 'Kalpavriksh', together with the Hindu College Nature Club, conducted a study in the Narmada Valley and pointed out some 'serious inadequacies and distortions in the information base' (Kalpavriksh and Hindu College 1986: 4).[14] Around that time, ARCH-Vahini, an NGO working on issues of health and environment, approached the World Bank, drawing attention to the plight of those to be evicted by the project. As already mentioned, the World Bank commissioned a study in 1983 on the relocation component of the Sardar Sarovar project which pointed out several inadequacies in the rehabilitation proposals. International NGOs such as Oxfam and Survival International in the UK and the Environmental Defence Fund in the USA, who obtained access to the report, started lobbying with the World Bank even as the latter negotiated the loan proposal with the Government of India and the riparian states for the SSP.[15]

On 8 March 1984, oustees of 14 adivasi villages in Gujarat and nine in Maharashtra marched from Vadagam village to the project headquarters at Kevadia Colony in Gujarat. In this first public demonstration they demanded a thorough revision of the Gujarat government's resettlement policy as outlined in its resolution of 11 June 1979, which offered compensation only to those with revenue landholdings. Although the Narmada Waters Dispute Tribunal (NWDT) made no provision for compensating 'encroachers' of wasteland and forest land, it at least spelt out the requirement that families and not landholdings should be treated as the compensation unit. Furthermore, while it had prescribed what Gujarat should do with oustees from Maharashtra and MP, it made no direct observation on oustees from Gujarat itself. Gujarat NGOs, e.g. ARCH-Vahini and Rajpipla Social Service Society, made the Tribunal's prescription on resettlement their reference point, demanding its incorporation in government policies.

At the same time, demands were voiced for stakeholder status for those left out by the Tribunal. In a memorandum to the government of Gujarat, the oustees and their representing NGOs demanded compensation for livelihood losses of landless people and 'encroachers' who were dependent on forest land, and resources and provisions for their adult sons as was the case with adult sons of people holding land titles.

The struggle for better resettlement intensified in the following years. In Gujarat, ARCH-Vahini spearheaded the boycott of project authorities; blockades to stop work on the rockfill dykes;[16] a writ petition in the Gujarat High Court and later in the Supreme Court of India; and intensification of the international campaign with the support of NGOs abroad, e.g. Oxfam, Survival International. Later, the Narmada Asargrasta Samiti (NAS)[17] was formed as a platform for the 5,000-odd people who had been displaced during the construction of SSP headquarters at Kevadia Colony back in 1960.

In Maharashtra, activists[18] from the Society for Social Knowledge and Action (SETU), an NGO based in Ahmedabad, initiated work among adivasi villages by helping them to form village-level committees.[19] In April 1986, the Narmada Dharangrasta Samiti (NDS) was formed in Dhulia, a committee comprising activists from SETU and representatives of the affected villages under the leadership of Medha Patkar.[20] In a memorandum submitted to the government of Maharashtra that month, the NDS demanded the release of degraded forest land for the purpose of rehabilitation if large quantities of revenue land were unavailable (cited in ARCH-Vahini 1991). A few months later, Maharashtra witnessed its first major demonstration in Bombay, organised by a Left-led 'Committee of Dam and Project Evictees' with the slogan, 'first rehabilitation then the dam' (Omvedt 1993). Two distinguishing features marked this phase of the struggle. First, demands for better provisions for resettlement and rehabilitation; second, adivasi villages in both Gujarat and Maharashtra were mobilised for the first time.

In the state of MP, the situation was more fluid. In central MP initial stirrings were around the Narmada Sagar Project (NSP) which was to be constructed at Punasa in Khandwa district.[21] Here, local NGOs expressed their concern regarding the social and environmental impact of the NSP.[22] This was shaped by accumulated experiences with the Tawa and Bargi projects[23] in MP, and a commitment to values of 'appropriate technology'.[24] As early as 1967 social action groups in MP, under the banner of NGNS, had opposed large dams on the Narmada on the grounds that they were dangerous and destructive.

In western MP, particularly in Jhabua district, several organisations and activist groups had been working among the tribal villages. The Khedut Mazdoor Chetna Sangath (Association for Awareness among Peasants and Workers, hereafter KMCS) was one such organisation. Working in over 100 villages in the Alirajpur *tehsil* from which its membership is drawn, it had, since its formation in 1983, launched several struggles against the state forest department over access to and control of forest resources. The coming of the SSP affected 26 villages in which the KMCS worked. In 1986, the KMCS facilitated a survey in these villages conducted by the Multiple Action Research Group (MARG), an NGO for research and policy advocacy based in Delhi.[25] The objective of the survey was to find out 'what kind of rehabilitation the affected people wanted, what rehabilitation they were offered and the extent to which it was satisfactory' (MARG 1986: 1).[26] The findings of the MARG study were revealing:

> In none of the 26 villages, had the government of MP given any notices under Section 4 or Section 6 of the Land Acquisition Act (1894).[27] Only in some villages and that too to a few farmers, notices under Section 9 of the Act were issued. Even these were not read out or explained and so generally people remained ignorant about their rights. In most villages, the initial information about the dam came from Central Water Commission's personnel, who marked the reservoir level. Information also came from varied sources including wandering *sadhus* (sanyasis). In no cases have the district authorities in MP informed the villagers about the dam. In only 9 or 10 villages (state) government officials have held meetings to inform villagers about their displacement (MARG 1986: 1–2).

Unlike its counterparts in central MP who opposed the NSP, the KMCS focused on the problems of displacement and resettlement in the initial years, disseminating information on the likely impact of the SSP, the extent of forest loss, the number of villages facing submergence, resettlement provisions as per the Tribunal award, the World Bank's credit agreements, and government resolutions and people's entitlements under those provisions.

The conditional environmental clearance accorded to the SSP and NSP in June 1987 activated a chain reaction.[28] Environmentalists and NGOs outside the valley strongly protested the decision of the government to approve the projects when sufficient studies on their environmental and social impact had not been initiated, and those started had not yet been completed. A month earlier, in May 1987, Medha Patkar, then leader of the NDS, had written to the Environment Defence Fund (EDF), an international NGO in the USA, that the environmental clearance to the SSP was expected despite incomplete studies, but that the NDS would nonetheless intensify its struggle for rehabilitation.[29] Patkar also mentioned that she had persuaded NGOs in MP, which were arguing for a 'no dam' position, to mobilise and organise tribals to put up joint rehabilitation demands (cited in Patel 1995). Patkar's initiative revitalised the Narmada Ghati Nav Nirman Samiti (NGNS) in MP with the participation of KMCS activists. Together with the NDS, the NGNS made several demands for resettlement and rehabilitation in 1987. These included the right to information on the technical aspects of the dam, the extent and schedule of submergence, land availability including amount, place, quality and legal status of the land selected for compensation; fresh land surveys to include those areas excluded from earlier surveys, the extension of rehabilitation benefits to those affected by project headquarters at Kevadia, the canal network in Gujarat and the compensatory afforestation programmes; and the assertion of the rights of those affected to settle in their own states under guidelines laid down by the NWDT award.

From Better Resettlement to Total Opposition

On 23 December 1987, the government of Gujarat announced substantial modifications to its Resettlement and Rehabilitation (R&R) package through government resolutions of 4, 14 and 17 December 1987. This followed pressures brought to bear by NGOs for R&R in Gujarat, MP and Maharashtra, and by a series of Bank missions, particularly in April and November/December that year. The major features of the new R&R policies were: (a) landed oustees were eligible to a minimum of 2 ha of land of their choice; the difference between the compensation paid by the government and the market price of 2 ha of land chosen by the displaced was to be borne by the government through an ex-gratia payment; (*b*) cultivators of government wasteland and forest land as well as the landless were to be accorded the same benefits as in (*a*). It should be mentioned that the Government of India still has no national policy on R&R and the approach of the states towards relocation has been extremely ad-hoc and inconsistent. Under these circumstances, the government of Gujarat rightly claimed that its R&R policy in the SSP was a 'revolutionary' step.

The policy announcements of December 1987 resulted in a split in the NGO movement that had so far spearheaded the agitation. NGOs in Gujarat, notably ARCH-Vahini, endorsed the new policies and offered critical support to the government in their implementation. The Vahini claimed that the implementation of R&R policies required 'objective, fair and continuous watch dogging' (ARCH-Vahini 1991:14).

In Maharashtra and MP, however, matters took a different turn. In November 1987, in a joint memorandum to the Narmada Control Authority, the NGNS (MP) and NDS (Maharashtra) put forward a list of 38 demands related to rehabilitation. The memorandum warned that if a clear decision was not taken by 15 December, a movement would be launched to get the demands fulfilled. On 5 December 1987, a meeting of NGO activists, environmentalists and intellectuals was convened in New Delhi to discuss 'social and environmental aspects of the Narmada Projects'; to 'produce relevant study material on the projects related to involuntary resettlement and environmental aspects which have been ignored by the government'; and to 'widen the network and to help those groups already working' (proceedings, T. Kochari, Narmada Action Plan, 5 December 1987, IIPA, New Delhi). Most of the NGOs active in the valley—NDS, ARCH-Vahini, NGNS, as well as Oxfam, Participatory Research in Asia, and Bombay Natural History Society—were represented at the meeting. Although the viability of the project was questioned by some participants, the consensus was to study in depth the various dimensions of the SSP and to initiate action plans involving research, mobilisation, monitoring, documentation, media exposure and fund raising so as to keep the 'pot boiling' (ibid.). The meeting marked the formation of what was called the

'Narmada Action Plan', largely governed by the feeling that 'it would be very difficult (but not impossible) to stop SSP at that stage' (correspondence, A. Kothari with A. Mehta, 16 September 1988), but that a demand for a complete, reliable appraisal of the project was justified and needed to be strongly pursued.

The proposed R&R policy changes by the Gujarat government on 15 December were not received well by activist groups outside Gujarat. Doubts were raised about the government's capabilities and will to implement the policies,[30] as well as about the availability of land for rehabilitation. International NGOs stepped-up their campaign against the project: in June 1988 two organisations, EDF and Friends of the Earth, testified before the Subcommittee on Foreign Operations, Committee on Appropriations, US Senate (EDF 1990), regarding the inadequate environmental impact assessments and cost-benefit analysis undertaken by project authorities, about the inadequacies in R&R policies of the government of MP and Maharashtra, and the unavailability of quality land for rehabilitation. In July 1988, Gandhian social worker Baba Amte organised a meeting of social workers and environmentalists, the consequence of which was the adoption of the 'Anandwan Declaration against Large Dams'.[31] In August 1988, the NDS and NGNS announced total opposition to the SSP on environmental, social and economic grounds, preferring to 'be drowned by the rising water of the dam if the government insists on building the dam, rather than giving tacit approval to these destructive schemes by agreeing to shift' (press release, NDS 1988). Thus was born the slogan *dubenge par hatenge nahin!* (We Shall Drown, but Not Move!) The rationale for opposing the entire project was, first, that proper rehabilitation of all those to be displaced was impossible since the governments had no real idea of the extent and impact of displacement; second, the extremely high environmental costs of the SSP had neither been assessed nor properly accounted for in the cost-benefit analysis and the governments had no action plans to undertake mitigative measures in this regard. A year later (in 1989), radical environmental opposition, the Narmada Bachao Andolan (Save Narmada Movement, hereafter NBA), was forged.

The Radical Opposition: Rationale and Responses

'Why Do We Oppose'? The NBA's Evaluation of the SSP

The resolve for total opposition to the construction of the SSP, according to the movement, rested on 'definite information and fundamental principles' (NDS undated: 1). Opposition was justi-fied on the grounds that 'even the most preliminary information regarding the number of families and villages affected, the ex-tent of the areas to be submerged, the number of hamlets likely to be displaced ... was not available with the governments... leave aside the detailed plan for rehabilitation.... Whatever promises the government may make on paper, the organisations of the oustees have come to the painful conclusion after full discus-sions, deliberations, studies and investigation that the government will never be in a position to give "land for land" for all 245 project affected villages' (ibid.: 3–4). Sustainable rehabilitation of all the oustees was therefore considered an impossible task.

However, the opposition was not merely over the question of inadequate R&R measures and impracticable policies; it was also over a wider issue that involved a major sacrifice by some, for an ill-defined, unestablished 'national good'. The wider is-sue involved the exercise of a democratic right to information with regard to all aspects of the SSP in particular, and develop-ment projects in general, and the obligation of the government to establish clearly the 'public interest' of such projects. Some questions were raised: (a) Whose development? (b) At whose cost? (c) What are the quantifiable and non-quantifiable costs and do they outweigh the benefits? (d) Is this development sus-tainable and just? (e) Is the project in the national interest? (f) Who are the people being asked to make sacrifices in the national interest? (g) Can their community life and resource base ever be compensated? (h) Are the decisions taken after com-plete and comprehensive investigations? (i) Since people depend on natural resources that are being affected by the project, are their rights to decide on harnessing and utilisation of such re-sources being recognised? (see NBA 1992: 3).

These questions brought to the fore issues related to the dis-tributional effects of development projects, issues related to their planning, execution and scale, their environmental costs, the financial implications of such investments, as well as issues of democratic governance and human rights. In the NBA's understanding, the SSP implied 'unprecedented displacement,[32] violation of the right to life and livelihood of people, the degradation of land, water and forest resources,[33] the untenability of benefits,[34] the staggering financial burden[35] and the consequent international debt trap...' (NBA 1992: 4).[36] The displacement had been planned without the affected population being informed (as required by law) and without any scope for their meaningful participation in the project, denying them the right to information and participation. The SSP thus stood for a 'faulty, non-viable, unjust and destructive project' (NBA 1994).

Against a Development Model: Self-definition of a Movement

As the struggle against the SSP began to crystallise, the movement was seen to be one against a development model manifested in the SSP and similar projects. An extract from a campaign letter entitled 'We want Development not Destruction', is illustrative:

> Since independence a preference for giganticism has come to dominate our development paradigm. Our planners, politicians and experts have opted wholesale for large dams and gigantic industrial units, and have dug mines and exploited forests in pursuit of their elitist vision of progress and development. The cumulative ill-effects of all this 'development' are now assuming disastrous proportions for a large section of the population, particularly for its most depressed strata—the tribals, the peasants and labourers—along with the already depleting natural resource base and our scarce financial resources (Action Committee for National Rally Against Destructive Development 1989:1).

At a national convention organised by the NBA a few years later on 'Development, Planning and Mega Projects', it reiterated its critique of gigantic development projects and expressed opposition to 'the human rights violation entailed in forcible displacement, the unsustainability of large-scale environmental disruption, the lack of public accountability of decision-makers, the absence of any genuine peoples' participation in development planning and the neo-imperialism of multi-lateral financial agencies' (NBA 1992).[37]

In the emerging critical discussion on state-led development projects and the 'elitist vision of giganticism', the movement linked 'large dams, the green revolution package, and the unmindful industrial-urban package', to the edifice of 'capital intensive technology and western indicators of development'. To the NBA leadership, the development model entailed 'increasing centralisation, capitalistic tendencies and vulgar consumerism', whilst causing the 'degradation of land, water and forests, increasing socio-economical deprivation and inequality and erosion of basic human rights' (NBA 1992: 1–2).

The NBA considered itself a challenge to this 'larger reality', defining itself as 'one of the major struggles in post-independence era to save the land, forest, the people and their resources, which are being inequitably consumed and destroyed by a few in the name of public purpose [and] development' (NBA 1992:1). As Kothari (1995: 428) noted, movement leaders argued that what they were doing was 'nothing short of challenging the fundamental structures of power and patronage, received categories and ideologies as well as representative processes that discriminate against the primary victims of economic development.'

Critical Responses of State and Civil Society to the NBA

The politics of radical opposition has had to confront governmental and also non-governmental actors. Whereas the SSP constitutes the NBA's major struggle arena, in Gujarat it is viewed as the only solution to drought and the shortage of water. Throughout the 1980s, in conditions of persistent drought, the government mobilised public support for the SSP, projecting it as the 'pride of Gujarat'. Industrial interests and farmers' associations backed the project, expecting to benefit from power, irrigation and industrial water. As might be expected, the announcement of total opposition drew sharp reaction from the government of Gujarat. It dubbed the movement as anti-Gujarat and anti-development, and clamped restrictions on it. In 1988, the dam site, project headquarters at Kevadia and 12 adjacent villages were subjected to the Official Secrets Act, 1923. Nearby

Bharuch district was declared a 'prohibited area'. A few years later, the NBA office in Gujarat was ransacked. In adivasi villages in MP and Maharashtra, where the movement was strong and where villagers had held back survey teams, the government used strong-arm tactics; in 1993, police repression in these villages was reported.

To counter criticism, the Gujarat government organised rallies, festivals and exhibitions throughout the state to highlight the benefits of, and to organise support for, the SSP. Within a short period, all major political parties, a large number of Gujarati NGOs, the chambers of commerce, farmers' associations and even Gandhian social activists in Gujarat extended their support. The project became popular as the 'real lifeline of Gujarat'.

With radical opposition from the NBA, the NGO movement for better resettlement in the valley had split into those focused on resettlement issues and those integrating it with a wider set of issues. The views of ARCH-Vahini were clearly shaped by the conditions within which it was embedded. Its immediate struggle was for fair compensation for affected villages in Gujarat. It therefore deemed the Gujarat government's policy announcements as a major achievement for the struggle, 'realising fully its responsibility to ensure the implementation of the policies' (interview, Anil Patel, Director, ARCH-Vahini, 12 February 1996). The next logical step was to demand similar policy modifications from the governments of MP and Maharashtra, and to work towards their implementation (ARCH-Vahini 1988).

ARCH-Vahini responded to the opposition by dismissing it as a 'lofty ideal', demanding to know 'if those who are making this radical shift will ...really ask the oustees to drown themselves in the rising water...'. It called for delinking the issue of rehabilitation from 'the battle on the wider front'. It did not 'share the strategic perceptions of those who are wittingly and unwittingly using the issue of rehabilitation of oustees in the cause of the fight against the dam', as this was 'not responsible activism' (ARCH-Vahini 1988:14). For the Vahini, the argument 'rehabilitation is impossible' was based on the 'alleged fact that enough land is not available'. It also called to question the NBA's arguments that 'oustees should not be asked or encouraged to identify the land they would prefer or that high prices of the land should not be given to the land sellers' (ARCH-Vahini 1988a: 8), because these arguments went against the interests of affected people seeking quality land for resettlement.

In Maharashtra, critical responses came from those groups which should have been the NBA's natural allies. The Committee of Dam and Project Evictees (CDPE), left-wing and the Shramik Mukti Dal (SMD), found the NBA's total opposition couched in rhetoric and romanticism. While the romantic image could capture media attention, it had no real strategy with which to gain mass backing (Omvedt 1993: 269). The appeal of its leadership to sentiments and idealism was considered no substitute for the strategy of mass-based politics. Omvedt mentions that the NBA's politics cut little ice with the popular farmers' movement in Maharashtra led by the Shetkari Sangathana. The Sangathana ignored the NBA's radical critique on the grounds that it divided the peasantry into losers and beneficiaries. For the Sangathana, the peasantry constituted the mainstay of rural India and struggles against the exploitation and appropriation by urban-industrial interests. Therefore, it could not support the NBA's division of the peasantry into rich cash-crop farmers of Gujarat pitched against poor farmers in MP and Maharashtra.[38]

Some of these early critical responses spurred the NBA to intensify mobilisation in the Narmada Valley, as well as in the national and global arenas. Between 1990 and 1993, the NBA made dramatic gains in its support base, assuming the shape of a multi-level network. As it drew upon different constituents and discourses in the network, its language of protest became more syncretic.

Network of Actors: Protest Events and Interest Articulation

Mass Mobilisation in the Narmada Valley

The NBA's success in mobilising affected people is evident from several protest activities organised in the Narmada Valley. The periodic setting-up of road blocks at strategic points, demonstrations and rallies, blockades of project authorities (Narmada

Control Authority and the World Bank) and political party leaders at different levels, were combined with specific events such
as obstructing the construction of bridges across the Narmada,
uprooting stone markers from the proposed submergence areas
and dumping them outside the Legislative Assembly in Bhopal,
and uprooting planted saplings which were a part of the project
afforestation scheme in order to draw public attention to the
monoculture of species. Less frequently, the NBA targeted
specific groups such as women and adivasis for mobilisation; of
considerable significance was a 1,500-strong all-women's rally
in January 1993 in the town of Badwani which demonstrated
the NBA's ability to mobilise women in the submergence zone.

Tools of protest such as satyagrahas, *jal samarpan* (sacrificial
drowning) and hunger strikes were combined with strategies of
rasta roko (road blockades) and *gaonbandi* (refusing state officials' entry into villages). While the former set of tools symbolically
highlighted the suffering and pain of affected people, the latter
strategies suggested symbolic delinking from centres of power
and non-cooperation with state agencies. Two major protests
organised by the NBA in the Narmada Valley are worth noting
in some detail.

The first was the Harsud Rally of 28 September 1989 held at
an NSP-affected township in MP.[39] The rally called for the adoption of a socially just and ecologically sustainable pattern of
development and an end to all projects affecting the environment and destroying people's livelihoods. The rally marked the
beginning of the syncretic protest language in which the struggle
against the SSP was considered part of a wider struggle against
a development model that benefited a few at the cost of a large
majority of people and their environmental resources. The NBA's
campaign newsletter described the rally as follows:

> People struggling against past or proposed displacement
> and environmental degradation by massive irrigation and
> power projects such as Sardar Sarovar and Narmada Sagar,
> Bhopalpatnam-Inchampalli and Koel Karo, defence projects
> such as Baliapal, nuclear power projects such as Kaiga,
> came together in an unprecedented show of strength. [The]
> defiant message to the politicians and planners was that
> people are no longer prepared to watch in mute desperation

as project after destructive project is heaped on them in the name of development and progress (Narmada 1990a: 4).

The 20,000 who gathered at Harsud, including affected people and representatives of NGOs and activist groups from different parts of the country, put the NBA at the centre of the environmental movement in India. Its campaign against the SSP received a tremendous boost (ibid: 10–11)[40] and its activists emerged as the accepted spokespersons of the affected people in the Narmada Valley. While the rally received solidarity support from at least a hundred NGOs abroad, in India it brought together several civic organisations and groups on a common platform for the first time. The direct outcome was the formation of the Jan Vikas Andolan (Movement for People's Development), a broad alliance of a 'wide range of movements, organisations and individuals, with its roots in a variety of struggles taking place in the country...' (Narmada 1990b: 25).

The second major event, the Jan Vikas Sangharsh Yatra (Struggle March for People's Development), was organised a year later. Considered to be 'the first move in the "final phase" of the anti-SSP movement ... its stated objective was to physically stop work on the dam, by offering satyagraha at the dam site and thereby pressurise the government to comprehensively review the SSP' (Narmada 1991: 3). The Sangharsh Yatra was a test case for the NBA's support base, both in the valley and outside, and the participation of more than 8,000 people in the six-day march to the project site bolstered its claim of steadily increasing support. The march (on foot) covered a distance of about 200 km before it was stopped at the MP–Gujarat border by the Gujarat government.[41] The Sangharsh Yatra pitched camp at the border where it stayed for a month. To pressurise the Gujarat government seven marchers, including Medha Patkar, went on an indefinite hunger strike. As the Gujarat government did not relent, the NBA decided to withdraw from the border 22 days into the strike. Baba Amte, then leader of the NBA, who had been allowed to camp on the Gujarat side of the border, returned to the Sangharsh Gaon[42] on 30 January to declare: 'Gandhism has died in Gujarat, and on the day Gandhi died, I return to the valley where Gandhi's ideal still lives' (Narmada 1991: 15). On 31 January 1991, the Yatra withdrew from the border with a pledge to take the struggle back to the villages under the slogan *hamare gaone mein hamara raj* (Our Village, Our Rule).

Translated into policies and actions, the slogan implied non-cooperation with an unresponsive government and the development of self-reliant institutions and actions in the villages. The resolve was that villages would henceforth boycott government activities like census operations and oppose all survey work related to resettlement. They would also take up reconstruction activities such as soil conservation, irrigation works, health training and adult education. The NBA newsletter described this as a 'gigantic social experiment... [that] can offer crucial insights into exploring alternative systems of governance and development' (Narmada 1991: 24).

From 1991 to 1996, the NBA organised monsoon satyagrahas each year. In the first satyagraha in 1991, groups of s*amarpit dal* (drowning squads) were formed who voluntarily drowned themselves in the rising waters of the river. The rallying call of the action was *dubenge par hatenge nahin* (We will Drown but Not Move). Amidst criticism that it was promoting collective suicide, the NBA reasoned that the people in the Valley were only honouring their pledge, *koi nahin hatega bandh nahin banega* (No One Will Move, the Dam Will Not be Built). Although the monsoon water did not rise high enough to engulf *Narmadayi* (a hut constructed to house the *samarpit dal* in one of the affected villages), the satyagraha gave the NBA impressive press coverage and a wave of 'solidarity' support from different parts of the country and abroad.[43]

Subsequently, up to 1997, monsoon satyagrahas in selected affected villages took place with varying degrees of success. While the 1992 satyagraha was less dramatic than the previous one,[44] the 1993 satyagraha saw the threat of *jal samarpan* reach new heights when the drowning squads, demanding a review of the project, went underground (amidst a police hunt) in their continuing resolve to drown. The Government of India yielded to the pressure and set up an independent team to discuss issues with the NBA. After 1993, the satyagrahas were less dramatic.

Apart from mobilising affected people, the NBA has gained support from a wide range of national and global NGOs, citizen and action groups. National and international attention has been drawn to the violation of human rights of the people in the Valley.

By questioning the projected benefits and the financial implications of the project, the NBA has framed the SSP as a sunk investment that squandered scarce resources coming from international and national tax-payers. In fact, the Narmada movement increasingly activated global and national networks in pursuit of a broader articulation of forces than those involving only local affected people.

Globalising Protest: Articulating Support, Achieving Ends

The loan agreement reached between the World Bank and the central and state governments in India provided the impetus for globalising protest against the SSP. International NGOs had opposed the loan agreement on social and environmental grounds. The Bank on its part periodically revised its policies and operational guidelines on these aspects and sought their implementation in the SSP. Its consultants on displacement and resettlement were apprehensive that proper rehabilitation of the displaced would not be possible unless the project authorities substantially modified policies and improved implementation mechanisms.[45] This was fertile ground for the NBA's international campaign. It directed attention towards the World Bank, demanding its withdrawal from the project.

Between 1990 and 1993, the NBA mobilised significant international support. Three of its activists testified at a special hearing of the US Congress Sub-Committee on Natural Resources, Agricultural Resources and Environment,[46] which followed up the hearing by urging the Bank to reconsider its involvement in the SSP. In May 1990, 120 members of the Finnish Parliament wrote to the Bank stating that the Narmada projects 'should not receive any Bank funding before alternatives have been thoroughly considered and before the R&R problems have either been solved or at least re-evaluated'.[47] In June that year, the Japanese government which had earlier sanctioned soft loans under the Overseas Economic Cooperation Fund (OECF) for turbine generators for the riverbed powerhouse of the SSP (the supply orders having been obtained by the Japanese companies Sumitomo, Hitachi and Toshiba), announced the cancellation of a $150 million loan at a meeting of Bank donors in Paris, stating that the Bank's assessments of social and environmental costs were grossly inadequate.[48]

International NGO support crystallised with the formation of the 'Narmada Action Committee', representing organisations from 15 countries. Active political and financial support (lobbying, advocacy and occasional solidarity demonstrations abroad)[49] by this network culminated in the setting up of the Independent Review Mission (IRM) by the World Bank. The Bank management, under pressure from influential sections of its Board of Directors, took this unprecedented step in June 1991. The IRM was appointed 'to assess the implementation of the resettlement and rehabilitation of the population displaced/affected and the amelioration of the environmental impact of all aspects of the projects' (IRM 1992: 359).

After initially opposing the terms of reference of the IRM, the NBA extended the necessary support and facilitated field visits by the mission.[50] The mission report unequivocally endorsed the NBA's position on the SSP. In a 1992 (July) communication to the World Bank, President Lewis Preston, chief of the IRM, wrote:

> We think that the Sardar Sarovar Projects as they stand are flawed, that resettlement and rehabilitation of all those displaced by the projects is not possible under prevailing circumstances and that the environmental impacts of the projects have not been properly considered or adequately addressed. Moreover we believe that the Bank shares the responsibility with the borrower for the situation that has developed.... If essential data were available, if impacts were known, if basic steps had been taken, it would be possible to know what recommendations to make. But we cannot put together a list of recommendations... when in so many areas no adequate measures are being taken on the ground or are even under consideration. Important assumptions upon which the projects are based are questionable or known to be unfounded.... Assertions have been substituted for analysis... [T]he wisest course would be for the Bank to step back from the Projects and consider them afresh (IRM 1992: xii–xxv).

To the NBA, the IRM report was an independent validation of its standpoint and a corroboration of 'everything that those opposed to the project have been saying' (Narmada 1992b: 9). It was termed a 'blockbuster report which laid bare all the pretences which either the Bank or the Indian authorities had regarding the SSP'. The Bank was asked to withdraw from the project within a month or face intensified opposition to its presence in India.

Meanwhile, NBA activists canvassed in the USA, Japan and European countries, meeting international NGOs, the media and members of the Japanese Diet. In July 1992 the European Parliament passed a resolution on the 'Narmada Dam' calling 'on all member states to ... urge their executive directors to vote against further World Bank support for the project', and 'on the World Bank to withdraw from the project, pay compensation to those who have suffered as a result of the SSPs and write off the US$ 250 million spent on building the dam if it is not completed' (EP 1992).

The Bank's management tried to salvage the situation, but on 29 March 1993 the Indian government announced its decision to terminate its contract with them. The victory of the Bank's 'withdrawal' saw widespread celebration in the Valley. But there were also fears that the state might become more repressive and less accountable in the absence of the World Bank. To 'prevent and document' state violations of human rights, the Narmada International Human Rights Panel (NIHRP) was formed, consisting of 43 environmental and human rights organisations from 16 countries.[51]

The support networks activated at the global level enabled the NBA to link the SSP to a much larger trend of dam building around the world. The involvement of the World Bank and other multilateral lending agencies in infrastructure projects on a world-wide scale, particularly in the developing world,[52] was a strong reason to globalise resistance against the SSP.[53] In July 1994, the NBA's international campaign resulted in the Manibeli Declaration, calling for a moratorium on World Bank funding of large dam projects all over the world. Within three months, 2,152 NGOs in 43 countries had signed the declaration.[54] Three years later, in March 1997, the international campaign against dams led to the Curitaba Declaration reiterating the need for an independent international commission to review all large dams financed and supported by international aid and credit agencies.[55]

The globalisation of protest also brought international recognition to the NBA and its leaders. In 1991, the 'Narmada Bachao

Andolan led by Medha Patkar and Baba Amte' received the Right Livelihood Award 'for their steadfast opposition to the ecologically and socially disastrous Narmada Dams—the largest river development project in the world—and their clear articulation of an alternative water and energy strategy that would benefit both the rural poor and the natural environment' (RLA 1991).[56] This was followed by the 1992 Goldman Environmental Prize for Medha Patkar 'in recognition of outstanding environmental achievement in Asia'.

Opportunities and Articulations at National and State Levels

The NBA constantly engaged project authorities, ministries at the national level, and the two state governments of Gujarat and MP, by seeking alliances and activating networks among various civil society groups. It must be stated at the outset, however, that state institutions played significant roles, both directly and indirectly, in providing opportune environments for fostering NBA's politics. The Central Ministry of Environment and Forests (hereafter MoEF) took a tough stand on the SSP. At various stages its reports were used by the NBA to bolster their critique of the project. More significant is the MP government's tacit support to the NBA. In two national review fora, the NBA and the MP government articulated similar positions on the SSP.

In the preceding chapter, we mentioned the role of the MoEF in demanding the full compliance of SSP and NSP authorities with its conditions. The MoEF periodically voiced concern over the lack of environmental impact assessment studies, detailed proposals for mitigative measures such as catchment area treatment and compensatory afforestation programmes, and a master plan on rehabilitation. Between 1985 and 1987, the MoEF withstood pressure from the World Bank and the Government of India for the mandatory environmental clearance to the SSP. Again, in December 1993, the MoEF asked for a halt in construction work as project authorities had failed to meet its conditionalities regarding environmental and resettlement measures. These actions of the MoEF helped the movement to rally support against the project.

The MP government's support of the NBA was expressed in two public reviews—the independent Five Member Review

Group (FMG) formed in 1993 and the Supreme Court of India. The FMG was the culmination of intense agitation; a 14-day hunger strike in Bombay in June 1993 followed by the threat of *jal samarpan* in the monsoon month of July. During the hunger strike in June, the Central Minister for Water Resources promised a review of the SSP.[57] In the absence of follow-up actions, however, the NBA's 'drowning squads' posed a serious dilemma for the government.[58] The FMG was the way out. An independent team of prominent citizens set up by the Ministry of Water Resources, Government of India, agreed to 'continue discussions with the NBA on all issues related to SSP'.[59] In its preliminary meeting, the FMG expressed deep concern 'to hear that the NBA proposes to proceed with its plan of *jal samarpan* ... [and was] anxious to prevent such an unfortunate occurrence which may have incalculable consequence' (FMG 1994). It appealed to the NBA to defer its *jal samarpan* as it was 'prepared to give careful consideration to any points or issues that the NBA may wish to raise' (ibid.). As a result the *samarpan* was called off.

The Gujarat government had opposed the formation of the FMG and refused to participate in the group's proceedings, as had the BJP government in MP. Only the government of Maharashtra participated. Towards the end of the FMG's term, however, a new Congress government in MP decided to participate in its review process. It requested a reduction in dam height by six metres, citing insurmountable problems of resettlement and environmental amelioration measures, and argued that a lower height would not alter the irrigation and water benefits for Gujarat.[60]

The NBA found an ally in the new MP government. In November 1994, when the NBA held the Bhopal agitation[61] to protest against the decision of the Gujarat government to accelerate construction on the dam, the MP government responded favourably. It formed two high-level committees: one comprising members of Madhya Pradesh's Legislative Assembly to look into the problems of MP's affected people resettled in Gujarat, and the other comprising Members of Parliament from MP to recommend measures to tackle the problems of Scheduled Tribes in the state.[62] Both committees recommended that the MP government should stall further work on the SSP and seek a reduction in the height of the dam.

The petition filed by the NBA before the Supreme Court challenged the construction of the SSP on 'social, environmental, technical, economic and financial grounds', arguing that the project as conceived was 'not in the national interest'.[63] In earlier years the NBA had taken legal recourse in local courts in Maharashtra and Gujarat and the High Court in Gujarat, raising a wide range of issues: seeking the status of project-affected people for those omitted from official lists, asking for changes in land titles for people who were given unsuitable land as compensation, requesting a stay on survey work undertaken by the project authorities, seeking detailed schedules of construction, submergence and rehabilitation, as well as seeking redress on cases of forcible eviction and police excess.[64] The fight in the courts pertained to specific cases of legal violation by the project authorities,[65] concern for the rule of law,[66] and adequate participation and access to information.[67] The Supreme Court petition, however, sought (*a*) to halt any further construction on the project, (*b*) to ensure the completion of all necessary studies on the project, (*c*) a comprehensive review of the project and, in addition, (*d*) a restriction on the height of the dam at 93 metres.

The Supreme Court asked the state governments to submit their responses to the FMG report 'uninhibited by any legal implications'.[68] This implied that the NWDT award on the Narmada projects, considered final and binding on all riparian states, could be renegotiated. The MP government submitted an affidavit to the Supreme Court seeking a reduction in the height of the Navagam dam,[69] whereas the Gujarat government strongly opposed any such reduction. The stalemate continued as the case became enmeshed in constitutional matters. The Gujarat government had raised the question of whether the Supreme Court could reopen a case that had been settled by a Water Disputes Tribunal Award. The matter was then moved from the 'division bench' to a 'constitutional bench' of the Supreme Court. In January 1995, however, by order of the Supreme Court, construction of the dam was stopped at 80 metres, but in 1998 the Court ordered further construction of the dam up to 85 metres.

These developments suggest that the NBA successfully used the conflicts among riparian states, as well as avenues of democratic institutions and practices, to its advantage. At the national level, activists of the NBA in general and Medha Patkar in particular addressed NGOs, activist groups, parliamentarians, trade unions, student and professional bodies, academic conferences, and Rotary and Lions Clubs. Academia and the press were actively pursued, ensuring on the one hand critical inputs towards reviewing the project components and, on the other, publicity for the NBA on an unprecedented scale. Moving beyond the parameters of the SSP, however, the forging of formal regional and national alliances around struggles related to development has been a crucial achievement of the NBA's civil society networking.

One such early alliance was the Jan Vikas Andolan (hereafter JVA), mentioned earlier in the chapter. According to its founders:

> [The JVA is a] movement against the development paradigm being practised in post-independence India whereby a narrow elite primarily benefits at the cost of a very large population that continues to be marginalised, displaced, and pauperised along with large scale plundering of our natural resource base. The movement... maintains that what today goes in the name of development is not genuine development but it is in fact socially disruptive, biologically and genetically homogenising and environmentally destructive (Narmada 1990: 25).

The formation of this broad front was of enormous help to the NBA in seeking support for its stand of total opposition to the SSP: it focused on a 'wider set of issues' rather than only resettlement and rehabilitation, i.e. the non-participatory nature of planning and implementation processes as well as the social and environmental costs of development projects. The JVA failed to make any significant impact however, and has since become defunct.

A few years later, the NBA became a part of the National Alliance for Peoples' Movement (NAPM), a consortium of a number of local peoples' movements across the country, functioning under a common minimum programme. Although established in 1992, it was only later that NAPM gained some credibility and publicity. Its professed objective is to 'challenge the current paradigm of development, oppose globalisation, privatisation and liberalisation', and to work towards 'a just, egalitarian, secular, non-violent and ecologically sustainable society' (NAPM 1996). In its view, while local communities are losing control over land, forests and water, the globalised control over technology, fertilisers,

seeds and water is rapidly destroying the self-sufficiency of agricultural communities and 'alienating them from their natural
habitat and resource base'. Hence the need to widen the struggle
frontier to oppose the 'intrusion of foreign multinationals and
their increasing grip over the economy and polity' (NAPM 1995).[70]
NAPM ideologues view the trend towards globalisation as part
and parcel of the present development model. In this sense,
globalisation strategies are extensions of state-led development
strategies. However, NAPM also perceives that the current 'economic reforms' pursued by the Indian state have led to dramatic
changes in the priorities of the state and its agencies. With concern, it notes that:

> Earlier the government used to plan projects for drinking
> water, irrigation, roads, education, health etc. The focus of
> government today is on projects like national highways,
> airports, modern sophisticated sea-ports, telecommunica-
> tions and electricity for big industries. In other words, [these
> are facilities] that foreign MNCs demand[71] (NAPM 1995).

The NBA is one of the leading actors in NAPM, and Medha
Patkar is one of its three convenors. Whereas both the JVA and
NAPM are national-level alliances, in western MP, an area with
a predominantly adivasi population, the NBA has activated the
Jan Mukti Morcha (Peoples' Liberation Front) along with three
other local organisations: KMCS in Alirajpur, the Adivasi Mukti
Sangathan in Sendhwa[72] and the Ekta Parishad in Dahi. The
Morcha has intensified the campaign against the liquor trade, and
against the exploitation of the adivasis, and supports the latter's
demand for self- rule (*swaraj*) in the region.

The Current Impasse

The lack of displacement impact studies and resettlement plans
had been a major plank of the NBA's anti-SSP posture. Despite
the fact that sufficient land was not available to resettle all affected people, some people from the reservoir-affected villages
had begun to accept resettlement. In the 19 affected villages in
Gujarat, ARCH-Vahini had mediated in the resettlement process,
helping oustees to locate land. In Maharashtra, the resettlement

process gathered momentum by 1992–93, with people from a majority of affected villages shifting to resettlement sites. In MP the number of people shifting to resettlement is negligible, largely because submergence is yet to take place. However, the number by itself was sizeable. More importantly, willingness to accept compensation and resettlement has spread across the Valley in adivasi villages as well as in the Nimad plains. In MP people have accepted compensation in cash despite the fact that this violates the NWDT Award.

The NBA faces a rather paradoxical situation. On the one hand, it has achieved success in temporarily halting the dam. On the other, there has been a decline in its mass following in the Valley. Together, these factors have brought a lull in NBA activities in the Valley. The monsoon satyagraha of 1995 in the villages of Jalsindhi in MP, and Domkhedi in Maharashtra, was very low-key. Manibeli village in Maharashtra, which had been the NBA's symbol of resistance in previous years, ceased to serve this function. By 1995, the few households that remained in Manibeli as part of the NBA showed signs of battle weariness. Some of them expressed willingness to accept resettlement (personal communication, Himanshu Thakkar, NBA).[73]

Such has been the loss of political momentum that at one of its rallies held in the Valley (in December 1996), the NBA had to rely on an extremely popular film personality to attract a crowd. The turnout was impressive, but only to see the film star; the moment he finished his address, the crowd quickly disappeared. Only a faithful hundred remained to fulfil the main purpose of the rally—to pledge to oppose displacement in the Valley (personal communication, Jai Sen, NBA Support Group).

The Supreme Court petition has preoccupied NBA interests, time and resources. The 'distorted and not factual' affidavits filed by the state governments have made the proceedings slow and cumbersome (NBA 1996). Proceedings in the Court have frequently been adjourned. Furthermore, the Supreme Court has sometimes expressed displeasure over the actions of the NBA. One such occasion is worth noting. While its petition was being heard by the Supreme Court, the NBA filed a petition in the National Human Rights Commission concerning the violation of human rights in the SSP. Taking strong exception to the duplication of adjudication, the three-member division bench of

the Supreme Court observed in one of the hearings: 'The peti-
tioners are ... enamoured to see their names [in the media]
every day.... The initial enthusiasm and the genuine feeling are
no more there' (*The Telegraph*, 6 May 1995). These observations
of the Supreme Court later led the NBA to withdraw its petition
from the National Human Rights Commission.

These reverses did not deter the NBA from remaining optimistic about its goals. During the monsoon satyagraha of 1996—
construction of the dam having been suspended for over a year—
it issued the following pledge:

> Now the people of the valley want to assert their right to
> live in the valley and the satyagraha will be an expression
> of this right. This will be done by the launching of many
> different programmes of construction and development
> (*nav nirmaan*) in the valley. It will be a celebration of liv-
> ing in the valley. Tree planting on a massive scale, hous-
> ing, bio-gas, soil and water conservation, small irrigation
> schemes, libraries, schools and so on. This is the people's
> way of asserting that they are determined to stay in the
> valley and will work to make this stay better, prosperous
> and bountiful (NBA 1996).

Several factors may have contributed to such optimism. The
MP government has reopened its old demand for a lower dam
height. It now functions as an ally of the NBA on the SSP. The
interests of a civil society-led movement and a state government seem to match. Also significant was the withdrawal of
ARCH-Vahini from the Gujarat Government Committee on
Resettlement in June 1995. Vahini expressed dissatisfaction
regarding the inadequate facilities accorded to numerous project
oustees from MP who came to resettle in Gujarat. The change
in ARCH-Vahini's position was justified thus:

> The policy promises made to us by the Gujarat govern-
> ment have not been kept. We accept that it was a failure
> on our part not to have managed to keep up the pressure.
> But we have not changed our fundamental position. We
> said no dam without proper rehabilitation then, we say it
> now (interview, Anil Patel, ARCH-Vahini, 14 March 1996).

On its part, the NBA viewed the shift in Vahini's position 'as the realisation of the mistakes of organisations like the ARCH-Vahini who had made the displacement as a fait accompli [*Sic*] while accepting all the claims of the government on the dam benefits and resettlement' (NBA 1996: 9).

Also, the incidence of some resettled people in Gujarat returning to their original villages has bolstered the NBA's claim of the 'impossibility of fair and just resettlement' of such a large population.[74] The NBA has recently started espousing the cause of the resettled population of Gujarat who have formed the 'Gujarat Narmada Vishthapit Sangharsh Samiti'.[75]

Finally, the World Bank's Project Completion Report on the SSP, published in March 1995, has also been an encouraging sign for the NBA. The report acknowledged major performance shortfalls and stated that the project was a 'lesson learning exercise' for the Bank. While maintaining that the 'basic rationale for the project is sound' (1995: 2), the report admitted that the World Bank had violated its own operational guidelines on R&R and environmental aspects, and that the robustness of economic rate of return calculations were clouded by many uncertainties (WB 1995: 4–5). For the NBA the report 'vindicate[d] almost all of the criticisms made by the NBA and the NGOs, criticisms that the Bank [had] deliberately ignored' (Letter to the EDs, NBA, 5 May 1995).

Notes

1. Satyagraha is a Gandhian tool of protest and implies political action based on truth and non-violence.
2. The Land Acquisition Act was used to acquire land (notice being served under Section 4 of the Act; see endnote 10 of Chapter 4); however, the TPC began construction even before land was legally acquired (Rodrigues 1984:191).
3. The Mulshi Satyagraha Committee had representatives from the Maharashtra Congress and local activists. It also had the tacit support of Mahatma Gandhi.
4. Pandurang Mahadeo Bapat (1884–1967) joined the Mulshi Satyagraha just before it commenced. He had earlier abandoned his studies in mechanical engineering at the University of Edinburgh and believed in revolutionary armed uprising rather than Gandhian passive

resistance. In Mulshi he committed himself to the principles of satyagraha, only to abandon them later. Local Malwas addressed him by the title 'Senapati', meaning commander.

5. Although the acquisition officer had located resettlement land in the Dindori *taluka* of Nasik district (150 km from Mulshi), not a single Malwa was interested in land compensation. The peasantry seemed inclined to accept compensation in cash.

6. The frequency of flood disasters led Mahatma Gandhi to write to Sir M. Visvesvaraya, the first renowned Indian engineer, to suggest ways of controlling them. Visvesvaraya suggested reservoir impoundment for flood control and drought prevention (Hart 1956: 120).

7. In 1933, there were 26 such gadjats spread over coastal and western Orissa, whose rulers were given the title of Raja (in 1874) by the British. Together with 14 gadjats in MP, they constituted the Eastern State Agency with the Governor General at Ranchi (Bihar) appointed as the agent of the gadjats.

8. Pattanaik et al. (1987: 53) translate the term *gauntias* simply as village headmen which conceals the clear connection of this 'class' of rural landlords in the then Central Province with a feudal order. Hart (1956:121) characterises this class as 'short-sighted spokesmen of Old India' and notes their resentment to the Hirakud project that 'promised abundant employment and relatively high wages which dam construction would offer to the cultivators hitherto held as serfs...' Hart's latter points seem to be at odds with his own illustrations of migrant labourers from all over India who were brought in by different contractors during project construction. However, his implied characterisation of the *gauntia pratha* as feudal is correct and has been supported by Chakrabarty (1987). The latter argues that projects like the Hirakud have challenged the existing feudal agrarian structure and to this extent have fulfilled historically progressive functions. Chakravarty also notes, albeit cautiously, the 'participation of big landowners and middle class farmers, religious institutions with large property and other feudal elements' in the anti-project movements in Tehri and Koel Karo (ibid.: 27).

9. For more details on the anti-Hirakud agitation see Baboo (1991) and Pattanaik et al. (1987). For an economic evaluation of benefits from the Hirakud project see Sovani and Rath (1960). A more recent evaluation that incorporates social and environmental aspects in its appraisal has been undertaken by D'Souza et al. (1994).

10. The list of studies is quite long. Some of the major contributions are CSE (1985), Fernandes & Thukral (1989), Cernea (1990; 1996; 1997), Scudder (1991), Das (1996).

11. The magnitude of displacement has inspired Gadgil & Guha (1995)
    to treat 'ecological refugees' as a distinct category in their three-
    fold classification of Indian society into omnivores, ecosystem people
    and ecological refugees. In their guestimate, however, as many as
    300 million people lead the life of ecological refugees.
12. Of the numerous environmental struggles that dot the Indian land-
    scape, the best known and most studied movement is the Chipko
    Andolan (see Guha 1989; Rangan 1993; Shiva 1991; Weber 1987).
    In many ways this marks a watershed in the Indian environmen-
    tal movement. It was one of the first struggles launched in the
    post-Independence era which drew world-wide attention to the dam-
    aging effects of commercial forestry and logging practices in the
    foothills of the Himalayas, both on the inhabitants and on the wider
    ecosystem of the Garhwal-Kumaon belt in the Uttarakhand region.
    It took place at a time when not much was known in India and the
    world about the significance of environmentally sustainable devel-
    opment. The movement derived its name from the collective action
    it espoused—the hugging of trees by local people to prevent forest
    officials and contractors from cutting them down. The word 'Chipko'
    in Garhwali means 'to adhere'. The movement inspired several simi-
    lar struggles in other parts of India, the Appiko Chaluvali struggle
    in the Uttara Kannada district in south India being one of them
    (Shiva 1991:117). But perhaps most importantly, the Chipko move-
    ment succeeded in achieving a 15-year ban on commercial logging
    in the region. The movement has been variously seen as a peasant
    movement, a women's movement, a Gandhian movement, and so
    on. For a critical reading of the movement see Rangan (1993, 1996).
13. In 1982, the Centre for Science and Environment published the
    First Citizens' Report on the State of India's Environment. Discussing
    large dams, the report read, 'Despite the impressive achievements
    the expected benefits in terms of the actual generation of electric-
    ity, irrigation and flood control have fallen short of the planned
    targets. If the costs of environmental degradation such as defores-
    tation in the catchment areas are included the price paid for these
    modern temples becomes truly staggering' (CSE 1982: 58). On the
    question of displacement and resettlement, the report said, the
    'government's rehabilitation programmes generally offer inadequate
    financial compensations. They fail to preserve and create the com-
    munity life of the displaced population.'
14. The study pointed out several flaws in the resettlement plan, cost-
    benefit analysis, as well as the magnitude of the project's impact
    on forests and wildlife in the Valley.
15. Some international NGOs lobbied hard with their respective gov-
    ernments and their executive directors in the Bank, pointing out

the environmental impact of the project and requesting them not to finance it (Letter, Minister of Finance, Government of Canada to Patricia Adams, Energy Probe, 2 March 1985).

16. Part of the construction process of the waterway system would lead from the reservoir to the main channel. Five villages in Gujarat were cleared from the area to make way for construction in the early 1980s.

17. Committee of Affected/Displaced People in the Narmada.

18. The activists were Achyut Yagnik and his colleague and employee at the time, Medha Patkar.

19. With the intention of helping the government include complete details while estimating compensation and resettlement entitlements, these committees would prepare comprehensive household data on land possessed, location and extent of submergence, quantum of produce from land, forest and river, data on the size of the house and amount of bamboo used (Parasuraman 1993).

20. The activists of the Narmada Dharanagrasta Samiti were employees of SETU. It was not until the end of 1987 that Medha Patkar parted ways with Yagnik, citing differences of perspective on resettlement and the 'need to break free from the bondage of foreign donations' (speech made by Medha Patkar, Support Group Conference, at Pune on 1 April 1995).

21. This project had been inaugurated on 23 October 1984 by Indira Gandhi, the then Prime Minister. According to official statistics, the NSP would irrigate a net cropped area of 1,41,000 ha while submerging 91,348 ha of land. The project would also generate power with an installed capacity of 1,000 MW.

22. (Personal communication, Avinash Deshpande, documentary filmmaker). Deshpande was then actively involved with Vidushak Karkhana, an activist group based in Shadol. He also produced the first documentary film on displacement titled *Narmada Puran*.

23. The environmental and social consequences of Tawa and Bargi projects were extremely adverse. The Tawa project caused acute waterlogging in the command area resulting in a decline in crop yields after irrigation, and sparked off the Miti Bachao Abhiyan. The displacement that occurred due to the Bargi project was much more than was initially claimed by the government. The compensation package offered was also extremely inadequate.

24. A noted social activist and Gandhian, Kashi Nath Trivedi, who was instrumental in forming the Narmada Ghati Navnirman Samiti (NGNS) back in 1967, opposed the Narmada projects on the grounds that big dams are dangerous and destructive and that they radically affect the *dharma* of the river (which is) to flow. The NGNS

was formed in 1967 to take up issues with the then MP government headed by Govind Narayan Singh regarding the displacement of boat people with the building of road bridges across the Narmada (interview, Anil Trivedi, NBA, 22 February 1996).

25. The Sangath had been made aware of the project's possible impact by ARCH-Vahini in 1983–84 (personal communication, Amit Bhatnagar, KMCS).

26. In later years, MARG conducted similar surveys in some villages in the districts of Khargone and Dhar (MARG 1987–88).

27. Under the Land Acquisition Act (1894), governments need to abide by the following procedures. In the first stage under Section 4, they have to issue preliminary notification that a particular patch of land is needed or may be needed for public purposes. Under Section 6, the government declares its intention to actually acquire land. Notice under Section 6 has to be issued within one year of the notice issued under Section 4. In the third stage the government invites concerned persons to raise objections, if any. After due valuation of the property that is to be acquired, notice under Section 9 is issued within two years of that under Section 6. The final award is made under Section 11 and land is acquired under Section 16 of the Act.

28. Subsequently in 1987, after the appraisal process was over, NDS activists pointed to limitations in the methods of displacement impact assessment and resettlement monitoring used by project authorities and suggested a reappraisal of the SSP through action research with a focus on environmental impact and human displacement. (Letter of 27 July 1987, Medha Patkar, Centre for Social Knowledge and Action and NDS to Chairman, Narmada Control Authority, with enclosure entitled: Some Comments on the Research Tools [survey-schedule] prepared for monitoring and evaluation of resettlement of Sardar Sarovar Oustees). The comments highlight inadequacies in the proposed survey: poor measuring standards of the lives of adivasi households without information on minor forest produce, labour-sharing, and fruit-growing; ignoring social life and institutions of adivasi communities; and failing to capture peoples' assessments of and experiences with the topography they inhabit (also see letter of 24 September 1987, Medha Patkar, NDS, to Shekhar Singh with enclosure entitled: Reappraisal of SSP with the Focus on Environmental Impact and Human Displacement). The proposal for reappraisal called for a 'scientific review of the studies and plans—related to environmental impact, human displacement, overall project costs and benefits, plans for environmental protections and rehabilitation of oustees ... [to bring out]

inadequacies in the premises, design, methodology and administration' (p. 5). The reappraisal also sought to analyse the decision-making process, assess the present status of resettlement planning and implementation, and bring forth the perception of oustees and their participation as envisaged in the plan and as enlisted in reality.

29. Although project work started in 1961 with the construction of infrastructure at Kevadia, it was only after 1985 that work on the headway of the canal was initiated.

30. While city-based environmental groups dismissed the policies as a 'mere piece of paper', some scholars argued that the policies of the Gujarat government were primarily aimed at dividing the NGO movement around the SSP (see for instance Parasuraman 1993).

31. The declaration was signed among others by noted activists like Sunderlal Bahuguna of the Chipko movement and the anti-Tehri Dam movement, the late Anil Agarwal, founder of the Centre for Science and Environment and publisher of the *Citizens' Report on India's Environment*, and B.D. Sharma, then Commissioner of Scheduled Castes and Scheduled Tribes and later leader of the Bharat Jan Andolan (Bharat Peoples' Movement). A portion of the declaration is worth noting: ' We... [are] all united by a common resolve to ensure that people are no longer denied their basic rights over natural resources. We affirm that the nation's rivers are the cradle of our civilisation and that they cannot be strangulated to meet the needs of the exploiting class within the society. The issues raised by the construction of big dams challenge the very concept of the pattern of the economic growth, unquestionably adopted by our planners. We appeal to the nation to halt all big dams here and now' (reproduced in *Social Action* 1988: 297).

32. According to the estimates of the NBA, the total number of affected people would amount to about 400,000 (writ petition [civil] No. 319 of 1994, Supreme Court of India: *NBA vs Union of India and Others*). However, it has been claimed elsewhere that the figure could be as high as one million (see ibid.: 1; Ram 1993:1). The figure of one million includes those affected by the reservoir, the canal network in the command and downstrea m, those affected by compensatory afforestation and catchment area treatment, tenants and labourers dependent on land that is being acquired by project authorities for compensation.

33. Land degradation would be in the form of waterlogging in the command area, salinity ingress near the coastal areas of the command and downstream in the district of Bharuch as well as resulting from catchment area treatment upstream. The project also submerges an estimated 13,500 ha of forest land (dense as well as degraded) in the reservoir.

34. According to the NBA, the major beneficiaries of irrigation and power from the SSP would be the already developed districts of Baroda and Ahmedabad, which consist of rich cash-crop farmers and industrial interests. Further, the NBA claims that irrigation, power and drinking water benefits are grossly exaggerated by the project authorities, and that there would be significant shortfall due to wrong initial estimates, the project design and inadequate financial resource allocation (NBA 1992: 14).

35. From an earlier estimated project cost of Rs 4,240 crore under- taken by the Tata Economic Consultancy Service in 1983 at 1981– 82 prices, the project cost rose to Rs 21,518 crore in 1995 at 1991–92 prices, as per the estimate of the Sardar Patel Institute of Eco- nomic and Social Research. NBA activists have argued that the Gujarat government can neither mobilise the required resources on a year-to-year basis, nor spend the mobilised amount in a timely manner. Also, given the lion's share of the SSP in the annual budget of Gujarat, it has resulted in the crowding-out of resources from other small projects that could possibly have been more beneficial to the drought-affected region in Gujarat.

36. For a more detailed review of the critique of the SSP see NBA (1992).

37. The convention was attended by eminent academicians, women's groups, trade unionists, NGOs and legal experts.

38. For an excellent analysis of Shetkari Sangathana see Dhanagare (1995).

39. Mooted by Baba Amte, a noted Gandhian social worker, at a meet- ing of activists at Hemalkasa in April that year, the possibility of a rally was followed up at the NBA meeting in Bombay in May and then at a meeting of representatives of over 60 organisations at Itarsi in August 1989.

40. Anticipating the events at Harsud, the Gujarat Legislative Assem- bly passed a unanimous resolution in support of the SSP three days before the rally.

41. The Gujarat government had organised a pro-dam rally in Chhota- Udaipur in Gujarat on 29 December to demonstrate support for the project. This rally marched to the Gujarat side of the border to prevent the anti-dam march from entering Gujarat.

42. The name given to the camp on the MP side of the border, where the people stayed for a month.

43. Bombay, Delhi, Bhopal and Baroda witnessed rallies and *dharnas* in support of the satyagraha.

44. In the months of March and April there were report of police ex- cesses and harassment at Manibeli. Different 'fact-finding teams' arrived at different conclusions. While the team for the People's Union for Democratic Rights reported violations of human rights,

a team from the Tata Institute of Social Sciences (the official monitoring and evaluation agency for R&R appointed by the Maharashtra government) reported no such violations.

45. See the several reports on the relocation component of the SSP by Thayer Scudder for the Bank and his letter to Paul Airman, Executive Director to the World Bank dated 11 April 1990 in which he mentions that 'the Bank disbursements to SSP should stop until the Government of Gujarat corrects—within a specified period of time—the various deficiencies relating to its own relocatees and to MP and Maharashtra relocatees. Should the required action not occur the World Bank should withdraw from the project.'

46. The activists were Medha Patkar of the NDS, Girish Patel of the Lok Adhikar Samiti in Gujarat, and Vijay Paranjpye, an economist who had undertaken a 'holistic' cost-benefit analysis of the SSP and the NSP. The testimony drew sharp reactions from various quarters and even sparked a debate within the NBA as to the validity and correctness of appealing to a foreign state in a matter which is internal to India. However, it resulted in several Congressmen writing to the World Bank urging it to reconsider its support to the project. The letter concluded with the following words: 'The continued World Bank involvement in the SSP sends a clear signal to borrower countries that the environmental and social conditions in the loan agreements are not enforced and bona fide established. In the light of the overwhelming evidence of the unsoundness of this project and its broader implications for the Bank, we believe it would be a gross misuse of public funds to consider an increased replenishment for an institution which has demonstrated its disregard for human rights and environmental concerns' (memorandum, US Congress sub-committee to the World Bank, 2 November 1990).

47. The letter was signed by members belonging to different political parties—Social Democrats, Conservatives, Centre Party, Democratic League, Swedish People's Party, Rural Party, Christian League, Greens and Liberals (Letter to Barber B. Conable, President, World Bank, 31 May 1990).

48. This was achieved through active lobbying by Japanese NGOs, notably Friends of Earth Japan.

49. The NBA has been extremely careful to create an image of local reliance in terms of funding as it has been constantly accused by pro-SSP forces of receiving foreign funding. In several of her speeches, Medha Patkar has made it a point to refute such allegations. For instance, the Goldman Prize money awarded to Medha Patkar was not brought into the country but was to be used for an 'international campaign against destructive, anti-people projects

in India' (Narmada 1992: 25 September). It is common knowledge, however, that the NBA regularly receives solidarity money from foreign-funded NGOs. For instance, as part of the National Alliance for Peoples' Movement, the NBA has welcomed foreign funding with the expected caveat that such funding should not have any conditionalities attached (proceedings, NAPM meeting, Kasravat, 4 March 1996). Local industrialists in Bombay have also become a major source of funding for the NAPM.

50. Although the NBA expressed total dissatisfaction with the terms of references of the IRM, which was limited to suggesting mitigative measures for environmental impact and improvements to R&R, *'the Mission members assured the Andolan that the former President of the World Bank had sent them a letter in which he agreed that the terms of reference could be expanded. The members also said that they would not hesitate to say whatever logically came into their findings.* Following this it was decided to extend the Andolan's cooperation to the Mission' (Narmada 1992:14 June, emphasis in original).

51. The terms of reference of the panel required members to reside in villages in the submergence zone of the Sardar Sarovar Dam and to be present at demonstrations and protests. They were, however, expected to restrict their activities to observation, writing, interviews and photography (See NIHRP 1992, Interim Report). The panel prepared two reports in 1992 and 1993.

52. By 1992, the World Bank had provided more than US$ 50 billion for the construction of more than 500 large dams in 92 countries, which have displaced an estimated 10 million people world-wide.

53. Under constant pressure multilateral donor agencies such as the World Bank and OECD have periodically reformulated and revised their operational guidelines on various aspects such as displacement, resettlement and environment (see OECD 1991; WB 1994a).

54. It is interesting to note that the 18th Congress of the International Committee on Large Dams (ICOLD) held that year in Durban strongly voiced the necessity for more large dams in developing countries to provide energy, water for domestic and industrial use, irrigation to improve agricultural productivity and flood control measures. In his inaugural address to the Congress, President Mandela said that 'no modern developing economy would be possible without large dams. There is opposition to the building of large dams, some valid. But in South Africa, we have no choice if we have to develop industry and feed our people' (Dansie 1994: 14).

55. The World Commission on Dams set up in February 1998 to review major dams in the world with adequate stakeholder participation is a direct outcome of the build-up in international pressure on dam-building regimes.

56. Extracts from the detailed text are suggestive of international opinion on the project and the movement: '... The movement has succeeded in generating a debate across the subcontinent which has encapsulated the conflict between two opposing styles of development: one massively destructive of people and the environment in quest for large scale industrialisation; the other consisting of replaceable small-scale activities harmoniously integrated with both local communities and nature.... The Narmada projects are the epitome of unsustainable development. NBA, under the inspiring leadership of Patkar and Amte, has ignited a historic debate of world-wide relevance especially in this year leading up to the Earth Summit. The victory of the NBA over the Narmada dams, Sardar Sarovar and Narmada Sagar would be a great symbolic victory for sustainability and a reprieve from homelessness and refugee status for several hundred thousand people.'

57. The issues discussed pertained to resettlement, environment, hydrology, drinking water supply to Saurashtra and Kutch, irrigation efficiency, distributive justice, benefits and costs, alternatives, information, human rights violation and project review.

58. The state responded by declaring Manibeli and adjacent villages as a prohibited area. In the face of a state-wide crackdown on NBA activists, the Samarpit Dal went underground with the public resolve to drown on 6 August 1993.

59. The FMG was headed by Jayant Patil, Member (Irrigation), Planning Commission.

60. The MP chief minister raised the matter of height reduction with the Prime Minister of India, requesting his intervention by calling a chief ministers' meeting for this purpose. The Gujarat government responded to these developments by ordering closure of the sluice gates of the dam two days later, marking the beginning of permanent submergence in the Valley.

61. The Bhopal Action was an indefinite fast by Medha Patkar and three representatives from the submerging villages.

62. The committee submitted its report in August 1995, recommending that the *adivasi gram sabha*, the body representing people in a tribal village or hamlet, should have wide-ranging powers over the natural resources used by adivasis. These included powers to safeguard rights relating to land, water, forest, minor forest produce; enforcement of customary rights over grazing and biomass collection; management, regulation and use of common property resources and maintenance of community assets.

63. The petition was filed 'on behalf of the tribal and other oustees of SSP ... other directly and indirectly affected persons from this project ... [and] on behalf of the people of India in general whose money is

going to be spent in excess of Rs 40,000 crores' (Writ Petition [Civil]
No. 319 of 1994, Supreme Court of India: *NBA vs Union of India
and Others*).

64. The legal framework governing the SSP consists of the Narmada
Water Disputes Tribunal Award of 1979, government legislation
such as Land Acquisition Act, Environment Protection Act and the
Forest Conservation Act, policies and resolutions as well as guidelines and decisions of various government and parastatal agencies
such as the National Commission on Scheduled Castes and Scheduled Tribes, Central Water Commission, Ministry of Environment
and Forests, Narmada Control Authority and its sub-groups on
Resettlement and Environment.

65. Civil Application No. 522, 1994, before the Gujarat High Court.

66. Civil Application of September 1993 before the Civil Judge, Senior
Division, Nandurbar: *Pukharaj Bora vs State of Maharashtra*.

67. Writ petition of June 1991 before the Delhi High Court: *Kisan Mehta
and Arvind Adarkar vs State of Maharashtra*, NCA, SSNNL, State
of Gujarat and the Union of India.

68. The FMG report, unlike the Morse Committee report, did not question the SSP but highlighted several problem areas. This argument
is developed in Chapter 7 by comparing the reports of the two review groups and their underlying methods and assumptions.

69. The new government of MP, in an affidavit-in-reply filed in the
Supreme Court, had asked for a reduction in the dam height from
FRL 136m to FRL 132m. In August 1996, at the intervention of the
Prime Minister, the three riparian states agreed that the dam height
should be 132m after which hydrological data would be collected
for five years before raising the height to 136m. The NBA condemned
this agreement and vowed to intensify the movement (*The Hindu*,
19 August 1996). However, MP government authorities later clarified that no such agreement was arrived at and that dam height
could be reduced to 88m.

70. The rallying call of the NAPM is *'jhute vikas se mila hey vinash,
visthapan, visamta aur gulami, hamey chahiye swadeshi,
swavlambhan, dharmnirapekhsta aur azadi.'* Translated it means
'The so-called development process has given us destruction, displacement, destitution and dependency, We want self-reliance, secu-
larism and independence.' Further, in the context of opposition to
the forces of globalisation, the NAPM's call for self-reliance is articulated in the slogan, *'hamara beej, hamara bhumi, hamara khad,
hamara pani!'* (Our seeds, our land, our fertilisers, our water).

71. Declaration of the NAPM at Sewagram meeting held on 16 March
1995. The NBA is a signatory to the declaration.

72. At the time of writing, a massive police operation took place against the Adivasi Mukti Sangathana (Adivasi Liberation Organisation) which had achieved enormous success in mobilisation in the Sendhwa region on issues of forests, forest lands and liquor trade. There were reports of activists being driven underground with arrest warrants issued in their names. It may be of interest to note that a similar form of police repression was experienced by the Chattisgarh Mukti Morcha (Chattisgarh Liberation Front) in eastern MP, which culminated in the murder of its leader, Shankar Guha Niyogi.

73. This necessitated a special meeting of NBA leadership with the people of Manibeli to convince them of the need to stay on. Manibeli, however, ceased to be a satyagraha site.

74. The problems in the resettlement sites are discussed in Chapter 6. It can be mentioned here that the return of oustees should not wholly be construed as an act resulting from dissatisfaction with the conditions of resettled villages. Many people who have received compensatory land have leased it to local farmers. Given the fact the submergence of large parts of the Valley has yet to occur, families who have received compensation have divided time and household labour between their original land in the submergence zone and the compensatory land in the resettlement site.

75. The Gujarat Narmada Displaced Struggle Committee. It may be of interest to note that, until farily recently, the NBA shunned any activism among the resettled as they were considered traitors to the movement, having weakened it by accepting R&R.

6

The Narmada Movement and the Significance of Risk Politics

Introduction

We now take the analysis of the Narmada movement a step further by situating the 'knowledge claims' which it made against those of the project authorities. The purpose is two-fold: first, it will show in greater detail how the movement questions the cost-benefit calculus in the SSP. In that sense, we extend here the arguments made in Chapter 4 on the appraisal of the SSP in order to show the ways in which the movement did not address major uncertainties and risks. Second, drawing on contemporary theories on environmental movements, this chapter brings out the significance of knowledge struggles in the SSP and the role of the Narmada movement in them (Eyerman & Jamison 1991; Jasanoff 1997). This second objective is to gain a deeper understanding of the practices of the movement, developing the arguments made in Chapter 5. A series of questions are addressed here, the central ones being: what are the claims about the SSP made by the Narmada movement, and, how have project authorities responded to such claims.

The primary data used are drawn mainly from relevant SSNNL and NCA documents; legal affidavits of the government of Gujarat; project completion reports of the World Bank and independent review groups; and relevant publications, unpublished documents, campaign material and legal affidavits of NBA and other allied NGOs and action groups in the Narmada movement.[1]

The NBA's Language of Protest

It is clear, then, that the NBA views its struggle to be as much against the SSP as against a 'larger reality'—a development model. It is not surprising, therefore, to observe that a variety of idioms, discourses, organisations and people are mobilised and linked in the movement's struggle over the SSP and over meanings of developmental means and ends. Issues featuring in the language of the Narmada movement include: (*i*) displacement and resettlement provisions; (*ii*) environmental impact, mitigative measures and sustainability; (*iii*) forceful evictions and violation of civil liberties; (*iv*) democratic rights such as access to information and participation in decision-making processes; (*v*) distributive justice and sharing of benefits; (*vi*) political decentralisation, cultural autonomy, indigenous knowledge and local control of resources and empowerment; (*vii*) institutional aspects governing river valley planning and water and power management; (*viii*) appraisal and evaluation techniques and financial implications of the project; and (*ix*) Western growth model and neo-imperialism vis-à-vis alternative development and appropriate technology.

The combination of these wide-ranging issues in the Narmada movement should be seen both as a consequence of and as a reason for the NBA's support networks consisting of diverse constituencies at the local, national and global levels: animal rights activists, adivasi and peasant organisations, action groups working among urban slum dwellers, civil liberty groups, trade unions, women's organisations, environmental action groups, developmental NGOs, media groups, retired government planners, development practitioners and academics. The hybridisation of actors and language allows the Narmada movement to *move* beyond the constituency of displaced people, their risks and losses and the language of compensation, to cover organisational, institutional and cultural dimensions around development interventions.

Meaning as Rhetoric or Meaning as Knowledge?

Again, we have already noted that the meaning of development constructed in the Narmada movement is synonymous with homogenisation, giganticism, capital- and resource-intensive technology and globalism. This meaning is then challenged by a vision of 'alternative development' in which the ideology of localism is combined with cultural and biological diversity, preference for micro projects and for ecologically and humanly benign technology, and demands for the democratisation of social life. The concrete manifestations of this 'alternative imaginary' as it takes shape in the SSP is discussed in Chapter 9. Here, our arguments concern the constructed nature of 'meanings' and the need to discount the visionary and revivalist implications of social movements in order to arrive at a deeper understanding of their politics.

To fight against a development project that is backed by interest groups pervading local and global levels requires a strategic syncretisation of idioms and the articulation of support at different levels with different constituencies. Thus, acquiring syncretic idioms (and deploying them selectively) is to be seen as a movement imperative. To unite different idioms and constituencies, movements often tend to construct what Melucci (1989) calls their 'highest meaning', i.e. a message that seeks to overlook heterogeneous elements and interests while underscoring unity. This is a larger frame that holds idioms and constituencies together in one movement, in the construction of which movement leaders and sympathetic scholars play a leading part. The framing of the Narmada movement as one for alternative or just and sustainable development is only one strong example (see Baviskar 1995; Gadgil & Guha 1995; NBA 1992; Omvedt 1993).

While it is perfectly legitimate to frame the Narmada movement as instigating just, sustainable and/or alternative development, this *highest meaning* should be seen as movement rhetoric. A movement's so-called highest meaning is first and foremost a reflection of its strategic predisposition. To hold together disparate constituencies, movement leaders construct a set of persuasive messages and tactics intended to appeal to the interests and values of diverse (and even conflicting) groups. The end product, i.e. the movement's highest meaning, is couched in a language that remains discursive and ambiguous so that each constituency finds space to plant its own interests.

Both sympathetic and critical studies of the Narmada movement consider this rhetoric of ultimate significance. The 'struggle over meanings' then becomes one of celebrating or critiquing this rhetoric. Celebrating the alternative rhetoric leads to eulogising local lifestyles, livelihood strategies and indigenous knowledge and culture, and conjures up images of self-contained village communities living in harmonious ecological utopias. It also encourages popular myths on rivers and cultures around them. The glossing-over of differentiation, poverty and every-day struggles in the real world of the locale is striking. Critics of the rhetoric use these celebratory features to seek the *a priori* dismissal of the movement as a backward-looking, romantic, *anti*-development force.

It is necessary to discount a movement's highest meaning in order to focus on its contemporary significance, a political force organising struggles around truth claims in the SSP. As we turn to an empirical demonstration of the knowledge struggles in the Narmada movement, we hope to have clarified that, far from having an anti-technology, anti-development predisposition (which opponents of the movement would have us believe and a viewpoint to which sympathisers of the movement inadvertently contribute), the issues raised in the Narmada movement pervade a scientific domain. Drawing from insights from Beck (1992; 1995) and Giddens (1990), we term this the 'risk politics' of environmental movements.

The Significance of Risk Politics

Central to the NBA's language of protest is the combating and contesting of official parameters of attributing risks and opportunities in the SSP. Official truth claims are subjected to systematic and rational interrogation, and fudging, inconsistencies, errors and concealment in assumptions, performance indicators, methods as well as inputs and outputs and costs and benefits are unravelled in the process. Defined broadly, 'risk politics' entails politicising conditions of uncertainty and lack of information on the one hand, and risks and lack of certainty on the other. Whereas the movement rhetoric (in its hard form) conveys a populist language that is anti-science, anti-technology in tenor, the language of risk politics remains within a scientised domain—referring to scientific data and technological knowledge claims.

We argue that it is this practice and the knowledge that is produced therein that give a movement like the NBA its significance.

Development interventions such as large dams entail uncertainties and risks at every step. The enormity of the scale of intervention—with engineering, technical, hydrological, economic, social, environmental, financial parameters—necessarily implies that project authorities work with an equally enormous set of probabilistic assumptions for each parameter. Their truth claims, i.e. how the dam will perform and what benefits and costs it will entail, rest on those assumptions.

It is an unstated but accepted practice for project authorities to overestimate benefits and overlook costs in their appraisals. Despite expressed concerns in policy discourses on mega dam projects, such problems of over- and underestimation have been the rule rather than the exception. In his pioneering work, however, Hirschman justified such a decision-making process and outlined the 'hiding hand principle', i.e. the errors of initial overestimation of resources and benefits are offset by an initial underestimation of alternative and remedial action:

> The river that is being tapped is frequently found not to have enough water for all the power, agricultural, industrial and urban uses that had been planned or that are staking claims, but the resulting shortage can then often be remedied by drawing on other sources that had not been within the horizon of the planners; ground water can be lifted by tube wells, the river flow can be better regulated by upstream dams, or water from more distant rivers can be diverted (1967: 10–11).

Contrary to Hirschman's principle which emphasises hidden solutions and benefits that offset project risks, contemporary environmental movements attempt to expose hidden uncertainties and risks buried in the probabilistic assumptions of project authorities. Whereas Hirschman recommended that project authorities should pursue the 'only feasible goal' of 'optimal rather than minimal uncertainty' (ibid.: 85) assuming that hidden solutions will enhance project benefits, risk politics in the Narmada movement bring to the fore hidden uncertainties and risks and their distributional implications. In the process, the movement deploys a discourse on 'minimal uncertainty'.

Risk politics question the trustworthiness of project agencies and institutions that handle uncertainties, attach probabilities, and calculate risks and liabilities. They highlight the fact that probabilistic assumptions of experts often tend to become political assertions about project performance. The data garnered in this process render official assumptions of risks and liabilities questionable, to say the least. What ensues from this form of politics is a counter-set of knowledge claims and assertions whereby projects such as the SSP become 'faulty, non-viable, unjust and destructive' (NBA 1994:10).

Two central arguments drawn from the works of Beck (1992; 1995), Buttel and Taylor (1994) and Giddens (1990) indicate the significance of risk politics in environment movements. The first is that assessing risks and assigning liabilities remain the prerogative of official experts who function in closed institutional domains and are mandated to alter the environment for productive transformation. To muster public acceptance of proposed interventions, Giddens (1990) has argued that experts tend to fudge or conceal the true nature of risks or even the fact that risks exist. He considers the circumstance more harmful 'where the full extent of a particular set of dangers and the risks associated with them is not realised by the experts. For in this case, what is in question is not only the limits of, or the gaps in, expert knowledge but an inadequacy which compromises the very idea of expertise.'[2]

The second argument is that modern social-environmental movements are prime vehicles of risk politics. According to Buttel and Taylor (1994: 223):

> Modern environmentalism, where the rubber meets the road, is increasingly an arena characterised by the deployment of scientific and technical knowledge, often in combat with rival data and knowledge claims that are set forth by their industrial, governmental and quasi-governmental adversaries in an attempt to deconstruct and delegitimate claims.

Giddens holds that the radical engagement and outlook of environmental movements are 'bound up with contestary action rather than a faith in rational analysis and discussion' (1990:137). Such movements question the assumptions and assertions of official experts regarding risks, and seek to mobilise societal opinion to 'reduce their impact or transcend them' (ibid.).

We deploy these two arguments in the specific context of the Narmada movement and list some claims and counter-claims in the SSP. The listing shows that 'expert' arguments are advanced in the movement which counter in some way or another virtually every claim made by project authorities.[3] The uncertainties and risks around the issues of hydrological estimates, assessments of environmental impact, project appraisal and project alternatives have been brought into the public-political arena for fuller interrogation.

Is the SSP Based on Sound Hydrological and Planning Assumptions?

How Much Water is there in the Narmada?

Hydrology, as already seen, was a major area of dispute between the Gujarat and MP governments, and had been resolved through political agreement between the heads of the two riparian states which was accepted by the NWDT. In 1979, the NWDT estimated the amount of annual utilisable water in the Narmada to be 28 million acre feet (MAF). Of this, 9 MAF (32%) was allotted to Gujarat, 18.25 MAF (65%) to MP, 0.25 MAF (1%) to Maharashtra, and 0.50 MAF (2%) to Rajasthan, based on the estimated annual flow of 27 MAF at 75% dependability.[4] The 75% dependability implies that a flow equal to or more than the assessed flow (27 MAF in case of the Narmada) would be available in 75 out of 100 years. This issue has been reopened by the Narmada movement to claim that there is insufficient water in the Narmada and that hydrological estimates, i.e. annual utilisable water flows at the Sardar Sarovar dam site, have been overestimated (NBA 1995b: Ram 1993).

The NBA has claimed that the annual amount of utilisable water at the dam site is not 28 MAF at 75% dependability as assumed by the NWDT, but is less than 23 MAF, i.e. about 17% less than planned (ibid.). This claim is supported by the Independent Review

Mission (IRM 1992), the MP government (GoMP 1994), and by two members of the FMG (FMG 1995).

The NBA attributes the erroneous figure of 28 MAF to an unreliable methodology, called hind-casting,[5] used by the project authorities in their calculations of water flow. The NBA has pointed out that studies conducted by the Central Water Commission, a Government of India undertaking, based on actual observable flow of the river, have yielded 23 MAF. From this the NBA concludes that the allocation of Narmada water to each of the riparian states should be proportionately downgraded. Since the actual flow is 23 MAF, Gujarat's share will have to be reduced to 7.2 MAF from an allotted 9 MAF. As a consequence, the proposed benefits of the SSP (planned on an assumption of 9 MAF) will not materialise.

Project authorities justify the hydrological estimates on the following grounds. First, the technique of hind-casting is an accepted and established international practice when observed data on actual river flow are insufficient. The Central Water Commission, i.e. the highest policy-making body in the country on this issue, has endorsed both the use of hind-casting and the accuracy of the hydrological estimate (CWC 1991).

Second, the NWDT considered the utilisable quantum of water at the Sardar Sarovar Dam site at 28 MAF on the basis of 75% dependability. This implies that 28 MAF or more would be available in 75 out of 100 years. However, the volume of water flowing in the Narmada at the Sardar Sarovar Dam site varies tremendously from year to year, ranging approximately from 5 to 61 MAF. The NWDT also made provisions for a proportional scaling-down of allocation among the four states during dry and water-deficit years.

Third, dependable availability is only one factor with which to determine the utilisable quantum of water. Total storage capacity in the river basin is an additional factor. Assuming the observed flow at 23 MAF, there is a case for increasing storage capacity in the river basin. This may well imply a higher dam for sufficient storage rather than a lower one as demanded by the Narmada movement.

Fourth, according to the schedule for development of the command area in Gujarat, the first 10 years will involve the development of one million hectares, requiring about 5 MAF of water.

Water demand in MP will increase more slowly. The problem in the initial 10 to 15 years will be excess and not inadequate water. By the time the full command of 1.8 million ha is developed (the World Bank estimates this to be in year 2025), the tribunal's water allocation will be ready for review. Observed data can then be used to alter allocations.

Estimating Utilisable Flow in the Narmada: Which Method is Correct?

The debate on utilisable flow is significant, connected as it is to the design of the SSP. Critics feel that to overestimate the water flow would affect the proposed benefits (see Ram 1993: 6). The inference is that less water than assumed in the Narmada will warrant the scaling down of the SSP. While the MP government and the NBA consider 27 MAF in the Narmada an overestimation, the Gujarat government finds the amount *sound* and *settled* (SSNNL 1994a).

Different agencies—governments, committees, commissions, tribunals and review groups, Bank consultants and experts—at different points in time using different methods have come to different conclusions regarding the 75% dependable annual run-off in the river. It is evident from Table 6.1 that a dispute over methods underlies the contention over estimates of water flows, i.e. whether hind-casting as a method is a reliable and valid adjunct to observed flow data in assessing utilisable flow. While the justification for hind-casting is based on the need to consider a longer time series of data, its reliability as a method is open to question as it is *derived* rather than actual data. In our view, the significance of the debate over water flows lies not so much in the technological virtuosity of hind-casting (even hydrological experts do not agree on this issue) as in the conflicting reading of its appropriateness.

The dispute over methods dates back to 1965. In its comments on the conclusions of the Khosla Committee, MP claimed that the committee's assessment of water availability at different sites was on the high side.[6] It argued that estimates of water flow should be based only on observed flow data. But MP simultaneously suggested that if a longer time series was indeed required, then hind-casting could begin from 1891 instead of 1911

Table 6.1 Assessments of Water Flow in the Narmada

Year of Study	Agency	Decision Methods	Estimate of 75% Dependability Flow	Remarks
1965	Narmada Water Resources Development Committee (known as Khosla Committee)	Hind-casting (1915–47) and Observed flows (1948–62)[7]	28.92 MAF	Stated that data collected from 1915 to 1930 could be unreliable because of inadequate density of rain-gauges.
1966	Official-level discussions among riparian states.	Negotiated agreement based on Khosla Report	27 MAF	Net utilisation flow estimated at 28 MAF (see note 5 in this chapter).
1974	Chief Executives of the riparian states	Agreement and appeal to NWDT	28 MAF	For planning purposes as stated above.
1991	Narmada Valley Development Authority (GoMP)	Hydro-meteorological (HM) cycle observation of the Narmada basin	27.64 to 27.89 MAF	It estimated a 79-year HM cycle of the basin; argued that run-off data were required for at least a 79-year period, hence justified hindcasting.
1991	(CWC) Central Water Commission, Note: for the Narmada Control Authority.	(a) Observed Flow (1948 to 1988) (b) Hind-casting (1891 to 1948) + (a).	(a) 23 MAF (b) 24.8 MAF	Stresses the probabilistic nature of hydrology and justifies 27–28 MAF for purpose of planning.
1992	Charles Howard and Associates (consultants commissioned by IRM)	Testing stream flow data.	Questioned quality of observed data and reliability of hindcasting.	More detailed analysis recommended.

(Table 6.1 Contd.)

(Table 6.1 Contd.)

Year of Study	Agency	Decision Methods	Estimate of 75% Dependability Flow	Remarks
1993	CWC	(a) Observed data from 1948–90. (b) hind-casting (1891–48) + (a).	(a) 23 MAF (b) 26.6 MAF	Argued for hind-casting since the required data length should be of 130 years given that the co-efficient of variation of annual flows is 0.35.[8]
1993	Wallingford Institute (consult-ant commissioned by the World Bank)	Testing primary flow records	26.6 MAF	Endorses CWC's position; rejects recommendation of Charles Howard and Associates.

Sources: NWRDC (1965); NWDT (1978a; 1979); IRM (1992); FMG (1994: 1995).

in order to offset the estimate by including some years of poor rainfall (NWRDC 1965: 207). Put simply, hind-casting as a method was not only accepted but its time series was lengthened (despite constraints in analysing rainfall data; see remark column on Khosla Committee in Table 6.1). Political negotiations facilitated by the central government in 1966 led to official agreement between the states on the 27 MAF quantum of utilisable flow at Navagam and its dependability. First in 1972 and again in 1974, the chief ministers of the four states reiterated their agreement that the quantity of water in Narmada available for 75% of the years should be assessed at 28 MAF and 'that the Tribunal in determining the disputes referred to it may proceed on the basis of this agreement' (see NWDT 1979, Vol-III: exhibit C-I). It was this *agreed* figure that the NWDT took as given in allocating water to the four states.

The CWC's note of 1991 to the Narmada Control Authority led to a fresh round of controversy, this time with the intervention of the NBA. The note was significant for the following reasons: It made a veiled and indirect justification for hind-casting by mentioning *difficulties in measuring actual water flow with desirable accuracy* due to constraints of equipment and facilities, especially in conditions of high flood, and argued that to rely solely on this data without proper consistency checks and simulation could be misleading.[9]

1. While noting that the observed data from 1948 to 1988 resulted in an annual yield of 22.9 MAF at 75% dependability, and that hind-casting from 1891 yielded a figure of 24.8 MAF, it sought to justify the figure of 27 MAF for planning purposes, using three arguments. First, that water planning should be such that the reservoir would have maximum possible storage so that even in the wettest years (flows in some wet years in the past have been over 40 MAF[10]) water can be used rather than wasted into the sea. Second, that statistical analyses of cyclical patterns of water flows indicate that a period of high annual flow can be anticipated in coming decades.[11] Third, that a new measure—the '90% probability of exceedence' factor—instead of 75% dependability factor, yields water flows between 28 MAF and 31 MAF.
2. Of qualitative significance was the opinion regarding hydrological uncertainty expressed in the concluding remarks, which

corresponded with Hirschman's plea of optimal uncertainty in development intervention. A fuller extract is illustrative:

> Uncertainties in engineering are often hidden by use of factors of safety, rigidly standardised procedures and conservative assumptions. In hydrological solutions, there is an extension of one's perception over large periods of time. With future realisations increasing the awareness of any phenomena, simple devices like factors of safety alone can not take care of future possibilities. In such an open ended situation as that obtaining in hydrologic analysis, the planner is forced to change his [sic] solutions. How to prosper under uncertainty is the question to be tackled...in water use planning and operation...' (CWC 1991: 8).

The Sardar Sarovar Project's water flow debate is a classic example of how project authorities, while negotiating limits of technical methods and assessments, seek the Hirschmanian optimal uncertainty. If indeed there is more water in the Narmada, the project will yield greater benefits (than if planned to harvest 23 MAF yield). The assessments made by the CWC indicate that there is a greater probability of this happening, hence the risk (of a bigger project) is worth taking.

In contrast, the position of the NBA on this issue is a minimum of uncertainty. Its main arguments are:

1. Observed data is more accurate than hind-casting. The latter cannot be used as an accurate measure given the inadequacies pointed out by the Khosla Committee of 1965. Observed data indicate a yield of 23 MAF.
2. If the observed data are difficult to estimate, as the CWC has pointed out, one cannot rely on hind-cast data which are derived from observed data.
3. The 75% dependability factor is an accepted water-use planning tool in India, hence the '90% probability of exceedance' is a clever statistical ploy to justify the (inflated) figure of 27 MAF.
4. Given the fact that yield estimate at 75% dependability is 23 MAF, the allocations to the four states will have to be scaled down. A scaling down of Gujarat's allocation from the current

figure of 9 MAF would require reduced storage, carryover and height of the dam (NBA 1993a).

In a submission to the Five Member Group, the NBA called for a 'total redesign' of the SSP, stating that data available today are 'of much better quality and quantity and this data shows that the water yield of the river was overestimated by NWDT and the states. This has immense consequences for the benefits and costs of the project' (NBA 1995). The NBA also points out that, from 1965 to 1995, all government agency studies[12] show observed data to be between 22 and 24 MAF.

Disputes Over Hydrological Estimates or Over Water Distribution?

The conflicting assessments of hydrological estimates illustrate an earlier argument that, while project authorities are inclined to take risks, environmental actors, through risk politics, highlight probable adverse outcomes regarding project benefits and costs. While the former tend to believe in 'probabilistic safety', the latter believe in 'super-safety' (Beck 1995). However, underlying the SSP dispute regarding methods of assessing water flows lie deeper objectives of the involved actors. For the Narmada movement, the hydrological estimate is one of the many unsound claims for which the project should be opposed. For the governments of Gujarat and MP it is a matter of who gets to utilise how much water from the Narmada.

Consider the following facts. At Navagam, the site of the Sardar Sarovar Dam, Gujarat's share (9 MAF) plus Rajasthan's share (0.50 MAF) will be available for use *provided upstream dams are constructed in MP*. If the dams are not constructed, MP will be unable to tap its allotted share of water (18.25 MAF), implying that more than the legal share of water for Gujarat and Rajasthan will flow past the Navagam site. Unless Gujarat's share is reduced legally, it is theoretically possible for the state to tap more water than its allotted share. As the IRM notes, 'the [Sardar Sarovar] dam and the canal, as presently designed, are theoretically capable of diverting more than 8 MAF during the *three and half* months monsoon' (1992: 242).

If we consider MP to be one among the least developed states in India,[13] it is most unlikely that it will be able to mobilise the necessary resources to tap its allotted share in the coming 50 years. In an office memorandum of the World Bank dated 14 March 1990, this fact was explicitly stated:

> It is unlikely that MP will be able to use its allocated share of 18.25 MAF within the remaining 35 years of the NWDT Award. Indeed it is probably technically impossible for MP ever to use this amount of water ... Gujarat could use and presumably will use for its expansive 1.9 mn.ha irrigation command area, rather more than its 9 MAF allocation (cited in IRM 1992: 252).

For the Gujarat government to consider the hydrological estimate to be sound and settled is therefore not at all surprising. As the NBA probes deeper into questions of uncertainties and risks around hydrology, the MP government gets a fresh handle with which to ensure that Gujarat does not corner the benefits from Narmada waters.

What Links does the SSP have with the Narmada Sagar Project?

The NWDT award states clearly that the performance of the SSP would depend critically on upstream projects, particularly the NSP (in Madhya Pradesh), for a regulated release of water. The bulk of the Narmada flows in the monsoon months and the Sardar Sarovar reservoir has live storage capacity for only 4.7 MAF (5,800 million cubic metres); Gujarat therefore depends largely on the NSP for its allotted 9 MAF of water. Given this critical link, the NWDT had stated that the NSP should be built simultaneously with or before the SSP (GOI 1979: 1431). The NWDT also estimated that without regulated release from the NSP, the power and irrigation benefits from the SSP would be reduced by about 17%. On the basis of this estimate, the NWDT asked the Government of Gujarat to contribute 17% of the total cost of the NSP (ibid.).

Although the Planning Commission gave investment clearance to the NSP in 1989, work has barely begun on the project. In 1990, the World Bank cancelled a loan to the NSP due to inadequate baseline data. In 1992 there were some renegotiations with the World Bank, and in 1994 there were reports of construction work on the coffer dams[14] at the NSP site. The time lag between the construction of the SSP and the NSP, however, is obvious. The Narmada movement claims that since the NSP has lagged behind in construction, and given the uncertainties that prevail over its construction,[15] the proposed irrigation and power benefits of the SSP will be drastically affected (NBA 1995b).

Authorities of the SSP have responded to this criticism by reiterating that the 'no NSP' scenario is only hypothetical since the project is being undertaken and will be completed. The worst scenario is one where construction on the NSP will be delayed (Patel 1991a; SSNNL 1994a). Although project authorities acknowledge that delays in the construction of the NSP will affect the longevity of the SSP (through increased siltation), they consider this problem to be manageable and a reason for hastening the completion of the NSP.

How will the Project Perform and what are its Benefits?

Will Water Reach Drought-prone Areas?

The SSP covers about 20% of total cultivable area in Gujarat. Its proposed project command of 1.8 million hectares spreads across 12 districts in Gujarat and two in south Rajasthan. Table 6.2 gives the district-wise break-up of the proposed irrigation benefits in Gujarat.

Claims advanced in the Narmada movement point to the untenability of water benefits to drought-prone areas in Gujarat, particularly Kutch, Saurashtra and north Gujarat, in whose interests the SSP is being constructed. It has been argued that, aside from hydrological factors and the NSP that are the cause of this non-availability, there are other reasons to believe that water will not reach these areas.

To begin with, large parts of these water-deficient areas of Gujarat do not feature in the proposed command area of the SSP. Only 1.6% of the cultivable area in Kutch, 8.5% of that in Saurashtra and 20% in north Gujarat are covered in the proposed SSP command (NBA 1995b; Ram 1993).

Table 6.2 District-wise Break-up for Gujarat of Existing and Projected Irrigation Targets

District Cultivable	Cultivable Area ('000 ha)	Area Covered Under Existing Major and Medium Schemes ('000 ha)	Area Under SSP ('000 ha)	% Area Under SSP
Bharuch	507.8	128.49	97.95	19.28
Baroda	593.0	123.69	340.15	57.27
Kheda	536.1	237.11	116.01	21.63
Gandhinagar	51.9	NIL	10.65	20.52
Ahmedabad	676.2	46.35	331.27	48.99
Panchmahal	547.8	112.12	9.68	1.77
Mehsana	753.3	68.55	150.19	19.55
Banaskantha	925.6	93.89	313.89	33.91
Surendranagar	782.5	23.38	303.73	38.81
Bhavanagar	703.1	94.28	48.27	6.87
Rajkot	810.0	98.90	34.12	4.21
Kutch	2363.1	24.52	37.85	1.60
Total	9250.4	1051.28	1793.85	19.39

Source: GoG (1991), Patel (1991a).

Second, Kutch, Saurashtra and north Gujarat lie at the tail end of the proposed command area. As with other irrigation projects in the country, beneficiaries at the head end, i.e. the already developed districts of Baroda, Bharuch and Ahmedabad, would receive water first and monopolise its use, and it would be very difficult to decrease water supply to them later. Farmers will switch to water-intensive cash crops such as sugarcane, as happened in the Ukai command in south Gujarat. Sugar factories have already been set up at the head end, anticipating sugarcane cultivation although it is not grown in the area, and nor does it feature as a major crop in the proposed cropping pattern. Other industries will become concentrated at the head end, expecting water supplies and agro-based raw materials. The major benefits of the project will therefore accrue to rich cash-crop farmers and industries (NBA 1992; 1995b).

Third, the irrigation efficiency[16] of 60% assumed by project authorities is extremely high and has not been achieved in any of the existing irrigation commands. Considering past performance of projects and practices in India, irrigation efficiency is likely to be between 35 and 40%.[17] Furthermore, the actual water requirements of proposed crops in the command are much

higher than has been assumed. Additionally, as the canal network spreads over thousands of kilometres, it will be extremely
difficult to control any unauthorised diversion of water. Hence,
per hectare estimates of water requirements should be increased
by 10%. All these factors imply that the project command will
be sizeably smaller than planned (NBA 1995b).

Fourth, the Gujarat government has made no specific allocation to reduce the adverse impact on downstream people and
their ecosystems. Although it had requested for this in order to
arrest salinity ingress and to maintain navigation, irrigation and
domestic use, the NWDT asked it to make downstream releases
from its allotted 9 MAF of water. If the Gujarat government
makes allotments to reduce the adverse downstream impact,
then the water allotted to the command area will have to be
proportionally reduced (NBA 1995b).

Taking all factors into consideration, the command area will
be less than 75–80% of the proposed 1.8 million hectares. In this
reduced command area, the tail end water-deficient regions of
Kutch, Saurashtra and north Gujarat will lose out on benefits.

In contrast to this assessment, project authorities reiterate
that water will reach these drought-prone areas and, more
generally, will be sufficiently available in the command. Their
assertions rest on two arguments.

First, the quantum of water that can be taken to Saurashtra,
Kutch and north Gujarat has been influenced by the NWDT's
decision. The allotment of 9 MAF (as against what Gujarat had
asked for), the dam height of 455 ft. (as against what Gujarat
had requested), the number of branch canals to these areas (the
Gujarat government had asked for three branch canals for Kutch;
the NWDT approved one), and the offtake point of the canal
(FSL 300 ft.), have all been fixed by the NWDT. These supply
constraints notwithstanding, irrigation has covered more than
12 districts, 62 sub-districts and 3,393 villages (Patel 1991a;
SSNNL 1994a).

Second, project authorities state that they have never claimed
that all areas in Kutch, Saurashtra and north Gujarat would
receive water (Patel 1991a: 37). The distribution of irrigation benefits is governed by several criteria: supply constraints, feasibility,
water scarcity and drought-proneness, as well as the ultimate irrigation potential of a district. Within these constraints, attempts

have been made to distribute benefits evenly (see Table 6.2) across the state. The major share goes to the districts of Baroda, Ahmedabad, Banaskantha and Surendranagar. The latter two lie in the water-scarce regions of north Gujarat and Saurashtra respectively. As far as the 'prosperous' district of Ahmedabad is concerned, 90% of the command is drought-prone.[18] In fact, according to the esti-mates of the Second Irrigation Commission (RIC 1972b), about 70% of the SSP command area is drought-prone.

Project authorities admit that the water allocated for irriga-tion to Gujarat is not enough even to irrigate 1.3 million ha of area under normal application (see Patel 1991a). The justifica-tion for a larger command, however, comes with the principle of extensive and protective irrigation adopted to ensure benefits for a maximum number of people through an ample spread. The command area of 1.8 million ha has therefore been planned to benefit 600,000 farmers in the command, around 50% of whom belong to the small and marginal category (SSNNL 1994a: 57).

The question of water shortage in the command caused by inefficiency and monopoly uses is countered with arguments on proposed innovations in water management. Water in the SSP will be controlled and supplied on a volumetric basis with pre-determined quota for each area (and sub-area) to ensure that farmers receive water in proportion to their landholdings. The supply system will rely on water users' associations and will be backed politically by appropriate legislation. The command area has been divided into various agro-climatic zones, and cropping patterns in these zones have been appropriately planned. Overall, the proposed crops consume less water but are comparatively remunerative. Water-intensive crops such as sugarcane and ba-nana do not feature in the cropping pattern as the water supply will be inadequate. Thus, only 0.5% of the crops in the command are planned as perennial. Farmers will be allowed to grow more remunerative crops such as sugarcane only when it is feasible for them to augment canal water with private investments in groundwater development. Project authorities claim that these governmental policy measures will not encourage sugarcane factories to be set up in the command (SSNNL 1994b).

To the charge of unauthorised and monopolistic uses of water in the SSP, the response of a member of the Narmada Planning Group (NPG) is worth noting:

the [Sardar Sarovar] project will behave as well or badly as society behaves. If the society behaves irresponsibly some groups could corner benefits as Harshad Mehta[19] did. The idea of distributing water to user associations instead of individuals is precisely to discourage unauthorised use. However if the whole village engages in theft [unauthorised use] then god help us (interview, 4 May 1995, D.T. Buch, NPG).

How Suitable is the Command Area for Irrigation?

This too has been the subject of conflicting claims. The Narmada movement argues that studies conducted by the Operation Research Group (ORG 1982) on the proposed zoning of the command according to agro-climatic conditions and irrigability class show that only 10% of the SSP command is eminently suitable for irrigation. The remaining areas are classified as moderately suitable (40%), suitable but with severe limitations (25%), marginal and unsuitable (20%) and unclassified (5%) (ORG 1982; Ram 1993). Large tracts of the proposed command area in Saurashtra and Kutch are classified either as having severe limitations or as unsuitable. It is therefore presumed that large areas in the command will suffer from serious waterlogging and salinisation (NBA 1995b).

Project authorities have responded to this debate by arguing that the problem of waterlogging in the command has been anticipated and control measures have been devised. The SSP advances the conjunctive use of surface and groundwater to increase overall water efficiency, as well as vertical, sub-surface drainage. Monitoring of the groundwater table is to be conducted throughout the command with 4,000 observation points. Surface drainage systems have been planned and will be implemented along with the development of the canal systems. Of the 13 agro-climatic zones only three (i.e. zones 4, 7, and 11 covering 20% of the total culturable command) pose drainage-intensity problems for which appropriate networks are planned (Patel 1991a: 78). In agro-climatic zones where soils are not suitable for sustained irrigation, the water dosage will be reduced (ibid.: 79).

Are Drinking Water Benefits Integral to the Project?

First, the drinking water component of the SSP has served as a moral, political and emotionally charged justification of the project. The NWDT had set aside 1.06 MAF (1,300 million cubic metres) of Gujarat's share of 9 MAF for non-agricultural use. Movement leaders point out that while this allocation has remained constant over the years, project authorities have increased the number of proposed beneficiaries from an initial estimate of 28 million in 4,720 villages and 131 urban centres to 30 million in 8,214 villages and 135 towns.[20]

Second, according to the NBA, no detailed plan, financial cost estimation and time schedules are available with regard to the drinking water component, although it has been listed and propagated as a major benefit of the SSP (Narmada 1993). As most of the proposed beneficiary villages lie outside the irrigation command area, the elaborate water supply system including canals, pumping pipes and purification, would require substantial resource investments (Ram 1992). In 1982, the Gujarat Water Supplies and Sewerage Board (GWSSB) estimated the cost of drinking water supply at Rs. 7,280 million (1981–82 prices). Since then, no proper estimate has been made except for vague references by project authorities that drinking water supply would cost 'several thousand crores of rupees' (NCA 1991).

Third, it is pointed out that the drinking water allocation to Kutch and Saurashtra (less than 0.35 MAF) is insufficient in view of the region's water scarcity and demands. Although large sections of the population are pastoral, there is no provision of water for livestock (NBA 1995b).

On their part, project authorities claim that drinking water benefits are integral to the project objective of drought-proofing. In the planned utilisation of about 1.06 MAF (1,300 million cubic metres) of water for M&I use, domestic and municipal supplies will involve 1,000 million cubic metres (0.86 MAF) and industrial water supply 300 million cubic metres (0.2 MAF). The M&I water supply has been planned by the project authorities (SSNNL) in consultation with the GWSSB as well as the Gujarat State Industries Department. Major offtake points for drinking water supply have been planned in consultation with the GWSSB (SSNNL 1994b).

Project authorities justify the increase in the number of villages and urban centres (from their earlier estimate in 1982) as a response to acute drought and consequent water shortages in Gujarat during the period 1985–88. Prolonged drought caused drastic alterations to the earlier plan: water will now be provided to 135 (instead of 131) urban centres and 8,215 (instead of 4,719) villages (SSNNL 1994b: 69). Although the total allocation for M&I use (1.06 MAF) has not changed, villages in the command area that were earlier included for domestic water supply have now been excluded (except where groundwater is saline or fluorine) to make way for water-scarce areas. The breakup of allocations of drinking water (0.86 MAF, the remaining 0.20 MAF is for industrial use) is as follows: Saurashtra (covering all six districts) 0.511 MAF; Kutch 0.049 MAF; north Gujarat (including Ahmedabad district) 0.301 MAF.

How Much Hydropower and at What Cost?

The claims of the hydropower benefits of the SSP have been challenged by the Narmada movement on the basis of low hydrological estimates and the NSP factor. It is held that since the dependable yield of river water is even less than 23 MAF rather than 28 MAF as assumed,[21] the power generation capacity of the SSP will be affected. Furthermore, a delay in completing the NSP upstream will cause a decrease of more than 25% in power generation at the SSP which depends on regulated release from the NSP (Dharamadhikary 1993; NBA 1995b).

Movement leaders advance two specific arguments. First, the figure of 1,450 MW associated with power benefits from the SSP hides more than it reveals. Although advertised by the project authorities, this refers to installed capacity. In the first phase of the project, actual power generation (firm power generated on average continuous basis) will in fact be only 439 MW (ibid.). In the final phase (with full development of irrigation and consumptive uses), generation will fall to 50 MW (ibid.).

Second, power generation in the project should have commenced when irrigation development was at an early stage in order to make efficient use of the RBPH, the main powerhouse. However, the construction of the RBPH has encountered severe technical and financial setbacks and may have to be abandoned altogether.[22] In the final phase, the RBPH, if built, will generate seasonal energy from monsoon spills and will essentially function as a peaking station. While providing power during peak demand hours, the peaking facility essentially involves a net loss of 30% of power (see Dharamadhikary 1993). Moreover, the SSP itself, with its lift irrigation, drainage and operation requirements, will consume substantial quantities of power, an aspect which has not been incorporated in the economic cost-benefit analysis (ibid.).

The cumulative impact of all these factors, according to movement leaders, is that power benefits will fall by 30 to 50%, which does not make the component economically viable (NBA 1995b). More importantly, the social viability of the power component is equally suspect as it entails a significant rise in displacement costs. The NWDT increased the height of the Sardar Sarovar Dam from 436 ft. to 455 ft. essentially for power benefits, primarily to compensate MP and Maharashtra for benefits foregone from the proposed Jalsindhi project (NWDT 1979). The additional height of 19 ft., however, entails a considerable increase in areas of submergence in MP. The NBA has argued that the additional height of 19 ft. increases the gross area of submergence in MP by 35%, cultivated area by 50% and the population affected by 70%. To the Narmada movement, therefore, power benefits are not only exaggerated but their social costs are enormously high (Dharamadhikary 1993).

Project authorities claim four hydropower benefits from the SSP. First, hydropower generation is economically cheap and environmentally clean compared to thermal, gas-based or nuclear generation. Second, the per capita annual consumption in India—an accepted indicator of development world-wide—equals 10 days of per capita consumption in the USA (NCA 1990; Patel 1991a: 87). Given rising demand and population growth, it would not be in the economic interests of the country to lose this opportunity for power generation. Third, the hydropower component of the SSP is the least-cost option in the Western Power Grid Expansion Programme. The supply shortfall in the western grid is one of the highest in the country; in addition the ratio of thermal to hydropower generation in the western region is heavily loaded in favour of thermal. Fourth, the economic benefits of hydropower can be gauged from the fact that in a 'power only' scenario, cost-benefit analysis yields a rate of return of above 13% (NCA 1990).

Have Adverse Effects in the SSP been Identified and Preventive Measures Planned?

Are Environmental Impact Assessments and Action Plans Adequate?

Controversies surround the likely environmental impact of the SSP. As discussed earlier, the MoEF had given conditional environmental clearance to the project on grounds that assessment studies and mitigative action plans should be completed by the end of 1989. Narmada movement activists stress that, several years after 1989, environmental impact assessments of catchment treatment, compensatory afforestation, flora and fauna, wildlife and fisheries, command area development (including drainage study), seismic studies, and health impact assessments remain grossly inadequate.

Studies in some areas are ongoing (e.g. related to seismicity) or have recently been commissioned (e.g. wildlife impact). In other aspects, the credibility of assessment studies leaves much to be desired. They are based on secondary data, have not been made public or properly peer reviewed (fisheries). In yet other areas, completed studies have not resulted in action plans (health aspects) (NBA 1993; 1995b).

Movement activists draw attention to risks and uncertainties on several of the project's environmental components. First is catchment area treatment, i.e. the need to undertake terracing, contour bunding, gully plugging and other soil conservation measures to minimise siltation and ensure sustainability of the project. In past projects the actual rate of siltation has been much higher than initially assumed (due to fragile and neglected catchment); this in turn adversely affects the lifespan of projects. In the SSP, a high-level, interdepartmental committee (M.L. Dewan Committee), set up at the behest of the Central Ministry of Agriculture in 1984,[23] estimated that the free draining area in which priority catchment treatment is required for the SSP was about 1.5 million hectares (cited by Paranjpye 1990:

238).[24] Progress in this area has been extremely tardy, however, as project authorities have not been inclined to accept responsibility for treating the priority. The siltation rate in the SSP is now expected to be higher than assumed and to affect the lifespan of the project. In fact, Paranjpye (1990) has argued that project life may be as low as 60 years.

Second, forest losses due to submergence and project activities and to the proposed compensatory afforestation programmes are cause for concern. Apart from the fact that forest losses have not been properly accounted for, the dependence of local people on forest resources seems to have by and large been ignored (NBA 1993). Although project authorities have claimed that compensatory afforestation will be double the extent of forests lost, the selection of species seems to meet commercial criteria rather than to fulfil material and ecological functions. Moreover, afforestation is not being undertaken in project-impact areas. Some members of the movement also feel that the species used for compensatory afforestation in the Rann of Kutch is highly unsuitable to the terrain.

Third, the seismic impact of the dam has not been properly assessed. Reservoir-induced seismicity is a well-known risk in large dams. The dam site is located near the Tapi-Narmada-Sone seismogenic lineament with poor geological foundations (NBA 1993; 1995b). The Sardar Sarovar Dam rests on a major geological fault that traverses the reservoir basin. Since these faults are active, there is the risk of tremors.[25]

Fourth, there will be an adverse impact on the downstream environment and population. Adequate studies and detailed plans on impact and mitigative measures are still lacking in major areas such as oceanic intrusion, effect on the biological cycle due to industrial and agricultural discharge, effects on estuarine fisheries, and loss of livelihood of the downstream population.

Fifth, although some studies on the command area have been conducted and others are ongoing, credible action plans for proper surface and vertical drainage are lacking. This implies that the command area will be subjected to flooding and waterlogging. In other areas of concern such as flora and fauna, fisheries, health-related issues and archaeological sites, adequate studies and action plans are non-existent despite promises made by the Government of Gujarat.

The most vociferous criticism in the Narmada movement is reserved for the *pari-passu* clause on environmental protection measures. Connoting simultaneous progress, this implies that environmental studies and mitigative measures must be undertaken at an equal pace with physical construction of the project. That is to say, the last mitigative measure would have to be in place concurrently with the last phase of construction. The *pari-passu* warranty by project authorities is regarded with much suspicion in the Narmada movement, which holds that the authorities use it as 'a license to go ahead with the construction, not caring about when and how the environmental studies were done. The *pari-passu* clause turned out to be nothing but the official creation of a loophole' (NBA 1995b: 57). The main contention of the NBA is that the *pari-passu* clause violates a fundamental principle of environmental planning which requires that impact assessments and action plans for preventive/mitigative measures should be completed during the planning stage of a project.

Since a fair number of environmental impact studies are incomplete, it is not possible to assess what the total costs of mitigative measures are likely to be (Thakar 1993). In the rough estimates of the NBA, it could be about Rs 43,000 million (NBA 1995b).

The claims and responses of project authorities hinge basically around their position that environmental impact assessment is an evolving field. They state that the generation of ground data on various environmental aspects of the SSP has taken time because 'in a country like India, there are problems of data availability and logistics for micro studies.... No systematic environmental parameters are available even in some of the developed countries' (Alagh et al. 1995: 4). A similar view is expressed by C.C. Patel, erstwhile chairperson of the SSNNL, to whom 'the SSP is the first large development project to be scrutinised for its environmental impact. It is therefore unrealistic to insist that the environmental standards and methodologies of developed countries be adopted by developing countries—especially in the initial stages of resurging environmental awareness' (Patel 1995: 82).

Project authorities argue that the environmental impact assessment of the SSP has been largely guided by the relatively recent guidelines prepared by various organisations like the World Bank, MoEF, and the Central Water Commission. Since the guidelines themselves are evolving, the various assessments of the SSP have also been undertaken in phases. The authorities claim, however, that despite an evolving environmental regime, SSP assessments have covered sustainability aspects regarding water management (allocation and distribution), soil classification in the command including waterlogging and salinity aspects, impact on flora and fauna, fisheries and health aspects of the population and the possible deterioration of agricultural runoffs and groundwater quality (Alagh et al. 1995; GoG 1991).

In defence of their position, project authorities advance the following arguments. First, as far as catchment area treatment is concerned, it is unfair to declare the entire cost of treatment as project costs. Catchment area treatment is a continuous and independent environmental measure. It has its own costs and benefits and has to be carried out with or without the SSP (FMG 1994). Thus, it is reasonable for project authorities to take responsibility for treating only the directly draining area of about 180,000 hectares.[26]

Second, with regard to siltation rate and reservoir sedimentation, project authorities argue that there is no risk to the lifespan of the project, which will serve for several hundred years. The siltation rate is based on 'actual, observed data—5.34 ha.m/10,000 ha/year'; hence, the question of underestimation does not arise. Taking the siltation rate in the Sardar Sarovar Reservoir into account, it is reasonable to expect a lifespan of between 180 years (if the NSP is delayed by 25 years) and 222 years (if the NSP is delayed by five years) (SSNNL 1994a: 59).

Third, forest losses are not as significant as they are made out to be. Virtually all the 11,000 ha of forests to be submerged by the SSP are degraded.[27] Surveys undertaken on flora and fauna suggest that 'there are no species present in the reservoir [area] that are not widely represented in the neighbouring areas' (Blinkhorn & Smith 1995:104). In Gujarat, the total forest area under submergence is about 4,500 ha, for which compensation of 4,650 ha of non-forest land in nine villages in Kutch district have been brought under afforestation. Degraded forests in other districts of Gujarat, both in the command and outside, have been treated as project costs. The claim is that the loss of about one million trees due to submergence is being covered by about 42 million trees (in the command, as part of compensatory afforestation,

in the dam vicinity and along the main canal) through the afforestation programmes (SSNNL 1994a: 34).

Fourth, the fear of earthquakes is misplaced. The Tapi-Narmada-
Sone fault is about 17 km north of the dam site and is of low intensity. Detailed studies on this aspect have been carried out by
specialised bodies and experts, both at the national and international levels, and have been endorsed by the intergovernmental
Dam Safety Panel. Based on the recommendations of these expert bodies, foundations in the fault zone are 100 ft. deeper than
the lowest riverbed and have been filled with reinforced concrete.

Fifth, the downstream impact will not only be adverse but
also beneficial. The impact of the wet season (four months of
monsoon, July–October) downstream will not be significant in
initial years. The decline in monsoon flows will be pronounced
only when fuller development occurs in the command area as
well as in the storage upstream in the Narmada basin. The major
impact during these months will be on the fish industry, particularly on the high-priced *hilsa* fish which migrate upriver to
spawn during the rains. The overall impact may not be very
significant. In the dry months, river flows could be higher and
less variable because of the dam. This implies the availability of
fresh water to communities downstream for a long time until
the command area is totally developed.[28]

Sixth, the risks of waterlogging and drainage in the command
area have been adequately studied. Water delivery in the SSP
command has been designed so that surface water input will be
minimum and efficient. In most parts of the command area, the
level of groundwater is extremely low and falling further. On
average, about 5% to 7% of the command area is susceptible to
waterlogging. Special care has been taken to ensure that these
areas are minimally serviced with provision for surface drainage. In the problem confronting the command area, the
bottomline is not that of preventing waterlogging, but also ar-
resting falling water tables.[29]

What is the Impact of Displacement and how are Associated Risks Addressed?

In the Narmada movement, the risks of displacement and resettlement is the major area of politicisation. While this is discussed

more fully in Chapter 8, mention must be made of the key problem areas identified in the movement, elaborating on claims
and counter-claims around the following issues: (a) displacement
impact is not fully known; (b) adivasi communities will be worst
affected; (c) the three state governments have different resettlement policies; (d) adequate land is not available for sustainable
resettlement; (e) conditions in resettlement sites are sub-standard;
(f) the displacement process entails violation of rights; and (g)
until recently, there was no resettlement master plan for a good
part of the displacement process.

1. According to the NBA, the scale and nature of displacement
   in the SSP have not yet been fully studied. Numbers of displaced people estimated during NWDT proceedings were
   based on outdated government records and are erroneous.
   Village-level surveys of project-affected people have not been
   undertaken. Except for people in the reservoir area, no others are recognised as project-affected people, whether due to
   the canal network, downstream impact, catchment treatment,
   compensatory afforestation, project site or colonies.[30]

   Even in the reservoir area, there are information gaps on
   many aspects regarding displacement. Data on people engaged
   in non-agricultural occupations, i.e. petty traders, shopkeepers and small business people such as tailors, carpenters and
   fisher folk are not available (ibid.). Similarly, the extent of
   dependency of adivasi communities on forestland and resources is unknown as governments have no up-to-date land
   records or individual and common property records. Adivasis
   in the project reservoir constitute 100% of the affected population in Maharashtra, 98% in Gujarat and 29% in MP.[31] These
   communities are characterised by a unique and distinct culture and economy. They are heavily dependent on the river,
   forests and land (jal, jungle, jamin) for their livelihood and
   subsistence. Their extended kinship networks (social capital)
   are resources of emotional and economic support. Rupturing
   this non-market, communitarian socio-economic fabric will
   involve severe consequences, especially in the absence of
   proper protective measures (NBA 1995b).
2. Although the NWDT clearly stated that a resettlement master plan should have been prepared by 1980, this was not

done. In 1987, this was categorically reiterated by the MoEF as part of the conditional clearance accorded to the SSP. It was not until 1995 that a master plan was prepared by the NCA. Fundamental issues such as the extent of displacement, the different categories of affected people and their specific needs and requirements, however, are still unknown. Therefore the master plan showing these details, together with time schedules of submergence, details of land availability, other amenities in resettlement sites, and the phasing of resettlement, can only be said to be complete after proper surveys have been done.

3. The resettlement process in the three states is being carried out on the basis of ad-hoc and discrepant policies. For instance, while Gujarat offers two hectares of land to the landless, Maharashtra offers one hectare and MP only cash compensation (see Appendix 3). Similarly, oustees from Maharashtra who have settled in Gujarat are expected to bear the cost of the (resettlement) land, whereas those settled in Maharashtra itself need not bear such costs. Unmarried adult daughters (18 years and above) are entitled to compensation only in Maharashtra and not in MP or Gujarat. Furthermore, whereas MP offers a minimum of four hectares of non-irrigable agricultural land to each of the project-affected people (PAP), there is no such provision in Maharashtra and Gujarat. This disparity in the NWDTs resettlement packages makes the choice offered to SSP oustees pointless.

4. Given the large-scale requirement of land (even if only for those in the reservoir area), the Gujarat government, which has major responsibility for resettlement, has not been able to identify and acquire the land needed. The government has depended on voluntary sales and purchases of land, most of which is not only outside the SSP command but also non-irrigable (NBA 1995b). The MP government too has shown no interest in identifying and acquiring land for its oustees who wish to settle there. In many instances, land identified by oustee families has not been allotted to them. There are several reports of less land being offered than is due, and/or of the land being of sub-standard quality (NBA, undated). Due to the lack of consolidated land availability, kin groups are being resettled in a widely dispersed manner (ibid.).

5. Apart from the inferior quality and inadequacy of land, infrastructural facilities in resettlement sites are poor. Research undertaken by independent agencies in many such sites note the lack of electricity, roads, schools, water and pastoral lands. Health problems (infant mortality, nutritional deficiency) have been reported (TISS 1993). The absence of health care services makes the adivasi population particularly vulnerable because in their original habitat they relied on traditional medicinal practices and remedies. Relations between oustees and host communities have been tense. Oustees periodically face varying forms of harassment; some are not allowed to cultivate land, others are not allowed to make use of common resources and of approach roads to fields (NBA 1995b). Women oustees have reported sexual harassment. Given the abysmal conditions in resettlement sites, many oustees who had accepted resettlement have returned to their villages. Some have even returned resettlement land deeds to project authorities (NBA 1995a).
6. The process of displacement and resettlement violates people's basic right to information and participation in the planning process. Displacement has been planned without the affected population being informed (as required by law) and without any scope for their meaningful participation in the project.

Legal procedures for acquiring land under the Land Acquisition Act have not been followed. Notices for land acquisition were not put up in some villages and were never publicly announced (given that the majority of the population is illiterate). The right of people to voice objections (under the Act) has never been respected, despite representations by some project oustees.

Although the NWDT award and Supreme Court orders categorically stated that the resettlement process should be completed at least six months before submergence, this condition has not been fulfilled: hundreds of oustees have faced temporary or permanent submergence without having been compensated or resettled.

Coercive methods have frequently been used to remove people from their habitat. The state has also used violent measures, e.g. firing, registering of false charges, arrest and torture, to suppress democratic means of protest (Narmada 1992; 1993, *Narmada Samachar* 1993–96).

The arguments advanced by project authorities in support of their position on displacement and resettlement are based on their adopted 'incremental approach' to displacement impact and resettlement provisions in the SSP. As C.C. Patel has observed, 'the government is open to reconsideration to solve genuine difficulties faced by affected people' (1991: 44). The so-called incremental approach has meant that project authorities have periodically, albeit under pressure from the movement, modified resettlement policies, improved entitlement packages and expanded their definition of affected stakeholders in the project. Some of their arguments are outlined here.

1. Studies on displacement impact have been undertaken at various levels by state governments. Independent monitoring and evaluation agencies have been mobilised to ensure that special attention is given to the cultural and economic vulnerability of adivasi communities.[32] Other academic and research institutes and expert consultants have been involved in this process and their findings have been considered in the planning and implementation of the resettlement process.

   On the basis of these studies, the number of villages affected in the three states, whether fully or partially, the extent of submergence of private revenue land, forest land and other land, the nature of submergence—temporary (due to flooding) or permanent (due to pondage)—and the phasing of submergence, have been assessed. Contrary to the claims of the movement, project authorities consider the displacement impact in the reservoir area to be much less. Out of 245 affected villages, only three face total submergence. Table 6.3 gives the details of submergence in the three states.

   Submergence in MP is not as significant as it may seem. The loss of private revenue land is less than 20 ha in 116 villages (each with an average of 350 ha of revenue land). Therefore, in only 77 MP villages can submergence be said to be substantial. The loss of houses in MP is much greater than the loss of agricultural land (see Table 6.4), implying that the resettlement package needs to prioritise the risk of homelessness rather than landlessness.

2. People affected by canal infrastructure are not compensated anywhere in the world and have not been compensated in

Table 6.3 Details of Submergence[33]

State	Private Ag. land under submergence (in ha)	Forest land under submergence (in ha)	Other land under submergence (in ha)	Total land under submergence (in ha)	Villages fully affected	Villages partially affected
Gujarat	1877	4523	1069	7469	3	16
Madhya Pradesh	7883	2737	10208	20828	1	192
Maharashtra	1519	3459	1592	6570	–	36
Total	11279	10719	12869	34867	3	245

Source: Patel (1991:46).

any other project in India. It is only in the SSP that compensation for this constituency of people has been provided. The demand that the compensation package for canal-affected people be similar to that of reservoir-affected people fails to recognise that the former continue to live in the command area and enjoy the benefits of irrigation for their remaining land (Patel 1992). In SSNNL estimates (1992), out of 169,500 landholders among canal-affected people almost 146,000 would lose less than 25% of their total holdings. Only 758 landholders would lose all their land and some more than 75% of their land. Canal-affected persons who lose their entire land will be compensated (Alagh et al. 1995). Special care is being taken to ensure replacement costs for small and marginal farmers who lose their land in the process.

3. The oustees of Kevadia colony were displaced in 1961–62 and were compensated in cash according to the prescribed and prevalent norms of the time. About 710 ha were acquired in six villages (with a total cultivable area of 1,110 ha) from 334 landholders. Apart from benefits of project construction that have accrued to these people, more than 660 people from these villages have been directly employed by the SSNNL. In 1992, the Government of Gujarat decided to further compensate the affected families by giving them cash compensation of Rs 30,000 per ha per landowner (with a ceiling of Rs 36,000). Major sons in these families are separately eligible for the same financial assistance.[34]

4. Studies to assess the downstream impact of the project have been carried out and plans are being considered to resettle

Table 6.4 Summary of Resettlement with Progress up to March 1995

State	Total no. of oustee families	Oustees eligible for house plots (H.P.)	Oustees eligible for agricultural land (A.L.)	No of oustees allotted H.P. up to March 1995	No. of oustees allotted A.L. up to March 1995	Oustees to be allotted H.P.	Oustees to be allotted A.L.
Gujarat Oustees	4600	4600	4577	4276	4346	324	231
Mah Oustees (in Gujarat)	999	999	999	594	656	405	343
Mah Oustees (in Mah)	2114*	2114	2114	841	1147	1273	967
Mah Oustees (Total)	3113	3113	3113	1435	1803	1687	1310
MP Oustees (in Gujarat)	14124@	14124	14124	2428	2535	11696	11589
MP Oustees (in MP)	18890	18890	830	672	NIL	18218	830
MP Oustees (Total)	33014	33014	14954	3100	2535	29914	12419
Grand Total	40727	40727	22644	8811	8684	31916	13960

Source: NCA (1995: 9–10).
Notes: *Number likely to increase after due verification; @Likely to undergo change.

some of the affected people in the reservoir area for fishing and other livelihoods (ibid.). However, the SSP reservoir will modify the pattern of river flows, particularly in the dry season, in effect implying year-round access to fresh water downstream of the dam, until irrigation and other uses of the river basin become fully developed.

5. The resettlement package offered to SSP oustees is the most progressive compared to other projects in India. Of the three states, Gujarat offers the most liberal package, but both Maharashtra and MP offer better deals to the oustees as compared to other projects in these states. Resettlement packages in Maharashtra have been revised periodically to incorporate landless labourers, encroachers of forest land, adult sons and adult unmarried daughters. The Gujarat government's package provides opportunities for landless and marginal farmers to own land (NCA 1995).

6. Considerable progress has been made on the resettlement front. Initial resistance to displacement has given way to increasing acceptance of resettlement. In the reservoir area, resistance to displacement comes not from adivasi villages but from non-tribal, rich farmers of the Nimad plains in MP.[35] Table 6.4 outlines the resettlement action plan and its achievement up to March 1995. From an original estimate of 12,072 families affected by the reservoir (based on the 1981 census), a later estimate shows 40,727 affected families (involving a population of 127,446). This increase in the number of project-affected people is not due to a lack of proper baseline studies as alleged, but to population increase and the liberalisation of resettlement policies. The inclusion of adult sons and unmarried adult daughters, the extension of cut-off dates for considering eligibility of adult sons, extending resettlement to those whose lands will become small islands (locally called *tapu*), the decision to extend resettlement benefits to encroachers, have all resulted in the upward revision (NCA 1995).

7. Efforts have been made to make the resettlement process participatory. Project-affected people have been taken to resettlement sites and shown different plots of land to elicit their choices. Only after they have chosen a particular site has land been allotted to them. This participatory approach was adopted for the first time in India under the directives of

the NWDT. Officials periodically visit the various resettlement sites in Gujarat. The cooperation of a number of NGOs has been sought and received in the resettlement process.

Are Investments Financially Sound and Socio-economically Justified?

Debates on the SSP cover financial risks as well as economic returns. The Narmada movement questions the financial viability of the SSP. It has pointed out that, like all major dams, the total cost of the SSP has increased rapidly. From an initial estimate of Rs. 42,000 million in 1982–83, the cost has shot up to about Rs. 146,000 million (1991–92 prices); a 130% increase over the costs approved by the Planning Commission in 1989, i.e. Rs. 64,000 million (at 1986–87 prices). It is argued that the total cost of the project on completion will be well over Rs. 300,000 million (see Thakar 1993). Up to 1995, Rs. 38,000 million had been spent (about 15% of the total financial cost), thus showing the need for enormous financial resources for completion of the project (NBA 1995b).

The withdrawal of the World Bank and suspension of the loan by the Japanese government due to the project's adverse social and environmental impact have made it difficult for project authorities to mobilise the necessary finances. To fill this gap, the Gujarat government has been allocating almost 80% of its irrigation budget to this one project. The movement sees this heavy investment in one project as hiving-off resources from other, much-needed projects in drought-prone regions (Narmada 1993; NBA 1992; 1995b). At the same time, the heavy market borrowing by the project authorities through bonds and deposits promising very high interest rates is financially untenable (Narmada 1992).

The economic justification of the project has also been questioned. The variance between cost-benefit analyses conducted by diverse agencies at different points in time are highlighted. To reiterate some rate of return figures discussed in Chapter 4, Tata Economic Consultancy calculated a rate of return (IRR) of 18.5% in 1983, the World Bank 13% in 1985, scaled down to 12% in 1990. The movement argues that the latter estimate omits

several other costs to be incurred in the project: adequate compensation for canal-affected people, compensation for loss of
employment downstream, cost of M&I water supply, and costs
of catchment area treatment. Moreover, calculations are not
based on a realistic assumption of cropping patterns. If such
costs are accounted for and benefits properly downgraded, the
IRR would be between 7 and 8% (Paranjpye 1990).

A major argument of the Narmada movement is that economic appraisals of the SSP do not consider probable alternatives
which could achieve the same objectives in a cheaper, quicker,
equitable, participatory and sustainable manner (NBA 1995b).
The movement considers that problems of water shortage in
Kutch, Saurashtra and north Gujarat could be better solved by
tapping locally-available water through such techniques as
groundwater recharge and watershed development. It claims
that simple, labour-intensive rain harvesting schemes can tap
three to five times more water than the SSP can ever take to
such regions. Similarly, power benefits could accrue at a lower
cost compared to the SSP if appropriate demand management
and saving techniques were adopted. A fuller discussion on the
debates around alternatives is provided in Chapter 9.

The Narmada movement also propounds a politically feasible
alternative, claiming that the dam height could be lowered by
145 ft. (from 455 ft. to 310 ft.) without compromising benefits.
Considering that the FSL of the main canal is 300 ft., a dam
height of 310 ft. would be sufficient to take water from the reservoir to the canal, and would also incorporate most of the
structures constructed so far. The reduced height would dramatically reduce submergence in the reservoir,[36] irrigation
benefits would be maintained and some power benefits derived
by suitable modification of generating structures. The MP government could provide regulated upstream releases from its
projects to fulfil Gujarat's share.[37]

As might be expected, project authorities dispel doubts about
financial risks. The Planning Commission approved the project
at a cost of Rs 64,000 million (1986–87 prices). Gujarat's share
was to be Rs 49,000 million and the other three states were to
contribute the remaining Rs 15,000 million. Given the delay the
cost of the project is likely to climb to Rs 90,000 million by the
year 2000. Gujarat has made Rs 53,000 million available from

its budget allocations in the Eighth and Ninth Five Year Plans. Market borrowings—from bonds worth Rs 15,000 million and public deposits worth Rs 5,000 million—will be mobilised. SSNNL raised Rs 300 million in the first 18 months of the public deposit scheme. Furthermore, it has provisions for adequate fund management mechanisms and has not allowed mobilised funds to remain idle (SSNNL 1994b).

Project authorities argue that economic cost-benefit analyses indicate only that the project is beneficial. The last estimate of the World Bank shows a healthy rate of return of 12%. Project authorities say that this should be seen in the context of the recommendations of the Nitin Desai Committee appointed by the Planning Commission of India. The committee approved development projects with a rate of return of 9% and those serving drought-affected areas with 7%. In view of these recommendations, the SSP fulfils the required criteria for assessing benefits (GoG 1991). In a project such as the SSP, although there are immediate human and environmental disruptions, benefits are significant and sustained and substantially outweigh the social and environmental costs. The benefits of such major projects can be gauged from the fact that during the three-year drought in Gujarat, the Ukai–Kakarpar and the Mahi–Kadana projects, i.e. the two major reservoirs in Gujarat, were the main sources of water.

In the context of debates over alternatives, project authorities also claim, and rightly so, that the SSP is the most studied development project in India. First, alternatives to the project were explored by the NWDT for about a decade, during which permutations of number, height and locations were debated and discussed. Hence, it would be wrong to say that alternatives have not been considered (SSNNL 1994b). Second, the proposal to construct a number of small projects is simplistic. Small or large dams are not 'either...or' options; in a country such as India both options have been, and need to be, exercised. Comparatively speaking, however, small dams cannot match the scale of benefits of large dams; they are uneconomic in terms of the amount of land they inundate relative to the volume of water they store. Most importantly, they fail to fill in during the very (dry) years when they are most needed (Mistry 1991; Patel 1991a). The SSP will inundate only 1.6% of the area it will irrigate; even if it were technically possible to find enough small dams sites to store the same amount of water, the land lost could well be over a million hectares (Patel 1991a). Although some legitimate concerns are voiced by critics of large dams, the latter also fulfil legitimate needs. Third, with regard to power benefits, while suggestions have been made to shift focus to demand management and service improvement, energy conservation can only reduce short-term power demands. Economic development and population growth inevitably imply that long-term demands for electricity will rise; hence, supply provisions have to increase to match this long-term demand.

Project authorities also oppose the idea of reducing the dam height. They consider the height to be non-negotiable, fixed as it was by the NWDT after extensive deliberation. To reduce the dam height hypothetically to 310 ft. would imply that reservoir storage is not necessary. In that case, only monsoon waters could be diverted for the kharif (rain) crops. To be able to guarantee 100% protection of kharif crops in Kutch and Saurashtra, however, storage is necessary. Given that the river's major flow is during the monsoon, storage is necessary for rabi (winter) crops: current plans indicate provision of water in three out of four years. Project authorities ridicule the argument advanced by the movement that a reduced dam height would keep the benefits intact: 'On the one hand the movement claims at this height water will not reach Kutch, north Gujarat and Saurashtra. On the other hand they claim that at a substantially reduced height benefits will not be affected. Who are they fooling?' (D.T. Buch, NPG, 4 May 1995, interview).

Project authorities also dismiss the suggestion that regulated releases from upstream dams in MP can be provided. Those who advance such proposals do not mention the names of projects, nor other details such as whether there is adequate water for those projects to justify additional storage and whether the implications of such storage for downstream projects and benefits have been considered (see Patel 1991a).

Knowledge Struggles and the Pluriformity of 'Truth'

In the Narmada movement's assessments of the SSP, the project appears very vulnerable to tests of robustness on account of its design, scale, performance and costs. Considering the arguments developed around economic, environmental and social risks and uncertainties, the description of the SSP as a 'faulty, non-viable, unjust and destructive project' (NBA 1994) and the radical political opposition to it seem justified.

Conversely, if the arguments of project authorities are considered, the SSP seems to be a reasonably well-designed intervention. Compared to those undertaken in past projects, the scope and quality of studies of the SSP have significantly improved. Project authorities seem attentive to environmental and social costs (and criticisms around these issues), and the mitigative measures set new standards in the country. Most significantly, there are sufficient reasons to expect important long-term benefits from the project that will fulfil the interests of the region in particular and the state in general. While it may still be debated whether the project is 'the lifeline of Gujarat', as claimed by project authorities, its thorough interrogation by environmentalists makes it qualify, as one of its architects contemplatively opines, as 'a victim of times' (Patel 1995).

Given the contradicting nature of claims in the SSP it is tempting to ask which set of claims is true. Pertinent as this question might be, it is not productive at this stage to try to answer it. Rather than create yet another 'truth claim' we are more inclined to explore the significance of knowledge struggles in the SSP. Their significance is that they bring the project's evaluation onto a public platform where criteria, methods, data and findings are open to critical and democratic scrutiny. Whether questioning the hydrological data and associated technological uncertainties, or the costs of displacement and the associated social risks, the evaluation of the Narmada movement fosters the public recognition of risks and uncertainties in the project. To say the least, the movement increases what Giddens (1990:131) calls 'lay knowledge of modern risk environments'. The intervention process becomes more transparent as facts and figures of the project enter the public arena. Risks and uncertainties are subjected to wider debate, as is the 'expertise' of the state to control and compensate such risks. Paradoxically, the very fact that project authorities seem inclined to respond to every issue raised by the NBA and its allies, albeit unconvincingly at times, is testimony to the knowledge interests and generation capabilities of the movement. It demands controls over risks and uncertainties, more and better information, studies, assessments, mitigative measures, etc; and in politicising those risks and uncertainties, it stakes a counter-claim to expertise. The NBA believes that enough information, studies and assessments exist to prove that the project will *not* behave as claimed. There are three ways in which this claim is constructed.

First, by framing uncertainties in the project as ignorance. Examples relate to the 'uncertainties' regarding the total impact of the project on different categories of people, or for that matter aspects pertaining to environmental impact assessments. Through this mechanism the movement calls for recognition of hitherto ignored areas such as displacement and environment as causes for concern and demands fuller commitment to them.

Second, by challenging risk assessments where the odds are known. Project appraisal is a good example—here risk assessments require a set of assumptions concerning resource availability, agricultural yields and inputs, volume of energy produced, economic values attached to benefits, implementation speed, irrigation system efficiency, water availability, etc. By identifying information gaps, fudging and errors, and by drawing from past experiences, the movement challenges some of these assumptions and makes the odds appear significantly higher.

Third, by highlighting disagreements over the interpretation of risk definitions and liabilities among authorities and experts. The annual utilisable water flow (hydrology) may be cited as an example. The governments of MP and Gujarat have disagreed on this issue for more than three decades. Given that hydrological estimates are in essence probabilistic estimates, some discounting for political interests and negotiations is difficult to avoid. The movement exploits the political division over estimates while simultaneously questioning the political means by which parameters are settled around project uncertainties and risks.

The kind of 'environmentalism' we come across in the Narmada movement is one where the movement accumulates data pertaining to a broadly defined environment and uses it in its struggle. It is environmentalism that Buttel and Taylor (1994: 223) think is moving towards 'an increasingly thoroughly "scientised" *Weltanschauung*[38] and mode of social movement strategy'. A more poignant observation has been made by Castells (1997: 123):

Environmentalism is a science based movement. Sometimes it is bad science, but it nonetheless pretends to know what happens to nature and to humans, revealing the truth hidden by vested interests of capitalism, technocracy, bureaucracy. While criticising the domination of life by science, ecologists use science to oppose science on behalf of life. The advocated principle is not the negation of knowledge but superior knowledge.

As Giddens (1990:128) observes, however, assessments that attempt to assign or demand strict probabilities are bound to be controversial. An intervention of the scale of the SSP necessarily contains many unsettled aspects. The uncertainties and risks involved warrant innumerable assumptions over how various features of the project would behave. Each of these unsettled aspects can be used either to justify or to oppose the project. For example, the dispute over irrigation efficiency of the SSP. Project 'experts' claim a 60% irrigation efficiency whereas movement 'experts' claim no more than 35–40%. Both use past projects as yardsticks with which to measure the present. As experiences in the past were dismal, however, 'experts' in the movement have every reason to believe that the present can be no different. On the other hand, past failures and resource illiteracy offer project 'experts' important lessons that can be, and have been, incorporated in the present. They even cite efficiency of irrigation rates of some existing irrigation systems in Gujarat which are above 60%.[39]

Whether water will or will not reach the drought-prone areas, whether the NSP will or will not affect the benefits of the SSP, whether adivasis can or cannot be adequately compensated, there seem to be conflicting 'truths' in most, if not all, of these issues. Beck alludes to this pluriformity of truth when he states,

> as risks become politicised, everybody has to pick a number out of the hat.... At the same time, however, no one any longer has privileged access to the uniquely correct calculation; for risks are pregnant with interests and accordingly the ways of calculating them multiply like rabbits...[E]very one produces ever more contradictory results, with ever more meticulous methods (1995: 93).

Interests, power and values of actors underlie the conflicting evaluations of risks and opportunities of the SSP. Such evaluations cannot be separated from interests and ideology. Integral to assessing the worth of an intervention (and the associated claims of benefits and costs) are questions of who undertakes it, with what assumptions and for what purposes. Beneath the scientised struggle over risks and uncertainties are deeper political commitments to ideologies and cultures that incline actors to exaggerate and emphasise some uncertainties and to underestimate and downplay others. Discussing in Chapter 4 the justification of the SSP on grounds of favourable benefit-cost analyses, we analysed the political problem of 'objectivity' in evaluation. If developmentalism and state interests tend to exaggerate benefits and to underestimate costs, one can argue that actors committed to the values espoused in environmentalism and discourses on alternative development (and to social-political actions) may well be inspired to exaggerate risks and ignore opportunities. Similarly, values permeate the framing of environmental problems. Whereas project authorities see the problem as one of growing population, food shortages, water deficiency and drought which the SSP will counter, project opponents see environmental problems as resulting from the SSP's erosion of those resources upon which mostly the poor and underprivileged depend for their living and sustenance, and from the ensuing permanent alteration of a river valley ecosystem.

Conclusion

In this display of claims and counter-claims in the SSP, the problem of 'whose reality counts' remains unresolved. Conditions where risk assessments become controversial often call for 'independent' assessments. In the SSP such conditions have generated independent 'risk arbitrators', but in no way does this undermine the significance of risk politics in the Narmada movement as analysed here. In fact, the Narmada movement should be understood not as a 'peasant movement' opposing a 'state project'. It is not an organisation of an interest group, but rather a cognitive territory or space which generates struggles over knowledge, interests and power. In contrast to the prevalent theoretical understanding of environmental movements as opposing modern science and technology for more traditional, indigenous knowledge systems, the Narmada movement shows how scientific discourses are used and deployed. Risk politics is not a struggle between planners' rationality and that of the peasants, or between indigenous knowledge and scientific knowledge. Rather, it is a struggle to define and refine the limits of scientific rationality and knowledge, what Beck calls 'the radicalisation rather than negation of rationality'. It is through risk politics that the Narmada movement challenges existing institutions governing risks and liabilities, to seek transparency and democratic accountability in decision-making around the SSP.

Notes

1. The materials used in this section are derived from various reports of project authorities, minutes of meetings of the environmental and resettlement sub-groups of the NCA, published material of members of the Narmada Planning Group, Narmada Control Authority and Sardar Sarovar Narmada Nigam Ltd, government correspondence; and interviews and discussions with project authorities and NGOs supporting the dam. In some cases, arguments of the World Bank in support of the project are also used.
2. In advancing this argument Giddens (1990:131) refers to the works of Dowie & Lefrere (1980) and Jouhar (1984).
3. Two aspects warrant noting. First, that claims and counter-claims have emerged dialogically over a period of time. In many instances, project authorities seem to have responded positively to critical assessments, for example, by revising the compensation package, preparing resettlement master plans and commissioning assessment studies. Second, in the zealous pursuit of advancing claims and counter-claims, both sides have made statements that support rival claims. For example, when project authorities say that the SSP should be higher if the Narmada Sagar Project (NSP) is not built, it clearly supports the NBA's argument that the NSP will affect the SSP. On the contrary, the NBA's argument that without the NSP, the SSP will fall short in performance, can be used to support the quick construction of the NSP.
4. The 28 MAF was calculated using the 75% dependable yield of 27 MAF less the project estimated evaporation losses (4 MAF) plus regeneration (return flow from irrigation and canal leakage) (2

MAF) and carryover storage (3 MAF) (see NWDT 1978b, Chapter XI in FMG 1995 11:7).

5. Hind-casting is a hydrological method of assessing water flows in a river when discharge data are unavailable. The method takes available data of actual flow of the river and correlates it with other hydrological data such as rainfall to arrive at long-term river flow data (SSNNL 1994a: 35). When data on the actual flow are inadequate, hind-casting is used as a proxy method. In the Narmada case, since actual flow data were from the 1950s because the NWDT had only about 25 years of data, the hind-casting method became necessary. The dispute over river flow is discussed in greater detail later in this chapter.

6. This was a minor contention compared to major contentions over sharing of benefits, areas to be irrigated in each state, the level of the Navagam Dam (the then SSD) and the canal (see NWDT 1978a: 20).

7. Madhya Pradesh had objected to the period 1915 to 1962 as the basis for calculations, and suggested that the runoff should be based only on those years for which actual data were available (1948–62); if earlier periods were to be included, the period 1891–1911, which had some bad years, should also form part of the series (see NWRDC 1965: 207).

8. The CWC has argued that if the coefficient of variation of annual flow is 0.20, then statistically a 40-year data length should be adequate. Since variation in Narmada is 0.35, the data length has to be 130 years (cited in FMG 1995:11). As observed data are available only for 45 years, the need for hind-casting arises.

9. In a subsequent note to the FMG, the CWC, through the Ministry of Water Resources, Government of India, made it explicit: 'The longer the series the more reliable would be the series obtained through simulation. Therefore the idea of hind-casting the series by established techniques universally used for increasing the length of the runoff series needs to be given the seriousness it deserves' (D.O No. 6/9/93-PP, Ministry of Water Resources, GoI, 8 December 1993).

10. The maximum flow recorded in 1961–62 was 60.30 MAF.

11. According to the CWC, the last 40 years form only a low-flow part of a longer cycle.

12. Totalling nine in number. In a recent study undertaken by the Narmada Control Authority in 1995, the series covers 47 years from 1948–49 to 1994–95 and estimates 75% dependability at 23.04 MAF.

13. MP, together with Bihar, Rajasthan and Uttar Pradesh feature as the BIMARU states. The acronym comes from the first letters of the four states and means disease in Hindi.

14. Dams built to divert the flow of the river for constructing the main dam.
15. In Chapter 5 we mentioned that the NSP faced considerable opposition in central MP.
16. The ratio between supply and actual use. Important factors that affect efficiency are canal seepage, evaporation losses, farmers' use. With regard to farmers' use, the term 'on-farm application efficiency' refers to the ratio of volume of water stored in the root zone to the volume of water applied.
17. It is estimated that 1 per cent drop in efficiency can reduce the command by as much as 16,000 hectares.
18. The case of Baroda is justified by stating that the ultimate irrigation potential of the district is 76% of its total cropped area. Without the SSP the irrigation potential developed in Baroda is 28% of total cropped area. In Gujarat, out of the total cropped area of about 10 million hectares (area cropped under normal rainfall), the ultimate irrigation potential is estimated at 6.5 million hectares (6.5%). However, some experts believe that this is an exaggerated figure and that the ultimate potential will have to be downgraded to 4.5 million hectares (Alagh 1991; Hashim 1991).
19. Harshad Mehta, a millionaire stockbroker, was arrested for artificially inflating share prices, particularly in the Bombay Stock Exchange, through his connections with senior bankers and highly placed politicians.
20. While 948 villages in Kutch and 4,877 in Saurashtra are listed as beneficiaries (NCA 1992), it has been pointed out by Ram (1993) that the 1981 census shows that only 887 villages in Kutch and 4,727 in Saurashtra are actually inhabited. The limits of desk-top planning are striking.
21. It is interesting to note that the Central Electricity Authority, a Government of India undertaking, takes into account only the observed flow and not hind-casting for calculating power benefits (see NBA 1995b: 47).
22. The technical problems relate to design and construction of the underground tunnel. Financial problems pertain to the withdrawal of Japanese aid for the reversible turbines.
23. The M.L. Dewan Committee was set up to examine the conditions of the catchment area in the SSP and NSP and to suggest measures for treatment of the catchment.
24. This estimate was based on the assumption that upstream dams will arrest siltation for the SSP and will have their own treatment area. Paranjpye suggests that delays in the construction of these upstream projects imply that the draining area of the SSP will be much larger; hence, the need for treating a much larger area esti-

mated at 8 million hectares. Considering that the entire catchment
of the Narmada river is about 9.8 million hectares, Paranjpye's
figure of 8 million hectares seems exaggerated. The MoEF estimated the total area needing treatment in the SSP at about 0.75
million hectares.

25. Support for this stand of the movement comes from experts to whom
    probability of a dam failure can never be zero no matter what the
    engineering standards, safeguards and controls, and margin of safety
    are. The risks of tremors are therefore a permanent threat (see T.
    Shivaji Rao 1993, *The Hindu*, 18 April 1993, 'Hidden Hazards of
    Sardar Sarovar').

26. The directly draining area is that portion of the catchment from
    which water flows straight to the reservoir. The freely draining area
    is much larger.

27. Estimates of the Government of Gujarat show that forest density is
    less than 0.1, with about 216 trees per hectare (SSNNL 1994a: 32).

28. Blinkhorn & Smith (1995: 105)—staff members of the World Bank
    associated with the SSP—suggest that if in future there is a risk of
    the river drying up in the dry season, it may be necessary to take
    water from a nearby branch of the Sardar Sarovar Canal.

29. Similar claims of having adequately addressed problems pertain-
    ing to health-related issues and architectural losses are made (see
    Alagh et al. 1995: 179–219 and 287–303; Blinkhorn & Smith
    1995:105; Patel 1991a: 49; SSNNL 1994b: 37).

30. In the NBA's estimate the total number of partially and wholly
    affected people will 'not be less than 4 hundred thousand' (NBA
    1995b: 4).

31. In the state of MP, about 12% of the population belongs to Sched-
    uled Caste (Dalit) communities.

32. While admitting to the specific needs of this category of affected
    people, project authorities vehemently oppose the preservationist
    discourse on tribal people. They consider the preservation of tribal
    cultures to be synonymous with denial of the right to development.
    For arguments in support of this position see Joshi (1997: 168–74).

33. In the 1995 master plan for resettlement and rehabilitation pre-
    pared by the Narmada Control Authority, Government of India, the
    figures are different. Total land submergence is estimated at 37,590
    ha. Forest land submergence in Maharashtra almost doubles to
    6,288 ha and private agricultural land in MP goes down to 6,598 ha
    (NCA 1995: 12).

34. Five years after the announcement, in November 1997, the people
    of Kevadia colony in a month-long sit-in demanded that these prom-
    ises be fulfilled.

35. In MP, project authorities have claimed that primary data collection in Nimad villages has not been possible due to the villagers refusing to allow enumerators into the village.
36. A more feasible demand is reduction of the dam height by 19 ft. to 436 ft. (see Shah 1994; also Swami 1995). It is estimated that a height reduction of 19 ft. would save 10,000 hectares and about 7,000 affected families.
37. It is not clear whether the movement holds Gujarat's share to be 7.2 MAF and not 9 MAF given its claim that the total annual utilisable flow is 23 MAF and not 28 MAF.
38. Referring to a world-view or philosophy of life. The 'use of this term usually points to a certain imprecision in an argument and almost invariably indicates that data appropriate to the particular case are wanting' (Marshall 1994: 564).
39. The following irrigation projects in Gujarat show irrigation efficiency higher than 60% (calculations based on rabi season 1991): Aji-II (66.91); Demi-II (68.03) and Uben (71.32). (Government of India, Ministry of Water Resources, Letter No. 6/9/93-PP, Dt. 8th October 1993, New Delhi).

7

Independent Juries and Reviews: Views from Above

Introduction

At a very early stage in the Sardar Sarovar Project, juries had to be formed to resolve conflicting positions of the governments of Gujarat and MP on the shape of the project. In addition were the Khosla Committee and NWDT, which then presided over conflicts on benefit sharing between states. However, in response to the Narmada movement's periodic demand for independent evaluation of the SSP, the Independent Review Mission (IRM) and the Five Member Group (FMG) were formed.

Albeit with different degrees of emphasis, the IRM and FMG, and a division bench of the Supreme Court are deliberating on the SSP's adverse environmental and social costs, the problems involved in resettlement policies and implementation, benefit-related issues, and less damaging and/or environment friendly alternatives. In that sense, their existence is directly connected to the *locus standi* of actors in the Narmada movement. As independent reviews follow one another within a timespan of a few years, one rather obvious question that emerges is, what implications do their findings have for the SSP and for one another? Insofar as this chapter critically analyses these findings, it may be seen as a 'meta-evaluation' of the SSP. More importantly, it advances the arguments regarding conflicting assessments (in the preceding chapter) by showing how sharp and subtle differences exist among independent review groups concerning the conflicting claims over uncertainty and risks in the SSP.

The data are drawn from reports of the two independent review groups, submissions to them and affidavits before the Supreme Court.

Review Groups in the SSP

Evaluation theories sensitise us to the fact that different and contradictory results in evaluations of a single project are not only possible but even expected since factors such as the approaches and methods adopted, the values espoused and intended uses, all influence evaluation 'findings' and thus account for their divergence (Shadish et al. 1991). Review groups are constituted primarily to enable the non-partisan and independent evaluation of projects. They are first and foremost evaluating bodies, and hence cannot avoid the influence of these factors. Following House (1980: 23), it may be argued that approaches to assessments of review groups can be classified broadly into decision-making approaches, goal-free approaches, goal-based approaches, critical inquiry, quasi-legal reviews and professional reviews.[1] Although actual reviews may draw from some or all of these ideal-type models, it is useful to bear in mind that each such approach is associated with different criteria and methods, different reference groups and different outcome measures.

All evaluations are inherently political.[2] Patton (1997: 347) argues that the political nature of evaluation stems from several factors: the involvement of people; data interpretation which is value laden and perspective dependent; action-based; and the fact that evaluation involves information and knowledge. As a political activity, evaluation involves the interplay of actors, information, knowledge and power.

Independent reviews may be considered unique in the business of evaluation as they are equipped to anticipate and counter specific political intrusions within particular political environments (see Patton 1997). Review groups usually come into the picture in conditions of controversy and conflict. Thus, in terms of their political bearings, when a development project involves contending actors, priorities, and perspectives, review groups, by virtue of their independence and mediation, meet with conflicting expectations regarding their potential role and contribution. Their predicament in this regard may be gauged from the observation by Doornbos and Tehral (1993: 183) on independent policy research in general:

> [R]esistance movements may welcome social researchers as long as they can be counted upon to share and expound their version of the generation of conflict, but may raise reservations vis-à-vis researchers who are not ready to make such commitments. Governments or international organisations, likewise may resist social researchers enquiring independently and possibly critically into some of their policy areas, but may invite the same or like minded researchers to offer an independent perspective once the policy area comes under heavy fire and is in need of thorough reconsideration.

Considering these observations, one would expect the contributions of review groups on the SSP to vary significantly. Before we go on to compare the context, review process and contributions of the IRM and FMG, however, some introductory comparisons of the two review groups are warranted.

1. The IRM was an international review group, the FMG a national body. The former was appointed by the World Bank and the latter by the Central Ministry of Water Resources (MWR), Government of India.
2. Both were made up of 'experts' in the field. The IRM had three 'experts' who had earlier participated in the Mackenzie Valley pipeline inquiry in Canada. The FMG had three 'experts' on water resources who served or had served in the National Planning Commission or MWR.
3. The constitution of both review groups followed specific protest actions by the Narmada movement. Their composition reflected consultations with movement actors, although the participation of the NBA was far greater in the FMG than in the IRM.
4. The IRM was explicitly a review group, whereas the FMG was not. The terms of reference (ToR) of the IRM were to

assess the implementation of resettlement and ameliorative measures on the environment and to recommend measures by which to improve implementation. The FMG's ToR made it clear that it was *not* a review group and that the SSP was *not* under its review. Rather, it was to carry out 'review discussions' with different opinion groups in order to obtain representative views on issues of implementation and benefits and to offer suggestions for improvement. Yet, considering the FMG's interpretation of its ToR, the contents of its reports, and the reactions to those reports from different quarters, it would not be wrong to consider it as a review group.
5. The Narmada movement considered the ToRs of both bodies to be inadequate and too narrow. Whereas economic and technical aspects had been ignored in the ToR of the IRM, that of the FMG did not cover the 'very basics' of the project. Having aired these reservations, movement actors participated in the proceedings of both review groups, whose later reports did cover technical if not the economic aspects of the project.

The Independent Review Mission

The formation of the IRM was an unprecedented step taken by the World Bank, testifying to the international outreach of the Narmada campaign on environmental impact and human rights violations in displacement. Neither the Government of India nor the Bank management was inclined to form such a review group, but on the insistence of a powerful section of the Bank's executive board, the IRM was formed in June 1991.[3]

International support for a comprehensive review of the SSP coincided with the global NGO campaign demanding that the World Bank be reformed to incorporate environmental and social concerns in its lending practices and policies, and to be publicly accountable and transparent. As Lori Udall, a leading international campaigner in the anti-SSP movement and an activist of the Environment Defence Fund (USA), observed:

Central to the NGOs' criticism of the World Bank are the Bank's lack of accountability to people directly affected by its projects and its promotion of large-scale, centralised and out-dated development projects that often are not fully completed or do not pay for themselves..., [its] chronic failure to involve and consult with local people, the inaccessibility of its project and sector information, and neglect of environmental and social impact assessment...[T]he Bank neither considers viable alternatives nor properly appraises projects. It does not implement and enforce its own policies and loan agreements nor does it conduct realistic cost-benefit analyses (Udall 1995: 203).

The implication of this campaign for the review group was that the IRM's assessment of the SSP also involved assessing the implementation and enforcement of the World Bank's resettlement and environmental policies.

The IRM received logistic support from the Bank management, including a separate budget of over one million dollars and access to all mission reports and documents. The Government of India and the three riparian states also extended support to the IRM despite their initial reluctance. The IRM had seven months (later extended to nine months) in which to complete the review.

With the 'sole desire...*to find the truth* [emphasis mine] and report it',[4] some of the main questions the IRM addressed were: are the stated environmental impacts accurately assessed; are sufficient data available to quantify those aspects; will the proposed ameliorative measures be effective; are the right issues being addressed in a timely way; do the resettlement plans take into account the characteristics of the people being displaced; are they appropriate to such people; are they capable of realisation as stated? (IRM 1992: 11).

According to the IRM, while its mandate on these questions was to engage in a 'problem-solving' exercise, it preferred to approach them in a 'problem-raising' framework. It observed that 'implementation cannot be carried out if the assumptions that undergrid the policies are themselves flawed. A consideration of implementation therefore requires a consideration of these assumptions' (ibid).

The IRM's Findings on Resettlement[5]

The IRM believed that loan agreements were signed without adequate attention to displacement problems. Although it lauded the World Bank for adopting an 'innovative and bold approach to a difficult problem' in setting up an independent review of the SSP, it held the Bank responsible for flouting procedures and of disregarding the adverse social and environmental impact during its 1985 appraisal of the project:

> In 1985 when the credit and the loan agreements were signed, no basis for designing, implementing and assessing resettlement and rehabilitation was in place. The numbers of people to be affected by the project were not known; the range of likely impacts had never been considered; the canal had been overlooked. Nor had there been any consultation with those at risk. Nor were there benchmark data with which to assess success or failure. As a result, there was no adequate resettlement plan, with the result that human costs could not be included as part of the equation. Policies to mitigate these costs could not be designed in accord with people's actual needs (IRM 1992: xv–xvi).

In specific terms, it observed the main problems with the resettlement component of the SSP were:

1. The assessment of the human impact of the project was inadequate;
2. the understanding of the nature and scale of displacement–resettlement was inadequate;
3. the project had failed to adequately take account of the needs of the affected tribal people;
4. the provisions of resettlement did not address people's 'real' needs, in particular the distinct needs of landed and landless people on the one hand, and the general needs arising from loss of forest land for gathering household material and other forest produce, firewood, grazing land and loss of river resources, on the other;
5. resettlement had resulted in breaking up families, although this was partly because oustees chose to go and partly because of limited availability of land;
6. the downstream impact on people had not been examined;
7. the compensation paid to oustees of the Kevadia colony and of the rock-fill dykes was meagre.

8. canal-affected people, especially those who were marginal farmers and/or had become landless due to land acquisition, had no proper compensation package;
9. project authorities had failed to consult the affected people, hence their opposition.

The IRM also observed discrepancies in policies among the three states. Apart from Gujarat, the rights of encroachers of forest land in Maharashtra and MP were not protected. Maharashtra and MP also adopted a restrictive interpretation of 'major sons' by not making them eligible for land compensation of 2 ha, as intended by the NWDT. Similarly, the majority of encroachers in these two states were likely to be deprived of land-based compensation as land records showing encroachment were outdated and unreliable. The disparity in the policies of the states implied that an oustee's legal right to choose the state in which s/he wanted to settle was violated. The IRM noted that even if MP modified its resettlement policy to match that of Gujarat, it could not be implemented given the time needed to meet the requirements of the SSP.

Observing that India had a poor track record on resettlement, the IRM stated that the 'incremental strategies' adopted by the World Bank and the Government of India, implying, encouraging and justifying phased improvements in implementation problems, have for the most part failed', and that 'a further application of the same strategy, albeit in more determined or aggressive form, would also fail. As long as implementation continues in these ways, problems will be compounded rather than mitigated' (IRM 1992: 355).

Findings on the Environment

The IRM acknowledged that India's environmental regimes (legal and policy) were quite advanced. In the specific case of the SSP, however, it noted the lack of basic data on environmental impact. It also noted that the World Bank had approved the project despite the fact that the Department of Environment (Government of India) had refused clearance to the project as far back as 1983. The IRM's specific observations were:

1. There was no such thing as an environmental work plan with which it could assess the project.
2. There were significant discrepancies on hydrological data which indicated that the SSP would not perform as planned.[6]
3. The upstream impact of the SSP, together with the downstream impact of the NSP, would have cumulative effects much greater than of the SSP alone. Confining the definition of impact and formulation of preventive measures to the SSP alone is ecologically unsound and self-defeating.
4. Afforestation and catchment treatment upstream are unlikely to be successful as they were being carried out without consultation and participation of the people in affected areas.
5. The practice of re-planting marginal forest land to substitute for better land submerged implied that forests will lose value.
6. Downstream ecological implications had not been considered; even rudimentary information on linkages and interdependencies was unavailable.
7. No plan for drinking water delivery was available for review, despite the priority attached to this benefit in the SSP.
8. There was no comprehensive environmental assessment of canal and water delivery systems in the command area, leading the review group to believe that problems of waterlogging and salinity would be acute. There were, however, some good works at the head end of the command area.
9. The threat of water-borne diseases was serious, particularly malaria, as proper safeguards and vector control measures were lacking.

Like the failure of the incremental approach in resettlement, the *pari-passu* strategy on environment, where assessments and mitigative measures run concurrently and finish simultaneously with project construction, greatly undermined the prospects of achieving environmental protection. The IRM noted that adoption of *pari-passu* policies created 'fundamental compliance problems' in the SSP:

> The history of the environmental aspects of the SSP is a history of non-compliance.... The nature and magnitude of environmental problems remain elusive. This feeds the

controversy surrounding the project.... To complete our work we had to assemble basic ecological information to establish the likely effects of the project upstream, downstream and in the command area. This should have been done by others before the project was approved (ibid.:).

In this context, the IRM noted 'a fundamental difference in the meaning and practical value of *pari-passu* between environment assessment (i.e. what is to be affected, how, what can be done, etc.) and environmental mitigative measures, some of which could proceed concurrently with construction' (ibid.: 275). In some instances it found that 'mitigative measures were actually ahead of assessment' (ibid.).

Overall, the IRM noted 'an institutional numbness in the Bank and in India on environmental matters' and a tendency among project authorities 'to justify rather than to analyse; to react than to anticipate' (ibid.: 226).

Recommendations of the IRM

The 'truth' that the IRM found in its review of the SSP was the lack of proper impact assessments and of undisputed data:

The absence of proper impact assessment and the paucity of undisputed data, have limited our ability confidently to make project specific recommendations.... The SSP is beset with profound difficulties.... No one is sure about the impacts of the reservoir and the canal on either people or the land. Without knowing what impacts were likely to be we found it difficult to the point of impossibility to assess measures by which they might be mitigated. It seems to us that the matters we have raised are fundamental. It would be prudent if necessary studies were done and the data made available for informed decision making before further construction takes place (IRM 1992: 354–55).

The IRM recommended that the World Bank should 'step back' from the project so as to enable concerned authorities (*a*) to design proper studies of assessments, (*b*) to evaluate these assessments to see if the project needed modification, and then (*c*) to design resettlement policies and see how these could be implemented. It concluded its report by stressing the need for consulting, informing and involving the people affected by the project throughout the Narmada Valley, the command area and downstream of the dam. Stating that the failure of consultation led to opposition to the project and particularly hostility among people in the submergence area, and that this opposition itself became a bottleneck in designing and implementing mitigative measures, it called for 'finding ways to rebuild confidence' and 'to demonstrate goodwill' in the project.

Responses to the IRM's 'Truth'

The NBA and its allies welcomed the findings of the IRM and demanded that one of its recommendations that the Bank step back immediately, be complied with. It was the biggest achievement in their campaign against the SSP. However, the responses of those supporting the SSP—stage governments, the central government, World Bank management and NGOs in Gujarat—were dramatically different. Setting aside three decades of conflict over water sharing and the shape of the SSP, the three state governments—then belonging to three different political parties, Congress, BJP and Janata Dal—were swift to condemn the findings. Gujarat's Chief Minister Chimanbhai Patel rejected them as showing partiality. In MP, the Minister of Irrigation Shitla Sahay dubbed the findings as 'wilfully motivated' by 'vested interests' (*The Hindustan Times*, 24 June 1992; 10 August 1992). ARCH-Vahini, six months after the IRM report, observed:

> Almost all the major conclusions of the report are not supported by the facts presented in the text...or they don't follow from the premise. At crucial places the facts are wrong although there is ample evidence that Morse-Berger had correct facts at their disposal. The report is littered with serious methodological errors. Ambiguities abound. The ultimate causality is clarity, objectivity and truth (Arch-Vahini: *Economic Times*, 11 April 1993).[7]

The Government's response was recorded by the Secretary, Ministry of Water Resources, in a letter to the Director, India Country

Department of the World Bank, in which he refuted the major charges of dereliction made in the IRM report.[8] He concluded:

> [I]t must be said that we were much disappointed to read the report. We were looking forward to have the benefit of concrete recommendations for improving the R&R process and environmental management measures for which we ourselves feel greatly concerned, as can be clearly seen from the fact that the NCA's sub-groups on these subjects are headed by no less persons than the Union Secretaries of the concerned Ministries.[9] It was unfortunate that the IRM chose not to make any recommendations on these matters but chose to deride the project without a proper analysis of the facts. We feel that a great opportunity has been lost (D.O. Letter No. 21/1/92-PP dt. 7 August 1992; footnote added).

The response by the Bank's management to the IRM was summarised in its document entitled 'India: Sardar Sarovar (Narmada) Projects, Management Response' (SecM92-849: 23 June 1992: 2), the main contents of which are:

> Resettlement: (a) General agreement with the description of the resettlement situation in each of the three states and in the canal area; (b) acceptance of shortcomings in the preparation and appraisal of the project's resettlement aspects; (c) disagreement over the conclusion that even if MP's policies were modified they could not be implemented.
> Environment: (a) Acceptance of delays in the completion of mandated environmental studies, and of the need for more effective central management and coordination of impact studies, and mitigation measures; (b) acceptance of the need to accelerate downstream impact studies and health aspect studies; (c) disagreement on the conclusions that delay in studies would have severe environmental consequences.
> Remedial Action: Action underway or planned at the project level to permit satisfactory implementation of resettlement, and at the Bank level to review resettlement performances throughout its portfolio.

The argument of the Bank management was that its capacity to influence policy implementation was limited, though necessary and progressive. It pointed out 'the very real limits of the Bank's ability to make up for the weaknesses in the Borrower's project implementation through intensive supervision', and that 'the Bank is ultimately reliant on the Borrower's commitment and capacity for effective implementation'. It justified its involvement in the SSP on the grounds that, 'while it would be safe to restrict Bank involvement to those situations in which the Borrower had already established an exemplary track record, this would in practice mean forgoing opportunities for potentially important changes', and that without the Bank 'the project in question may still proceed but under much less favourable circumstances' (ibid.: 11–12).

The IRM's Impact on the SSP

In less than a year after the publication of the IRM report, the World Bank was out of the SSP. The Government's announcement on 30 March 1993 that it would raise its own money and forego the undisbursed $170 million of the $ 450 million World Bank loan for the SSP, was anticipated given the events that unfolded after the IRM report was out (see Appa 1992). In the interim, the Bank management had striven hard to continue its involvement with the project. Its internal 'Cox Mission' visited India in July 1992 and recommended an action programme on resettlement covering three areas: policy improvements, organisation and management, and implementation (see Appendix 3). Based on the responses of the states to the action plan, Preston, the then Bank President, submitted a report: 'Review of Current Status and Next Steps' (R92-168, 11 September 1992). As Appa (1992) notes, however, the IRM was intensely unhappy with 'Next Steps' for ignoring and misrepresenting its findings. This resulted in two meetings between the IRM Chairperson Morse and the Executive Directors, first of the G–10 group and then with a larger group. The voting on 23 October resulted in a narrow victory for continued support.[10] Tough conditions pertaining to resettlement and environment were set to be implemented within six months by project authorities in India, failing which the project was to be suspended. The Government's announcement came a day before these benchmarks were to be evaluated by the Bank.

Paradoxical as it may seem, the Bank's 'withdrawal' was welcomed by both the NBA and the project authorities. The NBA considered the withdrawal a double victory, against the SSP at home and the World Bank outside. Medha Patkar, leader of the Narmada movement, regarded the withdrawal as:

a triumph for thousands of struggling farmers and tribals in the Narmada Valley and a shot in the arm for the wider struggle against advancing neo-imperialism in the form of the World Bank–IMF sponsored structural adjusting programme and General Agreement on Tariff and Trade (GATT). It is a victory...for those opposed to 'development' projects and development planning funded and pushed ahead with the 'aid' of multilateral agencies against our own developmental priorities and goals.... We are happy that the fight [now] is within our nation-state even if this may bring more repression![11] (NBA press statement, 30 March 1993, footnote in original).

The response of the international NGO community which supported the struggle against the dam was summarised by Patrick McCully, then Associate Editor of *Ecologist*: 'The World Bank's con game stands exposed in the eyes of the tax payers in the First World and the people of the Third World. The Bank has never been interested in social or environmental issues. They want hard cash.[12] In any event they hardly walked away from SSP with good grace. They had to be beaten into submission by the force of international public opinion after the Morse report' (interview with Bittu Sehgal, *Sunday Observer*, 11–17 April 1993; footnote added).

Project authorities in India did not officially explain the decision to forgo the loan, apart from reiterating their commitment to 'the full implementation of the resettlement plans and the environmental standards'.[13] Yet it is widely held that the 'withdrawal' was imminent as the authorities were in no position to fulfil the stiff benchmarks. The decision seems to have stemmed from imposed conditionalities by the World Bank that were 'not in keeping with the country's self-respect'. P.R. Chari, former Vice-Chairman of the Narmada Valley Development Authority in MP, explained 'having to give one bank official or another a mass of data almost throughout the year was creating an impossible situation. They were even interfering in the administration and trying to influence the staffing pattern and the places where staff should be posted.'[14] The immediate fear of a financial crunch ensuing from the Bank's 'withdrawal' was allayed by the Gujarat government, which dismissed the aid amount as 'peanuts' and proposed that the power component be separated from the project so that the private sector could be tapped to finance implementation.

The Five-Member Group

A few months after the World Bank's withdrawal from the SSP, the threat of *jal samarpan* by NBA activists led to the formation of the Five-Member Group by the Central Ministry of Water Resources.[15] Compared to the IRM, the issues before the FMG were broader and more complex. It had been formed by the Government after mutual consultation with the NBA. The task for the group was to initiate discussion on all aspects of the dam which had been raised in the movement[16] and to submit a consensus report to the Government within three months.[17]

The Five-Member Group did not have the luxury of the IRM budget and had to manage with a skeletal secretarial staff. Gujarat, MP and Rajasthan opted out of the review process, leaving only Maharashtra to participate in the proceedings.[18] Their common position was that issues such as hydrology and dam height could not be subject to review due to the binding nature of the NWDT. Noting this position, the central government later framed the FMG as a 'listening post' for itself to get representative views, rather than a 'review group'. It asked the FMG to 'focus on such issues which can lead to improvement on implementation and benefits planned for SSP' (FMG 1993: 4) and anticipated 'useful suggestions on resettlement and environmental concerns' (ibid.: 5).

In its review, the FMG considered the criticisms around the SSP from three vantage points: the benefits of the project, including the hydrological aspects, its environmental impact and, finally, the resettlement component. Some major observations of the FMG on the various criticisms of the project are discussed in the next section. It must be mentioned, however, that at no place in its first report did the FMG refer to the findings of the IRM.

The FMG's Views on Project Benefits

1. Hydrology: The actual flow data of the Narmada is 23 MAF at the dam site, whereas hindcasting yields a figure of 27–28 MAF. There is difference of opinion among experts as to which of the two figures should be used for planning purposes. Other than claimed by the project's critics, however, the use of the figure 27 MAF of utilisable water at 75% dependability 'doesn't necessarily vitiate the project planning' (FMG 1994: 22).[19] Noting that the development of upstream uses in MP will be a lengthy process, the FMG argued that more water will flow past the dam site than the allocations to Gujarat and Rajasthan (although the flow will fluctuate in the absence of upstream regulation).[20] It is only with full development of upstream projects that problems of inadequacy may arise. The FMG recommended that this issue be quickly examined by the Government and settled once and for all.[21]

2. Irrigation: The criticism that the project would encourage cash-crop farming is misleading, as is the justification that the project will increase foodgrain production. Insofar as the state provides water in pre-determined quantities on a volumetric basis and charges a price that discourages wasteful use, cropping patterns will be influenced by market forces and relative prices. The issue in question is whether farmers would prefer water-intensive crops like sugarcane and whether this will lead to a cornering of benefits in the head end of the command. To counter this possibility, the FMG noted that the 'effective measures of water distribution as planned in SSP (automatic and semi-automatic control, bulk and volumetric supply, proper pricing) should be strictly adhered to so that the discipline of the system is maintained (ibid.: 30). It recommended the creation of vested rights for tail end farmers by declaring water shares of different regions in advance, allowing tail-enders to acquire political strength, and facilitating the eventual reduction of water to head end farmers through legislation.

With regard to the opinion that irrigation efficiency would be no more than 35–40%, the FMG considered an irrigation efficiency of 60% to be achievable, but added that actual achievement will depend on controlled distribution, sound water management and pricing policies, as well as on the design of the project.

The risks of waterlogging and salinity were reviewed in the light of policy measures adopted by the Government, which envisaged the planning of drainage, the re-sectioning of natural drains and the provision of additional drains where necessary. In calling for assurance that these plans will be translated into realities, the FMG held the position of anti-dam activists to be 'cautions to be kept in mind and not fundamental criticisms' (FMG 1993: 33). On the 'conjunctive use' of surface and groundwater, it observed that steps need to be taken to ensure that the intended number of tubewells (both private and public) are constructed in the command area through the allocation of funds and the relative pricing of surface and groundwater.

On the critical view held by the movement that water would not reach Kutch and Saurashtra, the FMG noted the consideration of the Gujarat government to increase allotment to Kutch as positive, and endorsed this as 'desirable to consider whether allocation to Saurashtra and Kutch, or at least at any rate to Kutch can be increased' (ibid.: 40). Seeing the conflicting views among these regions on the issue of water endowment (the NBA arguing that proper harnessing of local water will better meet the needs of Kutch and Saurashtra and the Gujarat government stating that water to this region will have to be brought from outside), the FMG argued that if it was technically and economically unfeasible to allocate more water to Kutch and Saurashtra, then the Gujarat government should consider 'adequate plans and provisions for harnessing the local water resource potential' in the area. It also did not share the movement's pessimism that the SSP was crowding-out the financial resources necessary for the development of water resources in the area. Rather, it noted 'the priority attached to these schemes by the Gujarat government' and said that the latter 'will no doubt find ways and means of augmenting the financial allocation' for local projects (ibid.: 42).

3. Drinking Water: The criticism that drinking water was an afterthought was not taken seriously: 'there is nothing wrong with after-thoughts provided they are sound', the

FMG observed (ibid.: 47). Judging that the SSP did not imply the installation and operation of water-supply systems,[22] but simply that its canals would carry the necessary water to off-take points, it made certain recommendations:

- publication of the list of all 8,214 villages and 135 urban centres;
- identification of agencies which will manage the water-supply system for each village and urban centre;
- publication of the physical planning of water-supply projects, including financial arrangements and time schedules;
- publication of interim measures to meet the drinking-water needs of villages and towns in question before the canal water reaches them in 10 to 15 years time, as well as during the annual maintenance shutdown (about two months) of the canals.

4. Hydroelectric Power: Three major criticisms of the hydro-power component, i.e. that the firm generation capacity is much less than projected installed capacity; that its internal consumption will be extremely high; and that benefits will fall given less water in the Narmada and in the absence of regulated release in the NSP, elicited the following response from the FMG. The fact that firm generation is less than installed capacity is an essential requirement for the project to offer peaking facility. The proper way of looking at the project is *not* to view it as an installation where power generation will gradually diminish, but essentially as a peaking facility which makes use of the initial years of water availability for continuous power generation. 'If the full irrigation development is likely to take many years, it makes sense to use the available flows during this period for power generation' (ibid.: 56).

Noting the multipurpose nature of the SSP, the Five-Member Group disagreed with the logic of 'internal consumption'. Even if the project were only an irrigation and water-supply project, it would still require power for its various components. Given its hydropower component, the project will use its own power, contribute to the general pool (western grid) and draw from it for certain uses. The FMG reiterated the multipurpose nature of the project against the criticism that its power benefits entailed enormous social and environmental costs.

The FMG also observed that if the utilisable flow in the Narmada was less than planned, hydropower generation would be affected. The delay in the construction of the NSP upstream, however, did not warrant an adverse conclusion about the SSP; '...if for some reason...NSP has been delayed, the authorities...will have to accept this and manage SSP under those given circumstances. If the best situation fails to materialise, the second best situation will have to be accepted' (ibid.: 58).

   5. <u>Alternatives</u>: The FMG took note of the various 'alternative' proposals that it received on irrigation and power. Considering them outside the scope of its terms of reference, it forwarded them to the Government. 'Coming as they are from knowledgeable persons', the FMG recommended that 'they should be given the most careful attention' (ibid.: 45).

The FMG on Environmental Aspects

Two important observations on environmental aspects set the FMG's views apart from the IRM. First, it said that the 'projects were caught in a change in the climate of opinion' (ibid.: 63). Characterising environmental concerns as 'a new idea', it took note of the problems of institutionalising such concerns and argued that the mechanisms of evolving proper procedures of assessment and monitoring and having them accepted by state agencies who devise and implement development schemes can only be gradual. Specific to the SSP, it observed with satisfaction the environmental monitoring mechanism, doubting 'whether any more effective mechanism could have been devised or made to work within the framework of our existing political and administrative structures, particularly in the context of a federal system' (ibid.: 67).

Second, it held that an intervention of the scale of the SSP does not easily yield an a *priori* knowledge of all environmental consequences. '...[R]oom for further studies and for an examination of impacts not earlier foreseen' (ibid.: 62) will exist in such a project, and the MoEF's conditional clearance to the SSP and the attached *pari-passu* clause does not imply a deliberate closure of the space—an observation that appeared directly to question the framing of this problem by the IRM as a sure way towards 'non-compliance'.

The FMG, however, noted that schedules had fallen behind in many of the studies, plans and activities. Given that project construction had itself slowed down, it said that the lost ground could be retrieved through accelerated action, particularly in some areas (ibid.: 69):

1. Catchment Area Treatment: The 0.7 million hectares earmarked for treatment posed some problems in the mobilisation of financial resources. Project authorities were reluctant to bear the entire cost of treatment as other government agencies had obligations to share responsibilities. Action on treatment, however, has fallen behind schedule. A time-bound Action Plan will need to include location-specific treatment measures such as agro-forestry, grass-growing, fodder cultivation, tree farming. Successful implementation would require 'innovative methods, practical approach and social motivation' and supportive actions of non-government actors. A similar strategy is required for the compensatory afforestation measures which 'seems to be making progress' (FMG 1992: 71).

2. Wildlife and Fisheries: The impact of the impoundment of water on these aspects was difficult to determine fully. The complexities involved would require alertness on the part of the central and state authorities. As the project would substantially alter existing fishing practices, the rehabilitation of downstream fisherfolk would require specific efforts. There was an urgent need to ensure that valuable plant species are not lost and wildlife not drowned in the rising water.

3. Downstream Studies: The full impact of the project downstream, particularly in relation to livelihoods, domestic and industrial uses, had not been adequately studied. This should be done on a top priority basis.

4. Command Area Measures: Problems of waterlogging and salinity in existing irrigation projects should be enough reason for authorities not to take for granted that the measures proposed in the SSP would be sufficient or would work as

per plan. This too called for the authorities to be alert to problems that could arise in the command. This effects of fertilisers and pesticides on water quality would also require adequate attention.

5. Public Health Measures: There had been a general shift in approach from curative measures to preventive measures. The measures taken might require specific attention from the concerned ministries.

The FMG's Views on Resettlement Aspects

In analysing the submissions made to the FMG by various actors on displacement and resettlement aspects of the project, it noted that the materials were of 'conflicting nature':

> Divergent accounts of certain matters were given in certain documents. The claims made by officials concerned with rehabilitation and the criticisms made by NBA were substantially at variance. Further, while the NBA went as far as to argue that rehabilitation was impossible and that the project must be stopped certain other voluntary agencies felt that the human and satisfactory rehabilitation of the PAPs was entirely possible with more active support from the state governments and a greater role for NGOs. The criticisms made in the report of the Tata Institute of Social Sciences were questioned in a rejoinder by the Maharashtra government. Similarly the NBA's criticisms were countered in replies given by the Ministry of Water Resources as well as...by ARCH Vahini...(ibid.: 79).

The FMG claimed to have no means at its disposal to verify the facts and establish the relative accuracy of conflicting statements, given the refusal by state governments (particularly Gujarat) and Gujarat NGOs to participate, and the consequent lack of field visits by its own members. While noting the enormously difficult task of resettlement in the SSP, however, it summarily dismissed the NBA's slogan—resettlement is impossible—as 'a counsel of despair' (ibid.: 82). It acknowledged that the non-availability of land was a limiting factor in resettlement and that other forms of resettlement needed to be considered.

The FMG observed that the various categories of people would need different packages, and that a uniform package for all oustees was not required. Neither was it necessary to standardise the resettlement packages among the three states; 'if the package offered by Gujarat is more favourable and this weights the free choice offered to the PAPs in favour of settlement in Gujarat this is not necessarily undesirable' (ibid.: 83).

The FMG recommended that a comprehensive Master Plan, which should have been ready a long time back, should be completed within six months with detailed time schedules. The database for this Plan should be based on 'a complete census of all categories, groups, communities and individuals affected in any manner whatsoever, including canal affected people, communities downstream of the dam, groups and individuals providing supplies and services...' (ibid.: 86).

It also strongly endorsed the principle that 'a share in the benefits should be provided as a prior right to those who bear the social cost of the project', and said that this should apply to all future projects. It asked for appropriate legislation in this regard which would impose restrictions on land sales in the project command.

Finally, the FMG stressed people's right to be informed fully and well in time of the steps taken or under consideration. Government machinery should meaningfully involve experienced voluntary agencies in implementation of the resettlement package.

Responses to the FMG Report

The reactions of the different actors to the report were varied, to say the least. The Madhya Pradesh government reverted its earlier decision not to participate in the proceedings of the FMG. In the final phase of the FMG review, MP had made a submission for reducing the dam height by 19 ft. The FMG, on its part, considered this demand to be outside the purview of its terms of reference and a 'matter that needed to be resolved between the Gujarat and Madhya Pradesh governments' (ibid.: 60). In Gujarat, a pro-dam NGO, Narmada Abhiyan, legally challenged the constitution of the FMG in the Gujarat High Court on the grounds that the NWDT award on the SSP, which was final and binding, could not be reviewed. The High Court ordered that the FMG report not be made public until further notice. It was with the Supreme Court's intervention a few months later that the report was made public and copies provided to all involved parties for comment.

Not surprisingly, the NBA and its allies in the Narmada movement expressed dissatisfaction with the FMG report, albeit in a restrained and subtle manner. For the NBA, the report had first and foremost 'shattered the government myth that all is well with the SSP' (NBA 1994: 6).

The report contains many recommendations which clearly show that the benefits of the SSP remain strongly questionable, that environmental impacts remain unstudied and environmental works are lagging behind. The lack of a rehabilitation master plan and the need to rehabilitate all classes of project affected people has been commented upon strongly. However the group has failed to...tie together the implications of its findings...[as to] whether the project should be allowed to proceed. Also the group seemed to accept government assertions rather uncritically specially while discussing benefits (ibid.: 1).

The NBA blamed the Government of India for adopting 'unfair and underhand tactics' and for pressurising the FMG by restricting its terms of reference. It also noted that the FMG had 'unfortunately succumbed to the pressure in an entirely unwanted manner' (NBA 1994: 1). A well-known activist in the movement, referring to the antecedents of the five members, stated: 'perhaps one cannot expect more from these retired bureaucrats as it is very difficult for them to criticise the very system which they have served.' In a national seminar organised by the NBA around the report, Medha Patkar drew parallels with a famous parable:

The report reminds us of the story in which the King's parrot had died but no courtier had the courage to tell the king the truth. Instead, the king was repeatedly told that the parrot had refused food, had stopped talking and was not in the best of health. The FMG report is saying enough to suggest that the project was in crisis but not saying enough to halt the dam (proceedings of the National Seminar, Delhi, 4 January 1995).

Impact of the Report

In what may be termed a direct consequence of the FMG report, the Supreme Court of India[23] asked the state governments and the NBA to file their rejoinders 'ignoring legal implications'. It referred back to the FMG three specific issues for its opinion: hydrology, height of the dam, and resettlement of oustees, including environmental concerns. It ordered the suspension of further construction work on the project.

It is interesting to note here that, in its supplementary report, the FMG was unable to come to unanimous decisions and its members differed in their positions on all three issues. For example opinion was divided as to whether hindcasting should or should not be used in the case of the SSP, and whether the dam height should or should not be reduced.

IRM and FMG: A Comparison

Table 7.1 gives a comparative picture of the two reviews according to three main factors: (a) assessment; (b) policy and implementation; and (c) orientation of the groups towards the SSP and the Narmada movement.

It is evident that the two reviews give conflicting stories on the SSP. One holds that the SSP is a flawed project, whereas the other sees scope for improvement. One believes that the claims generated in the Narmada movement against the SSP are exaggerated, the other holds the same view with regard to the claims of project authorities. One finds the existence of disputed data as reason to reject the project. The other acknowledges the difficulties of taking sides when data are disputed.

What do we make of these conflicting findings of two 'independent' review groups? How do we relate the conflicting reviews to the power and knowledge struggles in the SSP? Are they reflective of those power and knowledge struggles? Or are they borne out of different approaches to the same questions? These are more significant issues than simply to pose which of the two reviews is correct.

The fact is that neither of the two reviews violates the 'truth' in the SSP. Paradoxical it might be, but the findings of both

Table 7.1 Comparison between the IRM and FMG

Review Aspects	Independent Review Mission	Five-Member Group
Impact Assessment Studies	Studies not done as per requirement or as promised. Studies done but inadequate, incorrect or disputed. Those being done cannot be finalised in time.	Studies need to be done on priority basis. Studies have been done and others can be done. Should await the findings of those studies that are being done.
Environment policy	Advanced policy regimes but violated in the SSP. Lack of basic data at the project level.	An evolving field; policies settling down although still somewhat contested. Master Plan needed.
Resettlement policy	Discrepancies across states undesirable. Policies close options for affected people.	Discrepancies across states not undesirable. Vast improvements over past policies in all states.
Hydrology and Benefits	Hydrology data—the basis of the design and operating features of the SSP—suggest that the project will not operate as assumed. Benefits are overstated.	Hydrological data reasonable considering the uncertainties in estimation. Project has the potential of achieving the claimed benefits with disciplined implementation and appropriate institutions.
Timing and Capacity for Implementation	Impossible. Incapable and inadequate.	Possible. Strong emphasis on capacity and institution building.
Affected People's Involvement in the Project	Project shows no participation or consultation.	There is a need for participation and consultation as is being done in Gujarat.
Approach to Review[24]	Goal-based.	Decision-making.
Position vis-à-vis that of the Narmada Movement	Endorsement and extension.	Critical engagement.
Framing of Findings on the SSP	Extended pessimism.	Cautious optimism.

Sources: IRM (1992); FMG (1994; 1995).

reviews are conflicting and correct at the same time. The IRM set out to evaluate three aspects of the SSP: the extent to which policies have been complied with, the nature of the social and environmental impact of the SSP, and whether the ameliorative measures are appropriate. Working towards these objectives, the review group drew up a standard of what it expected to find in terms of studies, assessments and ameliorative measures in a large project like the SSP. It then considered the empirical situation against this standard, noting what it perceived to be gaps, deviations and lapses. Since the studies, assessments and ameliorative measures undertaken in the SSP did not match its standard, the IRM decided that the project was flawed.

In contrast, the FMG approached controversial aspects of the project, judging the criticisms with regard to their credibility. Moving through the gamut of controversies, it identified potential problem areas in the SSP and suggested remedial measures. It acknowledged commissions in the SSP as compared to past experiences with similar projects. At the same time it presented a series of recommendations aimed at improving the quality and standard of the project. To put it in a contrasting frame, unlike the IRM which moved from a standard to the specific case, the FMG moved from experiences in the SSP to a standard.

The differences attributed to these approaches cannot in any way be essentialised. Standards may be low to begin with in order to show a particular case in a good light. Similarly, although some practices and experiences can be better than others, they may fall woefully short of what is acceptable. This could well imply that although criticisms of the project may be exaggerated, the project may still be undesirable.

The point being made here is that evaluation approaches and instruments, being reliable and valid, may tell different and even conflicting stories on development. The SSP reviews exemplify this aspect in no uncertain terms. On the one hand, the SSP fails to meet IRM standards. In the FMG review, on the other hand, it fares well in terms of existing practices in a given context. The issue at stake here is not so much the validity and reliability of the methods (see King et al. 1987; Rossi & Freeman 1993), but the more complex aspects of the politics of criteria and standards, and the claims to and constructions of 'truth' and 'objectivity'.

The IRM review became an instant international hit. Advocacy groups and NGOs at a global level welcomed its findings, and it led to the Government's subsequent 'withdrawal' from the loan agreement with the World Bank. Yet, the review raises a fundamental question: Why did the review seek the withdrawal of the World Bank from the SSP instead of following its terms of reference to assess implementation and suggest improvements to R&R and amelioration of the environmental impact?

The IRM's answer is at first glance convincing. In its view, 'if essential data were available, if impacts were known, if basic steps had been taken, it would be possible to know what recommendations to make' (IRM 1992: xxiv). Since the IRM held that, in many areas of the project, no adequate measures were taken on the ground or were under consideration, it could not put together a list of recommendations.

According to the IRM, on many aspects of the project it had to assemble information with which to establish its likely effects, while arguing that project authorities should have done this work themselves. Now, having established for itself the likely impact in these areas, why did it not go ahead to suggest ameliorative/mitigative measures on the basis of this information? Why did it instead subscribe to a circular logic to conclude that basic studies to establish likely effects are lacking? What about the studies that the IRM itself conducted? Were they not indicative of the effects? Were they not sufficient? Could these studies not form the basis of at least some mitigative measures? Was the IRM not capable of suggesting remedial measures? Or was it not inclined to do so?

Two reasons are most plausible for this logic. First, it is likely that the IRM (through its studies and assessments) perceived the social and environmental impacts of the SSP to be so adverse that it could not think of any mitigative measures with which to counter those effects. If so, perhaps it should have stated this clearly in the report, rather than pointing out the lack of studies and assessments, when the review mission itself had undertaken some such studies. The second possible reason is political. To suggest possible mitigative measures may have been construed as according some degree of credibility and legitimacy to the SSP, which the IRM was not inclined to do.

Were mitigative measures in the SSP possible or not? This question left unanswered in the IRM review leaves us with very little indication as to whether the suffering and despoliation in mega projects like the SSP could be minimised, ameliorated or mitigated at all. The silence also alludes to the social and political construction of 'goodness' and 'badness'. It underscores our argument that under claims to truth and objectivity lie the political interests and values of actors who influence the criteria and standards with which they reach the truth. Although the IRM claimed to have been guided solely by the objective pursuit of 'truth and report it' (1992: xxv), the question is, is there a single truth in a project like the SSP and can it be recovered at all? Equally important is the question, should the IRM's claims to truth be treated any differently from other truth claims made in the SSP? How objective is the IRM and its findings?[25]

Reviews are political activities, their outcomes are influenced by the interplay of actors, information, power and knowledge. Review groups are actors in the action field. Their constitution, audience and outreach, the resources and information at their command, the standards, criteria and methods they use and/or endorse, and the power they can command vis-à-vis the decision-makers of the day, influence their actions and outcomes. Recognising that subjectivity is an integral aspect of the evaluation process, Patton (1987) persuasively argues that concerns about objectivity can be better understood and discussed as concerns about neutrality. He seeks 'to replace the traditional search for truth with a search for useful and balanced information and to replace the mandate to be objective with the mandate to be fair and conscientious in taking account of multiple perspectives, multiple interests and multiple possibilities' (ibid.: 167). Viewed from this angle, the IRM's emphasis on 'what has not been done' and 'what cannot be done' in the SSP seems to echo the perspective and interest of political factors opposed to the project. In contrast, the FMG's focus on 'what has been done' and 'what needs to be done' is an attempt to accommodate competing and conflicting perspectives, interests and possibilities in the SSP. It appears that the FMG had trodden more neutral ground in the Sardar Sarovar Project than did the IRM.

The IRM focuses on the non-viability of a project because of its uncertainties and risks. At a general level, perhaps all large dams globally bear this characteristic and, by this yardstick, are non-viable projects. Moreover, in the context of developing countries, problems pertaining to assessment and management of uncertainties and risks in mega projects become even more complex due to the systemic 'non-occurrence of expected outcomes'[26] and the 'occurrence of non-expected outcomes'[27] (Padaki & Prasad 1995: 34). Given these project-specific and context-specific constraints that bear on the SSP, the IRM review could be seen as a mechanism of information and knowledge generation intended to minimise and manage uncertainties and risks. But it could also be read as a politically supportive actor project that helps the Narmada movement in converting uncertainties into ignorance and raising the odds of risks in the SSP. In that respect, it appears more to be a risk-advocate than a neutral risk-arbitrator.

The Supreme Court of India: The Last Jury?

At the time of writing, the Supreme Court was set to judge whether or not the SSP is a worthwhile developmental project. It asked all opposing views to be placed before it, 'uninhibited by the terms of the NWDT Award', and initiated a process by which each adversary could comment on the submissions of the other. Different actors in the conflicts responded in different ways pursuing their respective interests in this situation.

The hurdle of the NWDT award having been removed, the MP government used the opportunity to request (*a*) that the hydrology matter be reviewed by an independent group; (*b*) that in its view lower availability of water at the dam site could entail that the height be reduced to FRL 436 and even to FRL 427 ft. (less by 28 ft.); (*c*) that the power component of the SSP should be subjected to fresh cost-benefit review; and (*d*) that MP was willing to forego its share of power benefits that would result from a lower dam. The Gujarat government on its part has argued that a height reduction by 19 ft. would involve 20% reduction of electricity (cheap and non-polluting), whereas savings in terms of area submerged would be minimal.

During its earlier proceedings, the Supreme Court reverted to the FMG to get its opinion on hydrology, dam height and resettlement aspects (areas in which the FMG has either asked for further studies or expressed its inability to take a position).

The supplementary FMG report[28] has since been submitted to the Supreme Court. While the FMG reiterates its earlier position on hydrology, one member dissents on the method underlying the hydrological estimates and does not feel that hindcasting is necessary in estimating Narmada waters. On the issue of dam height, the supplementary report endorses the phased construction of the dam, first to a height of 436 ft. and, pending further studies, progress in the construction of ancillary irrigation structures and satisfactory progress on resettlement, to a possible full height of 455 ft. On the displacement/resettlement problem, however, opinion is divided as to whether 'the problem is manageable and that the deficiencies which have come to light can be rectified through appropriate remedial measures' (FMG 1995: 22). One member argued that since a reduced dam height can potentially reduce social and environmental costs, hardships and pain to affected people, livestock and wildlife (through reduced submergence), due consideration should be given to a fresh appraisal of different heights for optimising benefits and minimising suffering and despoliation.

The lead organisation of the Narmada movement, the NBA, is of course the 'petitioner', claiming representation of all affected people in the Valley. Its petition, while calling for a comprehensive review of the SSP, argues that the dam height should be reduced to around 310 ft.

What will the Supreme Court do and how will its verdict affect the project? Difficult as this question is, and not withstanding our lack of familiarity with the intricacies of the legal terrain, some possible scenarios can well be imagined. Two scenarios can be imagined but are implausible. One, in which the Supreme Court orders that the dam is a matter of government policy (i.e. a matter for the 'administration' to look into) and refers it to the government, with some strictures/conditions/guidelines on environment and resettlement. In the past the Court has made similar observations in a petition on the Tehri dam project, but this scenario is unlikely in the SSP as the 'public interest' in the project is far more pronounced and the conflicts have a non-negotiable character. And second, in which the Supreme Court would ask for a fresh review of the project, with the participation of affected people and their representatives. Weighing against this scenario are the arguments made on behalf of potential beneficiaries, especially when the project is in such an advanced stage.

Among plausible scenarios, the first is an observation by the Supreme Court that the phased development of dam height is strictly to be linked to progress in the proper implementation of environmental and resettlement measures. In the event of this happening, the Court is likely to attach detailed conditions for ameliorative measures. The second scenario is one where the Supreme Court, having analysed and weighed the benefits and pains, would ask for (appropriate) reduction of the dam height on the grounds that the resettlement burden would be substantially reduced. This will provide some solace to the MP government and to the actors in the Narmada movement.

It is difficult to forecast which of the two options the Supreme Court is likely to choose. Among decision-makers and in the Narmada movement, however, the 'worst' scenarios have been visualised and allowed for. A representative view among dam authorities on the possible reduction in dam height is likely to be 'social issues have been given a very high priority over economic benefits' (interview, D.T. Buch, NPG, 5 May 1995). In the Narmada movement, the worst scenario is one in which the Court holds that the dam is a worthwhile project. The likely response to this situation was foretold by an activist:

> The Supreme Court is part of the same state machinery. Its thinking more or less reflects the same convictions and logic. Three people who are not experts (unlike the IRM), who have other tasks (unlike the IRM) and who have no first hand information (unlike the IRM), are dictating the fate of millions. But the Supreme Court's verdict is the law of the land. They carry more power and therefore for us almost a compulsion to take up the issue with them. Our future actions depend a lot on the final verdict of the Supreme Court.[29]

Concluding Remarks

Having looked at the independent reviews, we might have been expected to arrive at the 'truth' of the SSP, i.e. whether it is a good or a bad project. This chapter should have helped us realise the futility of this question, simply because there is no agreeable standard against which to measure the goodness or badness of a project like the SSP. Chapter 4 demonstrated the inadequacy of a method like cost-benefit analysis to accommodate diverse interests; this chapter has shown how standards of the movement do not fully convince the FMG, and how the standards in the IRM hotly contested. Thus, a set of obvious questions remains: How do affected people perceive the project? What yardsticks do they use in judging the project? How do they respond to situations of systematic suffering? Do they think that their interests have been represented and accommodated by the supporters and opponents of the SSP?

In their own distinct ways, both the IRM and FMG reports draw attention to problems associated with displacement and resettlement. There are also clear indications that displacement and resettlement aspects of the SSP will weigh heavily in the judgement of the Supreme Court.

Notes

1. According to House (1980) the goal-based approach takes the objectives of a project as given and collects evidence as to whether these goals are being/have been/can be achieved. In the goal-free approach the set goals are not considered as given. The entire series of likely outcomes is evaluated including unintended side effects, positive and negative. The decision-making approach sees decision-makers as the audience to whom evaluation is directed, by taking their concerns and criteria as significant. Critical inquiry relies on 'disclosure' and, when properly executed, increases awareness of the significant aspect of a specific situation or project. The necessity of familiarity with the situation/project is greatly emphasised and the goal is to expand perception and not to arrive at definite criticism. Thus, critical inquiry is not akin to negative appraisal. Quasi-legal reviews are used to address controversial issues and are based on the assumptions that the partisanship of opposing sides reveals vital evidence. Finally, professional reviews tend to focus on quality of research, quality of services, and contribution of the project to regional economy and the state to gauge whether the said project is professionally acceptable.
2. In fact, Patton (1997: 352) approvingly cites an extract from an anonymous entry in a 1988 essay competition on 'Politics of Evaluation' conducted by the American Evaluation Association, which

reads as follows: 'Evaluation is NOT political under the following conditions: no one cares about the program; no one knows about the program; no money is at stake; no power or authority is at stake; and no one in the program, making decisions about the program, or otherwise involved in, knowledgeable about, or attached to the program, is sexually active.'

3. The team consisted of Bradford Morse (chairperson), Thomas Berger (deputy chairperson), Donald Gamble (senior adviser, environment) and Hugh Brody (senior adviser, resettlement).

4. Letter by Bradford Morse and Thomas Berger (IRM) to Lewis Preston, President, World Bank, 18 July 1992.

5. The IRM's report is voluminous (over 350 pages). It is based on written submissions made by different people and agencies, governmental and non-governmental; commissioned research by consultants on displacement and resettlement impact and hydrological and environmental impact; field visits by members of the team to the Narmada command area, downstream, submergence area, resettlement sites and discussions with project officials, policy-makers of the state and central governments, and NGOs and action groups.

6. Hydrology-related issues did not feature in the ToR of the IRM. However, it scrutinised the hydrological data, considering it the 'basis' and 'origin' of the environmental impacts. To ignore the hydrological estimates in the review would, in its words, have been 'irresponsible' (IRM 1992: 252).

7. In ARCH-Vahini arguments, the IRM had overlooked the 1983 resettlement mission of the World Bank, which included Thayer Scudder's report of 1983 which at no stage mentions 'the problem of information' let alone a complaint that information was adequate'. The IRM also completely failed to recognise the gains of an incremental approach to resettlement. 'The incremental strategy...has continued to secure policy reforms not only for Gujarat but also in Maharashtra and MP. Morse-Berger...have not understood that the best hopes both theoretically and in practice lie in the adoption of this strategy. If this doesn't succeed, nothing ever will succeed...[T]his is not only true for India and other Third World countries, it is also true for the affluent countries from where they came' (ibid.). The IRM's assertions were also questioned by Arch-Vahini; encroachers in most villages in Maharashtra and MP were entitled to 2 ha of land as their encroachment—prior to 1978 in Maharashtra and prior to 1987 in MP—has been regularised and government resolutions in Maharashtra prior to the IRM extend land-based benefits to post-1978 encroachers (ARCH-Vahini, letter 19 September 1992 to David Jones, Associate Director OXFAM, UK).

8. The letter begins with the remarks that the IRM overstepped its terms of reference and failed to recognise development as the basic right of human individuals, including tribals. The arguments in the letter address and refute major issues in the report and cover resettlement policies; rights of encroachers; rights of major sons; resettlement capabilities; resettlement packages; resettlement implementation; cut-off dates for major sons; canal-affected people; downstream studies; as well as environmental issues of siltation, waterlogging, salinity and drinking water supply.

9. NCA has three sub-groups, on resettlement, environment and construction. The resettlement sub-group is headed by the Union Secretary of Welfare and the environmental sub-group is headed by the Union Secretary of Environment. Both sub-groups have officials from all four states and experts from research, academic and non-governmental institutions. The sub-groups meet periodically to monitor and assess progress and to sort out inter-state problems.

10. Those voting against the management proposal were directors from the US, Canada, Germany, Japan, Australia and the Nordic countries. Together with directors from the South, those from Austria, France, Netherlands, Italy and UK supported the proposal (Appa 1992: 2580).

11. There was apprehension in some quarters that the Bank's withdrawal could mean stepping up the efforts to complete the project, possibly by compromising standards set by the Bank and through unacceptable measures. For instance, see two newspaper articles, one by Damandeep Singh in *The Pioneer*, 1 April 1993, and another by Rajiv Shirali in *The Economic Times*, 18 April 1993.

12. Appa (1992: 2580) also implies this view of the Bank when he argues that, for the Bank, the SSP is a sound investment because it is money loaned to a reputable borrower, unlikely to default on repayment.

13. Statement by Bimal Jalan, then Executive Director to the World Bank, cited in Damandeep Singh (*The Pioneer*, 1 April 1993).

14. Cited by Rajiv Shirali, *The Economic Times*, 18 April 1993. A senior Gujarat government official, when asked about the Bank's withdrawal, was sarcastic in his analogy: 'Having married for twelve long years it started finding faults with the wife [the project]! Because the neighbours kept saying that the wife was flirting. When you have made up your mind to divorce, you start finding fault with the wife saying that she is not cooking well, wearing expensive sarees and so no. The Bank's attitude can only be called cowardice' D.T. Buch, Narmada Planning Group, SSNNL, interview 5 May 1995).

15. The members of the FMG were Jayant Patil, Member, Planning Commission; L.C. Jain, Former Member, Planning Commission; Vasant Gowarikar, Former Adviser (science and technology) to Prime Minister of India; Ramaswamy R. Iyer, Former Secretary, Ministry of Water Resources, and V.C. Kulandaiswami, Vice-Chancellor, Indira Gandhi National Open University.

16. Eleven items were listed in the issues discussed at a meeting between the Central Minister of Water Resources and the Narmada Bachao Andolan on 29–30 at June 1993. They were: (1) resettlement—information on submergence areas and number of people affected, resettlement packages and compensation payment, forcible evictions, resettlement plan and implementation, improvement in packages, baseline surveys; (2) environment—studies, conditions of clearance and *pari-passu* conditions, tribal culture, poverty and deprivation, afforestation, catchment area treatment and flora and fauna; (3) hydrology—availability and specific utilisation; (4) drinking water supply at Kutch and Saurashtra—basis for claims, diversion of funds for the SSP vis-à-vis water management funds in these areas; (5) irrigation efficiency—water-consuming crops in head reaches, equitable planning; (6) distributive justice—more water to central Gujarat and less to Saurashtra and Kutch, benefits to adivasis north of the Mahi river, implications of water use, poverty and unemployment considerations; (7) benefit-costs—exaggeration of benefits as per the NBA, misconceptions, considerations for benefits such as migration, drought, regional and national benefits; (8) alternatives—integration of major, medium and minor projects in a basin, master plan, suggestions of Ashwin Shah (discussed later in Chapter 8), reduction in displacement and submergence; (9) information sharing; (10) review—framework of the NWDT Award, needs of participating states, height of the dam, development versus displacement; (11) human rights violation—hearing people's views, police firing, arrests and beating, right to livelihood (Government of India, Ministry of Water Resources, Office memorandum F.No. 6/4/93-PP, 5 August 1993).

17. The report was also to be made public within one month of its receipt by the Government of India. The government reserved the right to refer these issues to the group for reconsideration.

18. Gujarat's responses were as follows: 'The consistent stand of the government of Gujarat has been that it will not participate in any deliberations regarding the Sardar Sarovar Project covering clearly settled issues and related matters. Further, the Award of the Narmada Water Disputes Tribunal is binding on all the parties and it does not provide for any review at any time till 45 years from the decision of the Tribunal' (letter of Chief Secretary Gujarat to

Jayant Patil, Convenor, FMG, D.O. No. NMD-1093-5-NN & WRD, 9 August 1993). Surprisingly enough, MP responded by stating that its representatives will not appear before the FMG as MP was bound by the provisions of the NWDT award (letter of Secretary, NVDA, Government of MP to Jayant Patil, Convenor FMG, D.O. No. 1138/ 2/143/27/93-1, 12 October 1993). Subsequently, during the last phase of the FMG proceedings, the MP government reverted its decision after a new Congress government came to power in the state (it was the BJP government in MP that had refused to participate in the FMG proceedings). It expressed the view that the hydrology and the height of the dam should be reopened for review.

19. The rationale for this conclusion has been dealt with in the previous chapter where we analysed the hydrological question in the context of uncertainties in project planning.

20. It should be noted here that an argument of similar nature (mentioned earlier while discussing the defense by project authorities of the hydrology question) was propounded by the World Bank although it used the idea of incremental development of the command area to support the logic that, for a long time (at least 15 years), the problem will actually be more water rather than less.

21. The Supreme Court reverted the issue to the FMG for its opinion on the issue. In its supplementary report, the FMG failed to arrive at a consensus. Of the four members (Jayant Patil, Convenor of the FMG had opted out of the group during the second deliberation) one member—Ramaswamy Iyer—was not convinced about the need to use the hindcasting method in the case of the SSP as data on actual observable flow were sufficient to merit consideration.

22. This involves 'off-take points, filtration and treatment to make the water potable, a delivery system (including pumping and pipelines), arrangements for pricing, billing and collections' (ibid.: 48).

23. The hearing of the public litigation petition filed by the NBA was before a three-member division bench of the Apex Court, comprising Justice Verma, Justice Bharucha and Justice Paripooran. Members of the bench have since periodically changed as senior judges have retired from services.

24. Based on the classification proposed in House (1980: 23). To that end both reviews may also be termed 'quasi-judicial' to the extent that they admit arguments for and against the project.

25. It is interesting to note here that except for Bradford Morse, the three other reviewers were part of the famous Mackenzie Valley pipeline inquiry in Canada—the Berger Inquiry report published in 1977 under the title 'Northern Frontier, Northern Homeland'. The proposed pipeline project in the early 1970s involved the building of a natural gas and energy corridor from the Canadian Article

to mid-continent. The inquiry report based on the testimony of native people was controversial for its 'pro-environment' and 'pro-native' recommendations, among which was a 10-year suspension of construction activities till land claims of ethnic minorities in the region were comprehensively negotiated. The recommendations went against powerful economic interests in the USA and Canada in the gas, oil and water resources of the Arctic region. The inquiry was preceded by struggles over land rights and was followed by a long-drawn-out negotiation between the Canadian federal government and various ethnic groups on land claims and 'aboriginal title' (some still ongoing). While this antecedent of the majority of IRM members may mean nothing to some readers, a question arises whether the 'pro-environment', 'pro-native' inclinations and values of the members played a role in their condemnation of the SSP.

26. When expected targets are not or cannot be achieved.
27. Pertains to dysfunctional iatrogenic effects, when benefits go to the wrong parties and when existing anomalies are aggravated by the intervention itself.
28. At the time of writing the report is still under consideration before the Supreme Court and is not yet in the public domain.
29. Informal discussion with NBA activists, NBA Delhi Office, 11 February 1996.

8

Displacement and Resettlement: The View from Below

Introduction

Reflecting on some displacement related risks in the SSP, this chapter addresses how affected people in the Narmada valley have responded to the marginalisation and suffering caused by economic and social losses. We ask whether a reworking of resettlement in the SSP is feasible, following the concerns raised in the Narmada movement as well as by the IRM and FMG. People's risk perceptions and actions in their economic and social contexts but within a dynamic environment of political protest and incremental policy changes are discussed. This provides the opportunity to understand the social and political contributions of risks, particularly how risk perceptions and actions can change with changes in the political environment of a project.

The principal questions raised are: (a) how do affected people perceive risks to themselves in displacement and resettlement; (b) why do some people resist displacement whereas others resist resettlement? and (c) have the different action groups in the Narmada movement represented local interests, marginalisation and suffering?

The primary data are drawn from informal discussions and interviews with local people and activists in the three sub-districts of Alirajpur, Kukshi and Barwani in Madhya Pradesh, which bear the major brunt of submergence effects. Field visits to 10 villages in this area and to five resettlement and rehabilitation sites in Gujarat augment the database.

Displacement Risks: Effects or Perceptions?

One of the most common approaches to displacement related problems is the 'risk model' propounded by Michael Cernea (1990; 1995; 1996) and popularised by the World Bank (1994a). The principal assumption in this model is that displacement (with no or poorly-handled resettlement) results in eight main risks of impoverishment: landlessness, joblessness, homelessness, marginalisation, increased morbidity, food insecurity, loss of access to common property, and social disarticulation. Cernea (1995: 252) observes that these are high-probability risks and 'will undoubtedly become real if unheeded, or can be avoided if anticipated and purposively counteracted'. If the warning served by the model is acted upon through proper policy measures, Cernea, after Merton (1957), considers that the risk model becomes a 'self-destroying prophecy'. In fact, the reversal of the risk model, i.e. countering landlessness through land-based re-settlement or homelessness through sound shelter programmes, helps in identifying exactly what needs to be done to avoid the risk of impoverishment.

The strength of the risk model is that it outlines the major adverse effects of displacement, 'reflecting the fact that displaced people lose natural capital, man-made capital, human capital and social capital' (Cernea 1995: 251). It also serves as a powerful tool with which to fill gaps in policy: 'the flames of resistance are often ignited not intrinsically by displacement's hardship itself, but because the policy vacuums and legal vacuums leave few alternatives to political struggle' (ibid.: 258). The Cernea model is a useful starting point from which to recognise major displacement risks to the people in the SSP reservoir. People's risk perceptions and actions depend on a number of factors. Oliver-Smith (1991) outlines some factors which influence people's risk perceptions—patterns of internal differentiation within communities, multifaceted relationships with the immediate environment and the state, the availability of local and non-local allies, and the quality of the resettlement process. While the influences of these factors are context-specific, the latter two, pertaining to the political representation of risks and policy reforms to minimise displacement risks, have interestingly acquired significance in a general sense.[1]

The active involvement of NGOs and social action groups in displacement issues has contributed towards giving displaced communities a voice, raising national and global awareness of their problem, and building a radical critique of the ways in which such projects are justified as being developmental. As a consequence of this increased politicisation, incremental gains have accrued to the displaced communities as project funding and executing states, as well as multilateral agencies, have come under increasing pressure to formulate policies or to modify existing policies on displacement and resettlement (GoI 1994; OECD 1991; WB 1994a). For a fuller understanding of people's risk perceptions and actions we thus need to (a) take cognisance of the fact that different people (men, women, rich farmers, land-less, indigenous people and oppressed castes) may perceive their risks differently; (b) incorporate political representation of risks, the roles of action groups, their definitions and politics around uncertainties and risks; and (c) incorporate incremental policy changes in resettlement and the dynamics that they produce for risk distribution. Incremental policy changes maintain risks for some, minimise risks for some and generate opportunities for others. They also distinguish risks from displacement and those from resettlement, resulting from implementation lags and failures.

The meaning of 'risk' in the Cernea model is 'danger', echoing the formulation of the term in Giddens (1990) and Beck (1992, 1995). Notwithstanding the cautionary note around this usage,[2] both these scholars equate risk with danger, see it as pervasive in modern society and social life, and analyse how it is publicly produced and how it relates to resources and rights. Following this usage, the Cernea model translates the meaning of risk into danger of impoverishment in displacement. As a 'warning model' it cautions policy actors about the dangers of adverse displacement effects. Yet this 'danger' or 'risk' is not a static and fixed set of experiences and knowledge for affected people, as is implied in the model. That is to say, the dangers of homelessness, landlessness and joblessness and therefore impoverishment, which people perceive and act upon, are dependent on knowledge, information and experiences of people.

Although accommodating concerns in policy gaps, the Cernea model equates risk with certainty; that is, there is every possibility that displacement will result in impoverishment, the experiential dimension of this 'risk' is far more nuanced and necessitates differentiation between not just 'risk' and 'certainty' but also 'risk' and 'uncertainty', given the degree of knowledge, information and experience that people have at their disposal at any given point in time in the displacement–resettlement process. We need to differentiate among several situations—ignorance, uncertainty, risks and certainty in terms of people's access to knowledge and information. From risk as effect in the Cernea model, we thus call attention to risk as perception that influences action.

In theoretical terms, ignorance is a stage of 'no information'. Dependent on the context, this stage may be of shifting duration. Democratic contexts may substantially reduce this duration compared to non-democratic contexts. The very presence of this phase also implies that project planning is often a non-participatory process. The same is true for 'uncertainty' conditions where people have some information about displacement without adequate information on the nature of impact and/or resettlement entitlements, if any. Uncertainty may drive people towards resistance, at least theoretically; that resistance could be intense as it is not possible to make probability calculations of losses and gains with inadequate information. It is under 'risk' situations that such calculations are possible. People weigh their chances and risk either resettlement or resistance. This risk situation may also be a dynamic one with action groups and policy actors defining 'risks' in terms of their interests and influencing perceptions around it (Figure 8.1).

We refer to Beck's insight on the differentiated nature of risk perception: 'every interested party attempts to defend itself with risk definitions...the urgency and existence of risks fluctuating with the variety of values and interests' (1992: 32). Beck also argues that calculations of losses and gains are influenced by cultural norms of acceptance and legal frameworks of assigning compensation (Beck 1995: 43). When people (or groups) perceive

Figure 8.1 The Displacement–Resettlement Arena—A Dynamic Environment

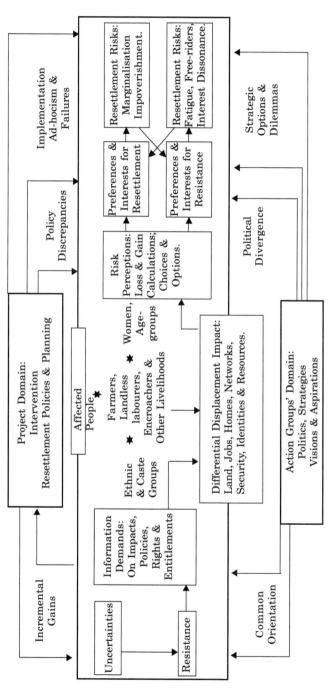

Source: Dwivedi 1999.

their risks to be more than what is culturally acceptable, or when action groups redefine what risks are acceptable, resistance to displacement will be acute, as also demands for change in public/state assignments of liabilities and compensation.

Sardar Sarovar: Differential Interventions, Differential Impact

As already seen, the adverse social impact of the SSP is spread across the reservoir area, the command area[3] and downstream of the dam.[4] We now turn to displacement in the reservoir area.

The reservoir area is spread over the states of MP, Maharashtra and Gujarat, where 80% of the population lives in rural areas and depends primarily on agriculture. A total of 245 villages in the basin are affected because of the reservoir and the Sardar Sarovar: 193 in MP, 33 in Maharashtra and 19 in Gujarat. The submergence area can be divided into two distinct socio-ecological settings. Villages in the lower hills of the river basin (upstream of the dam) spread over the three states are inhabited by adivasi groups, mainly Bhilalas, Vasavas, Bhils, Tadvis and Powras.[5] While affected villages in Maharashtra and Gujarat have a near-total adivasi population, approximately 65 villages in MP fall under this category. Adivasi villages are marked by a sparse population and subsistence-oriented livelihoods heavily dependent on rainfed agriculture, livestock and forests. Although land-lessness is extremely rare in these areas—almost every family owns some revenue land and 'encroached' forest land (locally called *nevad* or 'new')—the marketable surplus generated is negligible. The collection and sale of forest products (gum, bamboo and wood), fishing and seasonal migration to the Nimad plains in MP, and to Gujarat for wage employment, form an important part of the survival portfolio of these communities.

In the plains of Nimad, upstream, in Kukshi, Dharampuri, Barwani, and Thikri *tehsils*, the affected area consists of approximately 120 villages. Densely populated largely because of the fertile land, the Nimad plain is the 'bread bowl' of central India. Over the last ten years the region has undergone dramatic agrarian transformation, mainly because of a 'pump revolution' which has brought large parts under irrigated agriculture. Major crops include banana, sugarcane, cotton, papaya, chilli, soyabean and wheat. The Kanbi-Patidars, migrants from Gujarat, and in a few villages Jats, migrants from Rajasthan, are the big landowning castes in this area, their landholdings varying between 10 and 40 hectares. The middle peasantry, also largely of the Patidar caste (although Rajputs, Yadavs and Ahirs and in a few villages Bhilalas also belong to this category), own land ranging from 4 to 10 hectares. Landless farm labourers are mostly adivasis and Dalits, constituting about 50% of the population.[6]

Ignorance and Uncertainties

Although project appraisal began in the early 1980s, information about the nature and scope of displacement available to people in the reservoir area has been scanty. In Chapter 5 we noted the findings of a 1987 study conducted by MARG in 26 adivasi villages in MP which showed that in none of these had the MP government given any notice under Section 4 or Section 6 of the Land Acquisition Act (1894).[7] It was only in some villages and to a few farmers that the mandatory notices under Section 9 of the Act were issued. Even these were not read out or explained. People in the submergence zone who were to be affected remained largely ignorant about their entitlements and rights.

Paradoxically, while ignorance and uncertainty prevailed at the local level, policy advancements were taking place at the regional and national levels. By then, the NWDT had elaborately spelt out the provisions for resettlement, assigning liabilities and responsibilities of state governments as early as 1979. Compared to then existing practices, it had redefined established norms of resettlement: ruling out cash compensation, treating adult sons as separate families, seeking to offset major displacement risks pertaining to land, homes, health, education. In class terms, however, it addressed only the landed class who lost private revenue land in the reservoir area. Compensation for loss of jobs and other means of livelihood in the reservoir area were not explicitly considered by the NWDT. It was the involvement of the World Bank in the project and pressure brought on by affected people, activist groups, monitoring and evaluation organisations and consultants, which resulted in a compensation package for the landless and 'encroachers'.[8]

But as noted earlier, even by 1986–87 when construction work on the project was to begin, people at the local level did not have much information on land, forest, common property, job losses, i.e. who would be affected by the submergence and to what degree, and rights and entitlements as per the compensation schemes. The demands for information on extent and schedule of submergence, land availability, fresh land surveys and the assertion of the rights of those affected to settle in their own states prove this point. Major uncertainties prevailed in two areas:[9] (*a*) loss of lands—private and encroached forest and other government land; and (*b*) loss of jobs and livelihoods—landless agricultural labourers, other landless people, i.e. shopkeepers, fisherfolk, artisans, dairy farmers.

Information and Risks: Resettlement Policies and the Role of Action Groups

The early mobilisation of people by a few action groups—ARCH-Vahini and Rajpipla Social Service Society in Gujarat, Khedut Mazdoor Chetna Sangath (hereafter KMCS) in MP, and Narmada Dharanagrasta Samiti (hereafter NDS) in Maharashtra—yielded some information from project authorities on losses and rights among affected villages. As mobilisation intensified, further gains in resettlement policy accrued, although not uniformly across the three states. Moreover, their implementation was not trouble-free.

We noted earlier the substantially revised compensation package for landless labourers, which made them eligible for 2 ha. of irrigated land and other infrastructure amenities. As with landed households, adult sons were treated as separate families. Later, in 1992, the Maharashtra government modified its policies to recognise compensation of 1 hectare of land for adult sons, adult unmarried daughters and landless project-affected people, indirectly including 'encroachers'. The MP government did not undertake any such modification.[10] A major political obstacle in this regard from the MP government's point of view was that any revision of resettlement entitlements in the SSP would have a spiralling effect on all other projects. Implementation problems in resettlement also contributed to people's discomfort. While MP tried to compel people to move to Gujarat by offering substandard compensation, the Gujarat government ignored its policy obligation of 1987 to resettle those from MP who were interested in moving to Gujarat.[11]

The 1989 mission report by Thayer Scudder, while describing the Gujarat government's policies of 1987 as 'admirable' and 'excellent', noted 'innumerable and unacceptable implementation inadequacies', particularly with regard to Gujarat's efforts in preparing resettlement sites for MP oustees (Scudder 1989). Scudder noted problems in procedures of locating land of quality and quantity, lack of infrastructure amenities, high child mortality rates in resettlement sites, and lack of bureaucratic gestures of goodwill that would foster a favourable response among the people to policy changes (ibid.: 22). Far from creating goodwill, the Gujarat government has on occasion preferred to use force and violence to quash democratic forms of protest (see Srinivasan 1993; 1994). The two independent review groups, IRM and FMG, have both alluded to the deficiencies in resettlement, albeit in a different fashion.

Like the policy domain which showed gains, discrepancies and failures, the mediation of action groups was neither static nor uniform. We have discussed earlier how Vahini, KMCS and NDS, as part of the 'Narmada Action Plan', worked together up to 1987, and how the ensuing policy modifications of the Gujarat government led these groups to adopt different and conflicting politics. United in politicising uncertainty, the Narmada movement was divided over politicising risks.

Vahini directed its attention towards ensuring that oustees from the 19 adivasi villages in Gujarat were given land according to the policy gains of 1987. It assumed the role of 'careful watch-dogging of the implementation process' (Anil Patel, ARCH-Vahini, interview, 14 March 1996) and became actively involved in the implementation process, overseeing land acquisition through purchase and allotment in an attempt to minimise risks to those seeking resettlement.

In contrast, the NDS and KMCS combined to form a broad network, the NBA, declaring total opposition to the SSP. We have seen the several axes of NBA politics. Along one axis it highlighted the lack of surveys, inadequacies in policies and plans, non-availability of land for proper resettlement, terrible conditions on resettlement sites, and human rights violations through forceful evictions. A second axis expanded the definition of affected

people to include non-agriculturist occupational categories, i.e.
tailors, carpenters, shopkeepers, etc., in the reservoir zone as
well as people affected by the canal network and those downstream of the dam. Along a third axis it moved beyond displacement
risks to articulate the performance risks of the project, e.g.
whether the benefits proposed would accrue (Dharamadhikary
1993, Ram 1993); financial risks, e.g. whether money is available for the project (Thakar 1993); environmental risks, e.g.
whether reservoir-induced seismicity will affect the dam (NBA
1993); and distributional risks, e.g. whether benefits of the
project would be cornered by rich cash-crop farmers and industrialists (NBA 1992). Through the third axis, a fourth enables
the NBA to question all major large dam interventions (Manibeli
Declaration 1994, Declaration of Curitiba 1997).

Although working within the NBA network, KMCS focused
on risks to the adivasi people in Jhabua district of MP. While
recognising the specific nature of displacement related uncertainties and risks to these adivasis, KMCS has striven to integrate
these issues with other risks of material and cultural impoverishment faced by adivasis living in the area outside the submergence villages.

Risk Calculations: Some Scenarios

Two major displacement risks (using Cernea's terminology) are
plotted here: agricultural land losses and loss of livelihoods related to agriculture, in order to understand the differential impact
of displacement on different groups and to assess the ways in
and the extent to which resettlement policies (if properly and
uniformly implemented) can offset these losses. We shall also
give an indication of how people may subjectively calculate losses
and gains and how this may influence their actions.

The differential impact of the reservoir in MP on land loss
can be gauged from the estimates made by government agencies. Of the 193 villages affected, 49 villages lose more than 25%
of their land; 30 lose 26–50%; 14 lose 51–75%; 4 lose 76–90%; and
one will lose 100% of its agricultural land. In the remaining 144
villages, the loss of agricultural land is estimated to be less than
25% (NCA 1995). The break-up for these villages is as follows: 30
lose 10–25% of agricultural land, 54 between 1–10%. The remaining

60 villages lose homestead land and government wasteland (NCA 1995; WB 1995:17). The extent of submergence in the 75-odd adivasi villages (both in the hills and in the intermediary area) which lie closer to the dam is more than that of villages in the Nimad plains. According to estimates of the World Bank (1995), it is in these villages that the loss of land is more than 10%.

While these figures show the certainty of impact, they have been questioned on the grounds of inadequate land and village surveys. Although the figures (as information) form the basis of risk calculations, they are reason for suspicion in the Valley. For instance, in several villages there are cases where agricultural and homestead land situated at a higher altitude have been marked as being in the submergence area, whereas lowland areas and houses have been omitted from that area and therefore from the purview of compensation. Field visits gave the general impression that people did not trust the government's submergence figures. The more vocal and knowledgeable people pointed to the experiences of those affected by the Bargi project, where submergence was much more than initially claimed by the government.

Notwithstanding this general view, the impact of submergence is likely to be different across households within a particular village. Two rather obvious pointers are: (*a*) a household may or may not lose land; and (*b*) for those who do lose, the loss could be anywhere between 1 and 100%. Those who do not lose any land or lose only a part (less than 25%) face the prospect of having to remain in a truncated village or of owning unviable agricultural plots. In adivasi villages, the undulating terrain implies that hamlets in the lower reaches become submerged while others in higher areas are spared.

Government policies on the SSP consider those who lose more than 25% of their land to be eligible for compensation, with a minimum of 2 ha and a maximum equivalent to the prescribed land ceiling of the states, varying anywhere between 4 and 8 ha (see note 8). The compensation package is, therefore, unfavourable to those farm households which lose more than 8 ha of land. The incidence of such households is greater in the Nimad plains than in adivasi areas where a smaller percentage of households own more than 8 ha, even including 'encroached' *nevad* forest land. It is evident that those losing more than 8 ha (the upper peasantry) face the risk of marginalisation as ceiling limits disallow compensation beyond that amount. In a notional mapping, a household owning 20 ha of land will be left with 8 ha if losing 100%, 12 ha if losing 75% and 18 ha if losing 50%, after compensation. Furthermore, the latter two scenarios entail truncated holdings, one in the original village and the other in the resettlement site. For instance, a household losing 50% of a 20 ha plot is left with 10 ha in the original village (provided both plot and village remain viable) and will get 8 ha (the maximum compensation) in the resettlement site.

However, as many such households show a prevalence of joint ownership of land (a typical 'Indian undivided family'), the provision of land in the resettlement policy for adult sons (and unmarried adult daughters in Maharashtra) reduces the extent of marginalisation.

For those losing between 2 and 8 ha (middle and upper peasantry), the risk of marginalisation, in quantitative terms, is offset by compensation. Several factors enter the risk calculation of this category of losers. For instance, a Nimadi farmer losing 4 ha of fertile and irrigated land runs the risk of marginalisation as the quality of land currently possessed may be irreplaceable. A different problem arises for inhabitants of adivasi villages where the extent of dependence on, and loss of, *nevad* land is left out of the purview of compensation. In other words, an adivasi peasant dependent on 1 ha of revenue land and 3 ha of *nevad* is bound to face marginalisation. In all cases of 'encroachment', an adivasi displacee is entitled to only 2 ha as compensation in Gujarat and 1 ha in Maharashtra, no matter how much *nevad* land he/she loses under submergence. As in the earlier category, however, the provision of land for adult sons (and daughters) may reduce the extent of marginalisation.

For the landless and affected marginal farmers who own less than 2 ha (revenue and/or *nevad*), the loss of land, jobs and livelihood, common property resources and the security of the moral economy of the village, is offset by possible gains in land entitlement. If he/she moves to Gujarat, a landless labourer is eligible to 2 ha of agricultural land. An affected landless labourer wishing to stay in MP is entitled only to cash compensation.[12] The provision of compensation for adult sons (and daughters) is a critical variable in risk calculations for landless people.[13] The offsetting of livelihood loss through land-based compensation, however, might entail other forms of risk for landless people—the probability of their being able to mobilise resources to farming as a livelihood depends crucially on the quality of land and on adequate support services.

If losses in terms of agricultural land and jobs are given the stature of major risks facing people in the reservoir area, it becomes evident that the upper peasantry, particularly in Nimad, bears the greater risk; despite the compensation package, farmers in this category are the biggest losers. Pressure by activist groups and consequent policy changes may create opportunities for marginal farmers and landless agricultural labourers.

Nimad Plains: Risks and Responses[14]

The Landed People: Marginalisation as a Major Risk

Nimadi farmers have close associations and networks with the local administration and politicians, business and trade. The information at their disposal generates a certainty regarding loss of productive agricultural land and inadequate compensation. This certainty coexists, however, with uncertainty about the possible extent of their losses. Consider the following statement by a respondent in village Chikhalda:

> We have seen in earlier projects that the losses are much more than what the government initially claims. Look at Bargi dam. The submergence was three times the projected figures. The government does not have the slightest clue as to where the displaced have gone. It says that out of the 350 ha of land in this village only 100 ha will come under submergence. But we feel that the entire village is going to come under submergence, either because of the reservoir or because of the back water effects (interview, Nirmal Patodi, 27 February 1996).

Knowledge about the experiences of others generates doubt regarding the extent of loss. The perceived unfairness in the compensation deal also enters into risk calculations. The response of a farmer is illustrative:

The farmers cannot buy much land with the compensation as they are not given the right prices. Suppose I have 9 acres of land and depend on it with my father and three families of my brothers. According to the R&R policy [of Gujarat] we should be entitled to 20 acres of land. But the [MP] government is not talking about 20 acres. Instead it is offering cash compensation for 9 acres. Even if I get cash compensation for 20 acres I can still buy some land no matter how high the price is. The fact of compensating only 9 acres is nothing but just cheating us. We have no choice but to resist.

The marginalisation in holdings faced by the Nimadi farmers is evident from the default of the MP government. The option of settling in MP, a legal right derived from the Tribunal Award, is curbed because the MP government's policies are extremely inadequate. By maintaining sub-standard policies, it has impelled Nimadi farmers to seek land in Gujarat. Some resourceful farmers from MP have managed to acquire quality land in Gujarat. For instance, in Kukra-Rajghat village approximately 20 landowning households have accepted compensatory land and have settled in Gujarat. Others from the village, however, had to return because the costs of moving were higher than the benefits. As one farmer in the village explained:

Our family had 13 hectares of land here but in Gujarat we got 18 hectares as compensation. From 1993, I tried cultivating for four seasons yet the land there was so severely water logged that the crop output was abysmal. I had even invested Rs 20,000 on seeds. We were drowning here only during flood, but there one gets drowned immediately. If the land quality would have been good there was no reason for me to come back. Others have been lucky to get good quality land there (group discussion, village Kukra-Rajghat, 28 February 1996).

Two points of reference may be deduced. First, the Nimadi farmers' ideas about the unfair compensation deal cause them to resist displacement. These farmers constitute the major support base of the Narmada movement, a fact corroborated in earlier works by Baviskar (1995) and Dreze et al. (1997). Second, notions about a fair deal in compensation are not necessarily linked

to its form, i.e. whether land-based or in cash. Rather, the calculation is based on the extent to which the compensation can
offset the risk of marginalisation.

Cash compensation violates the norms laid down by the NWDT.
Its adverse impacts are also well-known. Yet Nimadi farmers today
function in a market economy, cultivate cash crops, transact deals
in cash, and may not be vulnerable to its ill effects. There even
seems to be a willingness among some farmers to accept cash
compensation as long as they see it as fair. In some instances in
MP, farmers have accepted compensation in cash.[15] Used fairly and
as an additional option, it is probable that cash compensation can
potentially offset the risk of marginalisation for the Nimadi farmer.

In fact, the MP government has exercised this 'option' but in
a manner that is questionable. Instead of systematic, purposeful and participatory estimation of the value of land loss (and of
other forms of capital), it has preferred to target specific households in the Nimad valley to demonstrate the people's acceptance
of cash compensation. Cash compensation has been paid to some
people in upstream villages, whereas those in the lower basin,
who will be the first to face submergence, have not received any
compensation. This strategy is not likely to yield any significant
results. In game theory terms, whereas the majority resist, some
people might accept compensation as there would be strategic
advantages to be derived from acceptance. But the dominant
strategy—the best individual response regardless of the other's
strategy—for Nimadi farmers will be to resist as long as they see
the compensation, both quantitatively and qualitatively, as unfair.

The essential core of their resistance is related to the issue of
land marginalisation. This may indeed be a tapered understanding of displacement risks. For example, one would expect Nimadi
women's reactions to displacement to be more conservative in view
of their dependence on established kinship and neighbourhood
networks, on the one hand, and on their immediate resource base
on the other. The risk of social disarticulation may seem to be
more crucial to them than land marginalisation.[16] Nevertheless,
while these issues are extremely important, they do not seem to
filter through explicitly in people's resistance. This is not to say
that Nimadi women have not participated in the resistance. In
fact, the NBA has mobilised them against the dam. What is being
argued here is that women's perceptions of risk have not been politically constructed or cultured by the NBA as a major resistance

factor, despite the fact that the leadership of the NBA is predominantly female.

Scholars of social movements have argued that resistance to nuclear plants, hydroelectric and irrigation dams the world over are often ignited by issues of land-ownership (see Beck 1995). If this argument is considered, the risk of land marginalisation of the Nimadi farmer forms the basis of articulation of different forms of risk—economic/material, environmental, cultural, financial, technological—by different interest groups. Moreover, that articulation could even suit the material interests of a Nimadi farmer to ward off the risk of marginalisation. Consider the critique of the SSP by a farmer:

> Look at the expenditure of the dam. First they said it would be Rs 4,000 crore, then Rs 7,000 crore, then Rs 9,000 crore and today the government says it will be Rs 20,000 crore. The command area in Gujarat will get waterlogged and water may not reach Kutch and Saurashtra. The officials say water will reach there in 25 years. In our view one can make immediate arrangements of the water needs there if we tap the rain water through small dams and check dams. The costs of SSP is enormous and the benefits too little.

Faulty cost-benefit analysis is only one among a range of issues that Nimadi farmers give as reasons for resisting the SSP. Their involvement in the movement enables them to express a meaning of resistance that encompasses a non-participatory development process, the plight of the adivasis who face displacement, the despoliation of nature and culture. As one respondent said:

> … prior to NBA we did not understand what environment is. But now we know that the dam will pollute the water which will be unsuitable for drinking, generate disease such as malaria and land will get waterlogged. In the land of *Sant Mahatma* [referring to Mahatma Gandhi], SSP will submerge *dharma* [justice], *prakriti* [nature] and *sanskriti* [culture] (Shobha Ram Jat, village Bagud, interview, 25 March 1996)

It is difficult to say whether these wide-ranging reasons for resistance centre around land marginalisation (and the associated uncertainty and/or loss of social status) or whether the Nimadi landed class shares the NBA's critique of development as its active constituent. This issue may become fertile ground for research in a situation where fair and adequate compensation is offered to Nimadi farmers. It would then be possible to examine whether the resistance results from what Beck (1995) calls the 'narrow-minded property interests' of the farmer, or from disenchantment with the language of compensation.

It should be added that the MP government's approach to resettlement prevents such a situation from being created. In our discussions with some affected people, however, the 'language of compensation' was rather forcefully prioritised. Our observation corroborates that of the NBA leadership which, in the past (and in public fora), expressed concern over the fact that in pre-election meetings to which contesting candidates of various political parties have been called to express their opinion on the SSP in Nimad, discussion revolves around compensation packages and rehabilitation demands. It could be that the Nimadi farmers' resistance to the SSP is due more to their having been given a raw (compensation) deal and less to their conviction about the espoused causes of the NBA. To expect the Nimadi farmer to think beyond issues of fair compensation is to place the burden of fighting the 'development dystopia' on people whose interests appear more immediate and who actually have benefited from this model of development.

Landless Agricultural Labourers: At Risk, Perceiving Gains?

The inclusion of landless labourers in the compensation policy was a major achievement of action groups in the Valley. The credit agreements between the World Bank and India in 1985 and the 1987 Gujarat R&R policies included landless labourers as project-affected people. In 1992 and 1993 more concrete proposals for their inclusion and exclusion and the nature of compensation were formulated.[17] The Action Plan of 1992 of the Gujarat and MP governments also observed that while the landed class has been resisting resettlement plans, it is the landless category who have made use of the benefits of the compensation package.

In its 1992 report, the IRM considered landless labourers (along with encroachers) to be 'groups most at risks' (IRM 1992:

196). It notes, 'many of them told us they...feel strongly attached
to the social structure that supports them' (ibid.: 177) and that
'peoples lives are not guided, especially in the tribal and scheduled caste communities [the landless labourers come from these
communities], by an attachment to *profit* rather than to place'
(ibid.: 182). These were the groups 'who expressed the most poignant fears...', to IRM (ibid.: 196). However, Baviskar noted in
her study that the landless labourers in Nimad maintain a distance from the NBA and protest activities, even though they
also lose their livelihood and community (1995: 220). How should
one view these rather contrasting findings?

Like the landed classes, the choices available to landless
labourers in Nimad are constricted. To offset the loss of jobs, the
security of the village economy and dependence on common property resources,[18] they have to move to Gujarat for 2 ha of land.
As landless labourers their powerlessness makes them vulnerable
to complex processes of negotiating with the Gujarat administration for adequate quality land of their choice, proper housing,
and other amenities to which they are legally entitled. The MP
government has so far offered cash compensation which is inadequate. Even more serious, however, is that landless labourers,
mostly poor Dalits and adivasis, are extremely vulnerable to the
well-known adverse effects, including alcoholism, cheating, unpro-
ductive spending, indebtedness, of this form of compensation.

For some landless labourers, however, resettlement in Gujarat
has been a risk worth taking. Consider the case of Magan Onkar
who settled in the Khanda resettlement site with 40 other landless households in 1993. Magan and his brother received 4 ha of
land of average quality. The water quality is good, and available
for both human beings and cattle. The initial allowance of
Rs 5,000 that each brother received for two pairs of bullocks
was pooled. Half was used to buy proper agricultural implements because those provided by the rehabilitation authorities
were of sub-standard quality. The household now grows cotton,
maize, sorghum and pulses. People in Khanda pointed out that
the first transitional year was very difficult; most households had
to survive on a grant-in-aid of Rs 3,000 each. Some households
could only cultivate sorghum during that year. 'But that phase
is now a thing of the past. We are happy here,' stated a woman
respondent.[19] Other households who had risked resettlement

preferred to return to their village, having leased their allotted land to farmers of the host community. The preference for an assured income, whether from land lease or from wage employment in Nimad, rather than trying to farm on the compensation land, highlights failures in implementation that do not offset the risks of resettlement. Land allotment is necessary but not sufficient to resettle the landless labourers as farmers.

Nevertheless, the expectation of owning a plot of land among landless people who face displacement can be sensed in the Nimad valley, although their position vis-à-vis the landed class may prevent them from unequivocal expression of their interest. Nimadi farmers understand the sensitive nature of the conflicting interests and strive to keep the landless as allies. 'The labourers are with us in the struggle. They have participated in rallies and demonstrations. In any case they are an integral part of a settled village, therefore they also feel alienated to go and settle in a strange setting. In Nimad villages we live as brothers', one respondent explained. Others alluded to the certainty of income from agricultural labour and the uncertainties of income from land. 'Don't be surprised if you find more silver ornaments and buffaloes in the house of my farm labourer than in my house', said a respondent during an informal discussion.[20]

But consider what emerges (Table 8.1) from an informal group discussion with landless labourers ('monitored' by a local activist and the landowner), who were taking time out from harvesting work in Kadmal village.

Table 8.1 Conversation with Landless Agricultural Workers in Kadmal Village

Q What do you think of the dam?
A The dam is being built, what else!
Q Has any one of you accepted resettlement so far?
A (By the local activist) No. No one from Kadmal and Kaparkheda.
Q Why not?
A Because the government has not invited us.
Q Will you go if the government invites you?
A Yes. When everybody goes we will go.
Q Have you been participating in rallies and demonstrations?
A (By the local activist) Yes, they have gone to Bhopal and to Delhi.
Q Why did you participate in the rallies and what issues did you raise?
A We are illiterate people. We remembered when we went there and forgot when we came back.

A few features in this brief 'public transcript' are striking. There is nothing that openly challenges the claims of the landed in Nimad. But is there a feigned ignorance and amnesia, an act of self-deprecation, an indifference to the movement against the SSP and a carefully concealed interest in compensation? To be able to 'introduce in muted or veiled form [interest or disinterest of the subordinate] into the public transcript' is what Scott referred to as the art of 'disguise' (1990:138). Considering the unequal power relationships and a perceivable conflict of interests between the landed and landless in Nimad,[21] one could perhaps expect a hidden set of transcripts of the landless (also of subordinate groups, e.g. women, and of dominant groups) that come into play 'off-stage' (ibid.: 4). But to ably recover them in the Narmada Valley would require an ethnographic approach to fieldwork.

Yet consider the language that the landless use in a safer haven—the resettlement sites. Perhaps, in Scott's terms, the resettlement site does not qualify as an ideal 'off-stage'. Neither can one characterise the language used there as a 'hidden transcript'. The closest we come to Scott's formulation is a condition where the subordinates engage in 'the open declaration of the public transcript' (ibid.: 208), but from the safe distance of the resettlement sites and without any element of personal risk.[22] If the opinion expressed by some landless labourers who have successfully resettled in Gujarat is considered, it gives a fair indication of how, in their risk calculation, the landless consider the benefits of a ruptured social fabric. 'What did we get there [in Nimad]?' asked a woman respondent, '... humiliation and a wage of Rs 15 per day.' The hostility towards the landed peasantry in Nimad was expressed in the following terms:

The Nimadi farmers kept declaring that their land is infertile. On a number of occasions they got their area declared as drought affected, so as to wrest all kinds of concessions and relief from the government. Now they are paying the price of perpetuating such blatant lies. In 1979, when Morarji Desai [then Prime Minister] gave clearance to SSP he was under the impression that the land in Nimad was infertile. Now, no matter how much you scream that the land produces diamond and gold, no one is going to listen to you (group interview, Khanda resettlement site, 20 March 1996).

What leaves little room for doubt is the distance maintained by the landless from protest actions in the Narmada Valley and their perceived interest in land compensation. In villages lying in the intermediary area (where class contradictions are not yet fully visible), people admitted that the exodus to resettlement sites has mostly included marginal farmers and the landless (group discussion, Amlali village, 26 February 1996). For the movement leadership this interest in compensatory land poses problems of representation. In constructing coalitions of interest groups, it is critical to balance different interests. When interests are not just different but conflicting, movements often tend to privilege certain interests over others (see Dhanagare 1995). In NBA politics, interest conflicts are not obvious. In claiming to represent all affected groups in the reservoir and command areas, downstream and in the catchment treatment areas, and in generating environmental sensibilities against mainstream development, the movement leadership has stayed clear of representing particular interest groups. Furthermore, as long as the government's resettlement policies, packages and implementation continue to remain inadequate and insensitive, the problem of class contradictions in the Valley will pose no immediate problem for the NBA. But its leadership is apprehensive of the effect on the movement of a probable exodus of landless labourers.

Some landless people find it perturbing that the NBA discourages them from accepting resettlement; those who attempt to move out are isolated and are left on their own to negotiate with government authorities for their entitlements. And those who have managed successfully to resettle, despite all odds, are 'invited' to return to their original villages. As one respondent stated:

Medha Behn [the leader of the NBA] had come to us here at the resettlement site. Her purpose was to convince us to go back to our villages in the valley. She said, 'When you come there, we will ensure that you get a formal welcome in Chhoti Kasravat.' If we decide to go back we will be welcomed and even garlanded by the activists in the welcome meeting. But after the meeting, when Medha Behn and the leaders leave, we will be told, 'bloody mother fuckers, you are back' (Magan Onkar, interview, 20 March 1996).

Despite this 'open declaration' from some quarters, there have been cases of landless labourers who have returned to their original villages and to the 'humiliation and a wage of Rs 15', albeit after leasing out their compensation land to local farmers. Yet, considering the interest among landless labourers for land compensation, a question does arise about the 'benefits' of a ruptured social fabric that can accrue to the most vulnerable sections among the ubiquitous category of 'people' in the Valley.

Adivasi Villages: Risks and Responses

An understanding of the perceptions of displacement risks in adivasi villages in the reservoir area needs to be situated within the characteristic life world of the adivasis.[23] The legacy of long-drawn-out struggles over forests and land resources amidst their systemic appropriation by the state, industries and non-adivasis, implies that for adivasis displacement has been historically pervasive and continuous, as has their resistance to it. In this context, displacement from the SSP goes much further. However, while the displacement impact is more severe, the policies and promises of resettlement temper risk perceptions to some degree.

The weighing of resettlement risks against displacement effects among adivasis is influenced by their experience of powerlessness vis-à-vis the state (notwithstanding their struggles against it), the differentiation and relative power within adivasi villages, and the role and orientation of their 'allies' in struggles. The last factor is particularly evident in the SSP as the three action groups, all claiming to represent adivasi interests, reflect different orientations.

Loss of *Nevad* Land

Unlike in Nimad, the problem of uncertainty was far more acute in adivasi villages. Apart from the lack of information on the displacement impact to which we referred earlier, uncertainities were compounded by the threat of displacement from the *nevad* economy, crucial to the adivasis' survival portfolio. In the affected adivasi villages, households tend to cultivate anywhere between 5 and 30 ha of *nevad* land (see Baviskar 1995:149). In some households, the *nevad* land may be more than revenue land by a factor of 20 to 80 per cent (IRM 1992:177). In a situation of population pressure, the *nevad* economy offers security as new 'forest' land can be brought under cultivation.

Nevad land has been contested terrain in almost all states in India (see Ambasta 1998; Guha 1989). The state labels the practice as 'encroachment'; it is followed up by forest department officials and the police with threats of eviction, fines, harassment and bribery. Adivasi access to forest resources is also curtailed in the name of conservation and protection.[24] In Alirajpur *tehsil* (one of the worst affected by the SSP), the KMCS leads the struggle over *nevad*. In the politics of KMCS, *nevad* is seen as 'both an assertion of customary ownership by the community and an economic compulsion' (Baviskar 1995:178). Its activists adopt the issue of *nevad* land as an instrument of politicisation and mobilisation through which adivasi people are made aware of the structures of exploitation that surround them:

> The knowledge of the modern, alien systems (courts, police, forest department, etc.) overarching and governing the adivasi life world is extremely essential for adivasi survival and struggles. The Sangath's task is to give information on the working of the system, so that it can be effectively resisted (Rahul Banerjee, KMCS, interview, 24 February, 1996).

Of the 90 KMCS member-villages in MP, 26 are affected by the SSP. For the KMCS,

> SSP is an issue through which one can explain how local party politicians respond, how industry agents and big farmers react, how contractors are powerful. Adivasi people clearly see for themselves the operation of these forces; hence SSP is important for awareness raising among these people in the long-term (Amit Bhatnagar & Jayshree, KMCS, interview, 7 March 1996).

It is this understanding that has led KMCS to join the NBA in resisting the Sardar Sarovar Project. Their fight over displacement around the SSP manifests broader struggles over resources:

> Big projects, big science and big technology should be opposed. Local adivasis cannot benefit from them. They increase our despair. We will oppose this science and technology manifested in dams, mines, forestry and even conservation. The direction of development has been one where resources have been siphoned out of our areas. What we need is a science that backs local use patterns and knowledge. Whether it is mining, water or electricity, local people should be the ones to derive benefits (Shankar Tadawla, former president, KMCS, interview, 4 March 1996).

Resettlement Risks

In Gujarat, 'encroachers' were entitled to 2 ha of irrigated land. Vahini's approach to interest representation was therefore different, focusing on resettlement risks after adivasis in the 19 Gujarat villages had begun to move to resettlement sites. The complex processes of resettlement involved identifying project-affected people, arranging suitable land of their choice, obtaining it (through purchase) and transferring it to their names, ensuring infrastructural amenities in resettlement sites—housing, water, power, access roads, schools and other service inputs—grant-in-aid, agricultural implements, grazing land, etc. Anticipating serious deficiencies in the transitional process, Vahini devoted itself to minimising the risks of non-implementation.[25]

The trend of moving to resettlement sites is more noticeable in Gujarat and Maharashtra than in MP. This should be seen in the context of policy changes in the two states, land availability, agency mediation and representation, and the immediacy of submergence as compared to MP villages. Gujarat adivasis face submergence first, a comparatively favourable policy package, and activist groups such as Vahini that are oriented towards resettlement. Resettlement figures published by the Gujarat government show that 4,572 project-affected people (about 95% of the total in Gujarat) have been allotted 8,549 ha of compensatory land in the state (NCA 1995: 199). In Maharashtra, where opposition to displacement led by the NBA was extremely strong, acceptance of resettlement increased after 1993. In September 1992, the Maharashtra government recognised an entitlement of 1 ha of agricultural land for major sons, major unmarried daughters, and landless project-affected people, indirectly including encroachers within these categories. 'The level of conflict subsided dramatically as a consequence, and more families accepted resettlement in a span of three months (between October and December 1992) than had accepted resettlement in the previous two years' (Gill 1995: 237). Government figures indicate that, of an estimated 3,113 project-affected people,[26] approximately 1,800 from Maharashtra have been allotted land in Maharashtra and Gujarat. The Maharashtra government has claimed that land necessary for all affected people willing to resettle in Maharashtra has already been acquired. So far, 690 affected people in Maharashtra have received 1,380 ha of land in Gujarat and 1,090 have received 1,702 ha of land in Maharashtra (NCA 1995: 206–10).

This acceptance of resettlement in Gujarat and Maharashtra cannot simply be construed as evidence that the adivasis are favourably inclined towards resettlement. Policy changes can only once be an important pull-factor. People may show a tendency to risk resettlement because of continued uncertainty and/ or resistance fatigue (e.g. due to state violence). Resistance fatigue and/or policy changes often split united villages into people who want resettlement and others who prefer to fight. Internal differentiation among adivasis is equally important. Marginal adivasi peasants may be more favourably inclined to accept resettlement in the hope of securing private land and thus ending the uncertainty of cultivating small plots of *nevad* amidst threats of eviction. Better-off adivasi households may be more inclined to stay where they are. Besides, they are also treated more favourably by the administration as 'opinion makers'. Acceptance of resettlement is also possible if action groups mediate in the issue of the risks involved.

However, it is only when adivasis perceive that their major losses (*nevad* land, cultural identity, labour and other social networks, forest resources such as bamboo, ropes, gum, fodder, fuelwood and timber, river resources such as drinking water and fishing) are offset or minimised that some success can be attributed to the resettlement process. To that extent, opinions about benefits and costs of resettlement will depend on the resources that adivasi households command in the resettlement area.

For those adivasi peasants who have received quality resources as compensation, life in the new environment is 'comfortable'.[27] Consider the family of Jam Singh Ugrania from Jalsindhi village in MP. Together with his father, five brothers and two sons, he received 18 ha of land in the resettlement site of Golagamdi, Gujarat, in 1991. Ugrania compared life in both worlds:

> In Jalsindhi, life was extremely difficult. When rains failed we did not get food for 12 months despite cultivating 16 ha of land. If you go there you will know how difficult life is. People go to Gujarat to toil in the fields of Patels when crops fail. Here we cultivate cotton, pulses and sorghum. In 1995, sale of cotton alone fetched us Rs 25,000. This place has electricity, school, street lights, good access road, and bus connection. Drinking water is available. The hand pump had been installed even before we were here. In the initial years of shifting we did face some hardship. We cultivated both lands for some time. The only problem here is fodder which was in plenty in Jalsindhi (Jam Singh Ugrania, interview, 12 March 1996).

The Ugrania household is perhaps an exception rather than the rule. The fact that Ugrania Senior was the *patel* (traditional village head) of Jalsindhi, had six sons and two adult grandsons (each counted separately as a project-affected person), and was allotted consolidated plots, are major factors to be considered. *Patel* households and their immediate kin groups in some adivasi villages in MP have shown a predisposition towards compensation.[28] Targeted by the state to influence community opinion on resettlement, such households received adequate resettlement facilities that offset the risks of displacement.

Among adivasi 'commoners' the situation has been different. Some project-affected people command better resources than others: land quality, irrigation facilities, brick houses, water supply, are unevenly distributed.[29] Hari Singh Rathwa from Jhandana village moved to the Tarswa resettlement site. Of the 6 ha of land allotted to him, 2 ha are absolutely useless and covered with wild grass (locally called *dab*).

The *dab* sucks all the fertilisers and virtually nothing grows in that plot. We hired tractor for Rs 2,000 to remove the grass yet it made no difference. Our expenses last year was so much that we could not break even. This year we have already spent Rs 9,000. We had petitioned to the Development Minister who had come here for changing that plot but nothing has happened so far (Hari Singh Rathwa, interview, 15 March 1996).

At the Tarswa site, where a few affected people from Jhandana and Kakrana have settled, almost 20 ha of land is of extremely inferior quality. Eight households who received this land have returned to their villages. The infrastructure in Tarswa is poor. The tin sheds (a common sight in resettlement sites) are barely habitable; in the rainy season water seeps in, summer makes them unbearably hot and winter extremely cold. Despite four hand-pumps, there are complaints about shortage of water for animals. In many adjacent sites, drinking water has to be supplied by tankers, often irregularly. The primary school had to close after two months because there were only five students.

Resistance Risks

The differential resettlement risks experienced by those who have moved throws up severe deficiencies in the implementation process. However, only Vahini[30] has played a mediating role in addressing resettlement risks. The NBA leadership has highlighted the adverse conditions of projected-affected people in resettlement sites but as vindication of its political stand (i.e. resettlement is not possible) rather than as a systematic identification of areas that need reworking. In fact, some local NBA activists brand those who seek resettlement as *dalals* (stooges) of the government. The state governments have indeed operated through middle-'men' (use intended) who are called *poonarvasat sathis* (resettlement friends); the *sathis* are paid monthly allowances by resettlement agencies and seek to motivate and encourage people to accept compensation. For adivasis who risked resettlement, however, the epithet *dalal* seems to hide more than it reveals. This is what Hari Singh Rathwa had to say:

Since the last three years, I have left NBA. I have no relationship with the local NBA leaders in the village and they blame me for splitting the village on this issue. I had been to Bombay and Delhi in the rallies of NBA. In the early days of the movement we did raise the issues of information and participation, and the destruction of our ancient homes. But it became clear that the government will go ahead with the project. The dam is a reality. Realising this, people also began to move. I moved here when I saw that people in Suryaguda who had settled there for two years had got excellent land and were happy. The fact that I have been cheated is a different story (interview, 15 March 1996).[31]

It is possible to interpret Rathwa's words as expressing a fear of displacement on the one hand and resistance fatigue on the other. What is clearly evident is that the success of a few households managing resettlement generates interest in resettlement among others. As will be seen later, this interest in resettlement has been a major factor polarising political activism in the Valley.

More forceful criticism of the NBA's non-representation of adivasis seeking resettlement has been voiced. Gill (1995) cites the change in response among people in Dhankhedi village in Maharashtra where initial resistance was so strong that it was impossible for government officials to enter the area. According to him, in December 1991 residents of Dhankhedi had stated their intention not to oppose resettlement. The explanation given by local leader Uday Singh Bonda was:

We have realised that we are expected to fight against the dam, remain naked and keep performing our traditional dances. We are being deliberately encouraged to remain like this, so that our photographs can convince the world to halt the dam. We don't care if the dam is built or not. We want a good deal for our children. We have fought for the activists for years, but have got nothing in return. We are with you only if we get everything that is listed in the resettlement policy as our right. Until you give us what is rightfully ours, not even one person from this village will move.[32]

The sharpness of this response—recorded by a bureaucrat involved in the resettlement process—does not warrant dismissal as a 'pro-SSP' viewpoint. Even among the most ardent supporters of the NBA, resistance surfaces because of expectations of better compensation:

> We do not recognise paper-work. The government has to give us good quality land and proper information about it: whether it is forest land, if people are already residing in it and about other facilities we are entitled to. Instead it is giving us assurances in paper which is useless (Bava Maharia, village Jalsindhi, interview, 1 March 1996).

Baviskar notes the gap between the aspirations of the movement leaders and activists to frame the problem of displacement within a wider development dystopia, and the perceptions of 'people in the valley, both adivasis and non-adivasis, who understand the issue of displacement in a much more particularistic way' (1995: 222). How does this particularistic way manifest itself, what tensions could this create in interest construction and representation, and can people's perceptions be ignored (as false consciousness) or form the basis for reworking the 'displacement-rehabilitation problem'. These are hard questions that need to be considered.

On a sombre note, it is plausible to attribute the (poor) adivasi peasant's 'particularistic ways' of understanding displacement to the uncertainties that characterise her/his life world where risk-aversive or safety-first behaviour is compulsively adopted to negotiate between two realities—survival and deprivation. Most evident in production decisions, e.g. the preference for diversification of plots and mixed cropping, is the adivasi peasant's 'survival algorithm', to borrow Lipton's phrase, which drives her/him to seek security. To the extent that resistance is deemed to fulfil this aspiration, i.e. so as to be secure either within the original village without displacement or in the resettlement site after displacement, adivasi people can be expected to be part of it. If the resistance project does not yield these results or appears to fulfil other purposes, disengagement is probable. In this act of disengagement, risks of resistance may be overcome (by accepting resettlement) but those of impoverishment remain in the absence of policy and implementation improvements and activist group mediation. For some (like the Ugrania household) risks can be offset totally and for others they will remain high. The worries of a KMCS activist are worth noting:

> As far as the dam is concerned perception gained ground among the local people that it cannot be stopped. So they started moving out. They realised that the movement was no longer in their interest.... Instead of attempting to broaden our mass-base, we have concentrated on a mobilisation strategy that seems to exclude local people. Our rallies have more of the so-called prominent citizens than local people (Rahul Banerjee, KMCS, interview, 24 February 1996).

Problems of a debilitating mass base in the adivasi area notwithstanding, people in some adivasi villages in Maharashtra and MP remain strongly associated with the resistance movement. And what have been the benefits of resisting displacement? 'Without the NBA you would not have seen a single adivasi in this place' one respondent stated.

> We were told that we will be displaced in 1990. It is 1996 now and we are still here. In these six years the World Bank was thrown out, our problem was known all over the country and the case is being heard in the Supreme Court. It is because of this struggle that some people have got quality land as compensation. And because of the struggle we expect a better compensation deal (Biharilal Dawar, Kakrana village, interview, 29 February 1996).

Comparison of Perspectives and Strategies of Local Action Groups

We have earlier alluded to the diverse mediation strategies of activist groups to the displacement-resettlement problem in the SSP. A fuller mapping of the different meanings of risks to the NBA, KMCS and Vahini, their convergence and divergence, and the implications of their approaches to the displacement-resettlement problem is attempted in Table 8.2.

Table 8.2 Comparing Action Groups' Orientation to Displacement[33]

Action Group	Risk Definitions	Representation of Interest Groups	Strategic Positions on Displacement	Observations and Implications
NBA: Network of individuals and organisations inside and outside the valley; Outside middle-class leadership.	SSP-related Displacement risks; Financial risks; Project risks; Technological risks; Environmental risks	Diffused: All affected by the SSP—reservoir, canal, downstream, catchment area, compensatory afforestation programmes, secondary displacements.	No dam position. Successful resettlement not possible. Supports the demand for lowering dam height. Close association with the MP government on this issue.	Outside support (national and global) more than the submergence area. Mass base stronger in Nimad than in adivasi villages. Building networks on displacement and alternative development in the country and abroad.
KMCS: Organisation with mass base in Jhabua district (MP); Operations both inside and outside the submergence area; local adivasi leadership.	Adivasi-related Displacement risks of the SSP; *Nevad* risks; Impoverishment risks of adivasis outside the submergence area.	Specific: Displaced adivasis in MP; Adivasis in Jhabua district. Diffused: Support to the NBA on the SSP	No dam position. Critical support to the NBA. Felt need to work with displaced adivasis who would move up to the hills instead of going to resettlement sites after submergence.	Strong local mass base. Small in size. Broadening struggles to include risks to adivasis outside the submergence area. Building networks among adivasis; articulating a demand for a separate adivasi state.
ARCH-Vahini: NGO in Gujarat; Operates both inside and outside the resettlement area; Facilitates community organisation among adivasis.	SSP & adivasi-related resettlement risks; bureaucratic risks of non-implementation.	Specific: Displaced adivasis of Gujarat. Also some displaced adivasis from MP Displacement threats to adivasis in Shoolpaneswar Sanctuary.	Close association with the Gujarat government in the resettlement process. Critical support to the dam 'as a necessary evil.'	Vulnerable to government mechanisation. Has recently articulated dissatisfaction with the Gujarat government.

The cursory mapping indicates points of divergence[34] among activist groups, i.e. their nature, meanings of risks and strategic preferences, implying different methods of reworking the displacement-resettlement problem. The NBA's current position holds the problem to be 'unworkable'; considered together with all other risks of the SSP, the solution lies in the dam not being built. In the NBA's view, the most glaring problems lie in the resettlement component; the need for its strategic reworking, however, is given very low priority. It views resistance to displacement and interest in resettlement as opposite poles. To narrow the gap between poles would take the NBA a step closer to 'developmental' resettlement. To widen the gap would take it a step closer to the struggles for alternative development.

ARCH Vahini, on its part, supports the SSP, although recently it has publicly expressed dissatisfaction with the tardy progress on resettlement in Gujarat. It has even deemed its involvement in the resettlement process 'a mistake' (Anil Patel, ARCH-Vahini, interview, 14 March 1996). Despite admitting the 'mistake', Vahini holds the view that displacement risks reincarnate as resettlement risks if gains in policies are not followed-up in the implementation phase. Vahini activists argue that the priority of action groups in MP should have been to seek policy modifications and reduce implementation risks, rather than opposing the SSP on wide-ranging issues.

Whether such a strategy could have achieved more for the affected people is debatable. On the one hand, it can forcefully be argued that the NBA's total opposition to the SSP placed the displacement problem on the national agenda. In similar vein, Vahini's expressed disenchantment with its close involvement in the process shows the limits and problems of strategic co-operation with governments. On the other hand, development-induced displacement will remain a reality (until the visions of alternative development are realised) and resettlement will continue to be a challenge in social engineering. The need to rework the displacement-resettlement problem can hardly be ignored, both in the context of the SSP and outside it. The close association of Vahini with the state administration in Gujarat may not place it on a high ethical ground to articulate this viewpoint; however, the viewpoint itself can hardly be ignored.

The KMCS no doubt considers the SSP as an archetype of 'destructive development'; its activists support broader struggles for resisting such projects. Its roots in the life world of the adivasis, however, burdens it with the responsibility to address the pervasive uncertainties and risks that adivasis confront in their daily lives. Perhaps these roots (and an adivasi leadership) explain why this organisation appears to reflect more on the displacement-resettlement problem than do others in the NBA network. In strategic terms, that orientation is manifested in the debates that the KMCS has initiated within the NBA to incorporate the risks of adivasis outside the reservoir area. In this regard, some KMCS activists appear to criticise the NBA for its single-minded campaign to stop the SSP. The problem of adivasis risking resettlement, despite policy and implementation inadequacies, is deemed to result from the failure of a resistance movement to perceive and adequately represent peoples' interests.

Displacement–Resettlement in the Sardar Sarovar

Displacement and resettlement aspects of the SSP (reservoir) are therefore beset with problems, both at the level of policy and implementation. A sizeable (and powerful) section of the local people face the risk of marginalisation, and it is their collective strength that has kept a long struggle (taking the Nimad Bachao Andolan of 1978 as the starting point) alive in the Narmada Valley.

Indeed, government policies have improved over time. In some supportive quarters, the Gujarat government's resettlement package is considered a model for the rest of the country. To that extent, even the MP government's package, which falls far short of Gujarat, has aroused popular interest. The incremental policy gains create opportunities for the rural poor in the Valley, and interest among them is noticeable. Overall, however, the SSP's displacement–resettlement component is clearly vulnerable to criticism on two major counts: (*a*) *insufficient and inappropriate compensation*, despite offering a better deal as compared to past projects; (*b*) *inadequate and ad hoc nature of implementation*, despite sustained political pressure and concomitant improvement in institutions, policies and delivery structures.

Insufficient and Inappropriate Compensation

It is evident that affected people in the Narmada Valley understand and use the language of compensation. Whether this understanding can be characterised as 'a particularistic way' of looking at displacement is worth pondering. In the risk calculations of affected people, however, material losses are paramount, and hence they show some interest in compensation provided it is perceived to sufficiently offset losses.

Clearly, the package offered to Nimadi farmers is insufficient. Both the quantity and quality of agricultural land owned by a well-off Nimadi farm household feature in its risk calculation, since what is offered in compensation (particularly under the MP government's policies) falls far short of that calculation. Continued protest action is a risk worth taking for this category of farmers. Under such circumstances, the only acceptable option before the government of MP is to negotiate a better compensation deal with this section of the people.

In studies on past dam projects, a frequent argument is that the well-off among the affected people are better endowed to wrest concessions from the government, whether legally (court battles), politically (through political pressure) or administratively (through connections in high places). The implication of this argument is that better-off farmers can represent and protect their own interests in the SSP, whereas policy must be attentive to the security of more vulnerable social groups. While agreeing with the general thrust of this argument, we also need to consider that unless the articulated interests of the Nimadi farmers are addressed, the project cannot go ahead through democratic means.

For marginal and landless people, Gujarat's resettlement packages offer gains and opportunities. They can be said to be adequate in Maharashtra but are insufficient and inappropriate in MP. Cash compensation for land and job losses to this category of people in MP is inappropriate for the very same reason that it is appropriate for well-off farmers. In all probability the cash received will not be put to productive investment by landless and marginal farmers. The same applies more or less to adivasi villages, although elsewhere we have pointed out instances where adivasi farmers have invested the received cash compensation gainfully in private pump irrigation (Dwivedi 1997b). This category of people is also vulnerable to 'bureaucratic risks' such as under-payment. As a matter of policy, therefore, affected marginal farmers and landless agricultural labourers (in the entire Valley) should be entitled to land-based compensation.

For the self-employed, artisanal and service groups the loss is mainly that of clientele and of markets for their goods and services. In the resettlement policies they bear the status of landless people (different from agricultural landless) and are entitled to cash compensation. Since most people in this category retain their productive assets and skills, compensation in cash is appropriate.

Overall, the policy realm requires some immediate overhauling with regard to the form and amount of compensation. This applies to the three states in general, but to MP in particular.

Inadequate and ad hoc Implementation

Government institutions and agencies set up for implementing and monitoring resettlement in the SSP have been the best in the country[35] and have improved over time thanks to pressures generated from 'below' and 'above'. The SSNNL in charge of project implementation had to give up the function of implementing the resettlement component, which is now carried out by an autonomous government agency, Sardar Sarovar Poonarvasat (Resettlement) Agency. Monitoring of the resettlement component is the responsibility of the NCA which has its in-house experts. Its sub-group on resettlement is headed by the Secretary, Central Ministry of Welfare, the highest administrative authority on resettlement in the country. The Authority meets frequently and has officials concerned with resettlement in the three states as well as NGO representatives. Apart from this, independent monitoring and evaluation of the resettlement process has been carried out by reputed research institutes such as the Tata Institute of Social Sciences (TISS) in Mumbai and the Centre for Social Studies (CSS) in Surat.

Notwithstanding advances in resettlement policies and institutions, problems in implementation have undermined policy gains and sensibilities. A fundamental bottleneck to successful implementation is the lack of trust among affected people towards government agencies and officials. Policy promises of land-based resettlement should be followed up with participatory land identification, timely allotment and sustained inputs for its gainful use.

Whereas administrative problems abound in these areas, a few that relate to delivery capacity may be mentioned here. First, people's knowledge of land availability and quality in the area can constitute an important input for land identification. Second, administrative personnel in charge of resettlement should be suitably qualified. There are several instances in the SSP where technocrats and engineers are in charge of resettlement at different levels. What engineering skills does one need to implement a resettlement programme? Rather, the attempt should be to mobilise the relevant professional expertise of sociologists, anthropologists and regional planners. Similar expertise is needed to tackle the specific problems of women resettlers. Inputs of NGOs and action groups in monitoring resettlement, disseminating information, eliciting participation and monitoring quality control are equally critical for successful implementation. In this context, one may also mention that it is not enough to appoint renowned research institutes as monitoring and evaluation bodies. Their findings need to periodically filter through the implementation process. Finally, each and every extension officer in the field must be made to understand that implementation of resettlement is a long-term project. One was pleasantly surprised to see government officials in two resettlement sites in Gujarat during fieldwork. Resettlers in these sites confirmed that these were no chance visits and that the officials came quite frequently. In many resettlement sites, however, people's problems have remained unaddressed, even after repeated complaints. Sustainable resettlement requires that people's problems in resettlement areas are addressed with utmost commitment, failing which the resettled will feel encouraged to return and those wanting to resettle will be discouraged from doing so.

Despite such teething problems, the fact that resettlement has been successfully implemented in some cases leads one to ask whether a more aggressive approach could offset administrative bottlenecks and make success more universal. Policy changes may earn some goodwill for the SSP in the Valley, but improvements in implementation is the sure way by which to regain and build trust.

Of course, it has to be borne in mind that the resettlement 'discourse' does not consider cultural or psychological losses, nor does it foresee the long-term impact of displacement on people's lifestyles. It also tends to deflate the social energy necessary to explore creative alternatives to development and displacement. In that sense the resettlement 'discourse' legitimises the development 'discourse'. If suffering and risk in displacement constitute the core of the crisis in the SSP, how should we address that problem? Should we accept the SSP and the displacement that comes with it and focus on resettlement? Or should we accept the movement's critique of the SSP as 'destructive development', take note of its environmental despoliation, inefficiency, non-participatory process, tied aid, and search for more efficient, sustainable and equitable alternatives in the water and power sectors? The next chapter turns our attention to the changes that the SSP crisis seems to have induced in all these spheres.

Notes

1. It can be argued that political actions and policy modifications are not common characteristics in projects so as to incorporate them in a general model. However, two counter-points can be advanced. First, the list of large dams that have faced displacement related resistance is rather long—historically and globally. Hence, political factors obviously exist. Second, a number of these projects are (co-)financed by multilateral and bilateral donor agencies; not only does this attract international and national NGOs, but pressures from them have resulted in policy modifications in a number of projects (see for details WB 1994a).

2. For instance, according to Douglas (1992: 48–9): 'Probability analysis arrives at politics in the form of a word "risk". The word gets its connection squeezed out of it and put to the same ...uses as any term for "danger"... [I]t arrives at the moment in which it cannot deliver what politics most wants from it ... certainty.' Elsewhere she mentions: 'Risk is the probability of an event combined with the magnitude of the losses and gains it will entail. However... from a complex attempt to reduce uncertainty it has become a decorative flourish on the word danger' (ibid.: 40).

3. It is estimated that between 60,000 and 85,000 ha of land will ultimately be required for the 40,000 km canal network. Official estimates indicate that the majority of affected farmers (estimated

between 145,000 and 170,000) will lose less than 25% of their land (see Alagh et al. 1995: 305). The argument is that the remaining 75% of their land will get irrigation and hence the risks they perceive would be qualitatively different from those in the reservoir area. The extent and nature of impact and terms of compensation have been a matter of dispute. For an overview of the disputes see IIM (1991), Patel (1992), and IRM (1992). Government policies on canal-affected people have since changed and they have been brought under the purview of compensation.

4. Downstream effects will mainly be deprivation of livelihoods, i.e. fishing and river transport. At least 5,000 fisher families will be deprived of their livelihood (see IRM 1992; Soni 1993: 7). The proposal to resettle a section of the affected people downstream (especially those dependent on ferry services and fishing) in the reservoir area is still at a preliminary stage and has not featured in a policy package.

5. The term Bhil is used as a general category to denote different adivasi groups in the region, including Bhilalas and Vasavas, particularly in government records. In the hills, however, Bhils are a distinct sub-group, different from other groups such as Bhilalas and Vasavas.

6. The distinction between caste-plain villages and adivasi-hill villages somewhat conceals an intermediary area. Inhabited mostly by Bhilalas, these few villages have a relatively heterogeneous composition compared to the adivasi villages in the hills, both in terms of class and ethnic groups. A small class of farmers at the upper end of the spectrum own 10 ha of land, suggesting economic differentiation. For over 10 years these villages have had access to irrigation through pipelines, as in the Nimad Plains. For the main arguments in this chapter, the intermediary area—west of Barwani in MP and across the river in Akrani *taluka* of Maharashtra—does not warrant specific attention.

7. Under the Land Acquisition Act (1894), governments need to abide by the following procedures. In the first stage under Section 4, they have to issue preliminary notification that a particular patch of land is needed or may be needed for public purposes. Under Section 6, the government declares its intention to actually acquire land. Notice under Section 6 has to be issued before the expiry of one year of the notice issued under Section 4. In the third stage the government invites concerned persons to raise objections, if any. After due valuation of the property that is to be acquired, notice under Section 9 is issued within two years of the notice under Section 6. The final award is made under Section 11 and land is acquired under Section 16 of the Act.

8. The 1985 World Bank loan agreement with the three states and the Indian government incorporated landless labourers as eligible for livelihood compensation in the agricultural and non-agricultural sectors.
9. Although activist groups did demand resettlement of villages as units, perceived risks of socio-cultural disarticulation were tempered by feuds and conflicts between and within villages, hamlets and kinship groups (see Baviskar 1995: 126; Joshi 1991: 36, 43, 51) and within households (Thayer Scudder, World Bank consultant on resettlement, personal communication), which partially inhibited local assertions over socio-cultural disarticulation.
10. MP has been reluctant to adopt policy modifications and instead has asked for a reduction in the dam height (since 1993) so that the displacement impact can be reduced. For an overview of the contention over dam height see Shah (1994) and FMG (1995).
11. On this aspect see Scudder (1989: 23).
12. In Maharashtra, the landless agricultural labourer is entitled to 1 ha of land 'if the oustee moves with others' in the village (NCA 1995).
13. In fact, compensation for adult sons (and daughters) seems to be a critical variable in influencing people's decision-making. There has been consistent demand for the revision of cut-off dates for eligible children. The dates have been revised at least twice; the cut-off date is 1 January 1987 at the time of writing; all those who were 18 on or before this date are eligible to be treated as separate project-affected people.
14. A few 'testimonies' recorded during informal discussions and interviews with affected people threatened with submergence, those who have accepted resettlement, and members of social action groups are featured in these sections. These subjective responses are a set of voices (among others) in the Narmada Valley and allow for an interpretative understanding of how people perceive displacement–resettlement problems.
15. In village Kukra-Rajghat a few farmers have accepted compensation in cash (Mahindra Singh Solanki, village Kukra-Rajghat, personal communication, 26 February 1997). Also, during informal discussions, references were made to the long queue of farmers in the Narmada Valley Development Authority Office for cash compensation.
16. Following these risk perceptions to resettlement, two additional points may be stated. First, cash compensation goes to men and puts decision-making power in their hands rather than female members of a household. In land-based resettlement land is often given to male members. Maharashtra's policy on the SSP is an exception and should be followed as a rule. Second, in land-based

compensation male farmers tend to give almost exclusive priority to land quality. Some of the respondents in this study mentioned that village teams visiting resettlement sites in Gujarat consisted of men who did not bother to find out if water and fuelwood were available in the vicinity of the sites. However, specific risks to women do not imply their disentanglement from land. On the contrary, feminist scholarship has highlighted the important role of agricultural land in women's empowerment (see Agarwal 1994).

17. For instance, non-agriculture-based landless people from MP are not entitled to land-based compensation either in Gujarat or in MP, the latter offering them a compensation package of Rs 25,000 (their major sons are also eligible for the same amount). In the case of landless agricultural labourers, both Gujarat and MP hold the view (as per the Action Plan of 1992) that since many Nimadi landless agricultural labourers are from marginally affected villages, they will be entitled to R&R in Gujarat only if (*i*) their house was submerged, and (*ii*) land on which they were working was to be acquired by the SSP.

18. Jodha's 1986 study on common property resources (CPR) in dry tropical areas spread over 21 districts across seven Indian states shows that between 85 and 100% of landless and small farm households depended on CPR for food, fuel, fodder and fibre (Jodha 1994: 158).

19. Khanda even has a properly functioning primary school with regular teachers.

20. Farm income in Nimad is subject to production uncertainties. As one farmer explained, 'per acre yield of a given crop could fluctuate anywhere between 8 and 28 quintals'. Fluctuations in crop prices also make crop choices a speculative venture.

21. For landless labourers, while irrigated farming in Nimad assures them regular employment, the wages they receive are (not surprisingly) low. The minimum wage for agricultural labour is Rs 25 in MP, whereas the wages they actually receive range from Rs 12 to Rs 18. This is a well-known fact and has also been highlighted by Baviskar (1995).

22. Scott implies that the open declaration of the public transcript is best used in situations where the subordinate blow the lid 'on-stage' in the presence of the dominant for personal release, satisfaction, pride and elation but with the associated risk of victimisation and sanctions for such acts of open defiance (ibid.: 208).

23. This is not to argue that the adivasi world is homogenous. Economic and social-cultural differentiation exists between and within ethnic groups and within villages.

24. The digging of cattle prevention trenches (CPTs) by the forest department around protected areas is the most recent manifestation

of this exclusionist regime. While the rationale of such activities derives from income generation (adivasis gain employment from digging CPTs) as well as conservation objectives, they limit access to forest resources (including land).

25. It has since identified 300 project-affected people who have been allotted sub-standard compensatory land and has also voiced concern over the non-availability and/or sub-standard civic amenities in resettlement sites (Anil Patel, ARCH-Vahini, interview, 14 March 1996). Vahini has also demanded denotification of Shoolpaneswar sanctuary in Gujarat, considering it an infringement of the usufructory of local Vasavas who live in about 100 villages in and around the area. The sanctuary has been proposed to ward off criticism of the environmental impact of the SSP, particularly on wildlife. The Shoolpaneswar sanctuary (600 sq.km) is in effect an extension of the 150 sq.km Dhomkal Sloth Bear sanctuary to the reservoir shoreline of the SSP. Local Vasavas have protested this move and have formed the Gujarat Vanavasi Sangathana to resist this project (for details on this struggle see Dwivedi 1997a).

26. Belonging to Tadvi, Vasava, Bhil and Pawra communities.

27. In her ethnography on the Vasavas in Gujarat, Hakim (1996) shows how Vasavas demonstrate adaptation in the resettlement sites, seeking a new identity and cultural ways that are distinct from their original villages in the hills.

28. One respondent, active in the NBA, named some of the villages from which *patels* have accepted compensation: 'Akadia's *patel* is gone, Jalsindhi's *patel* has gone, Chilakda's new *patel* is going (the old *patel* who refused to go has since died). In Sakarja and Kakadsila the *patels* have taken land and houses in the resettlement sites but are still living in their villages' (Bava Mahalia, village Jalsindhi, interview, 1 March 1996).

29. During a survey conducted by the KMCS in some of the resettlement sites, women respondents reported having good crops that season. The Khadgotra resettlement site now has irrigation facilities and adivasi peasants grow sunflower, wheat and maize (Vania Bhoura, President, KMCS, personal communication).

30. Other local NGOs in Gujarat were involved in the resettlement process. Among them two prominent NGOs, Rajpipla Social Service Society (one of the first NGOs to demand a better compensation deal for Gujarat adivasis) now supports the anti-SSP movement, and the Anand Niketan Ashram at Rangpur (which took an active interest in resettlement), is currently involved in controversy over disreputable activities in the Ashram.

31. During the interview Rathwa revealed that he and others interested in resettlement have sought help from Vahini.

32. As quoted in Gill (1995: 253–4). Gill was the additional collector in Dhule at the time and notes that several other villages in Maharashtra, which initially resisted resettlement, later accepted it. According to him, the resistance was directly related to the earlier experiences of those resettled and was an important factor in effecting R&R policy modifications in 1992. Official figures (1993 data) on Dhankhedi show that out of a total of 97 affected people (22 landowners, 40 landless and 35 major sons), 89 had moved and 72 had been allotted land in the R&R sites of Simamli and Choupadva in Gujarat.

33. The information is based on informal discussions and interviews with activists as well as published material of these organisations, such as newsletters, petitions and correspondence.

34. The points of divergence also bid for possible interconnections: For instance, the KMCS transcends its locality via the NBA, whereas the NBA manages a base in the adivasi area because of the KMCS; Vahini and KMCS strongly highlight the dependency and rights of adivasis on forest resources and their vulnerability, despite different approaches to displacement and resettlement; the NBA focuses on displacement risks, Vahini takes over from there, focusing on resettlement risks; the middle-class leadership of the NBA fosters national and international networks, the local leadership of the KMCS fosters a mass base. (For details on the politics of the action groups, particularly the NBA, see Dwivedi 1997b.)

35. However, the SSP lags far behind when compared to planning and implementation in projects outside India.

9

Conflict as Change-agent: Reform in the SSP

Introduction

The SSP crisis has inspired changes along several frontiers in the form of negotiation, dialogue and reform. What are the varied processes at work which, in different ways, aim to minimise conflicts in the SSP and other such large projects at different levels, both in the short and long term? What have been the responses by state and civil society actors to the crisis and conflicts in the SSP, and how do these compare with each other? These are some of the concerns which are dealt with in this chapter. As will become evident, state actors seem more inclined to reform in the resettlement aspects whereas civil society actors are predisposed to search for alternatives that fulfil the values of equity, sustainability, efficiency and community control. The discussion also draws attention to the dialogical processes initiated between these two positions at a global level around the issue of large dams.

The data in this chapter are drawn from published and unpublished material of researchers, submissions before the FMG, campaign material of global NGO networks, published documents related to policy debates by the Government of India, NGOs and the World Bank, and relevant materials and proceedings of the World Commission on Dams.

Displacement and Resettlement: Incremental versus Radical Perspectives

We saw earlier that discourse on development by planners justifies the inevitable and necessary evil of displacement for the greater public good, where a few bear the costs of development, the compensatory principle being that if gainers gain more than losers lose, the gainers can potentially compensate the losers in a 'socially desirable' developmental intervention. That losers are actually rarely compensated is neatly sidelined as 'externalities'. Notwithstanding the promises that are made to affected people before displacement, political and policy environments have been inadequate at best, and absent at worst, for compensating for the lost livelihoods of those evicted. Authorities tend to view relocation as a project bottleneck that needs to be removed rather than addressed. In the words of Ramaswamy Iyer, a member of the FMG:

> Faced with a problem, the first tendency of the administration is to say that the problem does not exist; after some time (which can mean some years) the problem is admitted; and after further delay some kind of a solution is found... (FMG 1995: 131) (parentheses in original).

This attitude in the management of displacement–resettlement is the root cause of the untold suffering of those affected, as is illustrated unambiguously in the case of the SSP. However, the SSP crisis seems also to point to two broad sets of correctives. The first, a radical perspective which denies the language of compensation and outrightly opposes displacement. Expressed by forces such as the NBA, this approach considers the rupture of people's social, economic and cultural fabric to be a violation of their rights to life and livelihood. No compensation package, no matter how informed, can offset such a loss. Here, displacement is the most obvious manifestation of an unequal relationship between the proponents (and beneficiaries) of development and those who bear the costs. Moreover, the resettlement process is rife with structural and institutional impediments that predict its quantitative and qualitative failure. In this radical perspective, the problem can be solved only through imaginative ways of rethinking development projects, e.g. in alternatives for large dams.

The second, in contrast, is the incremental perspective articulated by those such as ARCH-Vahini which calls for minimising the effects of displacement through development resettlement planning. The focus is on policy and institutional reforms as well as on effective implementation through more participatory and democratic procedures. To remain legitimate in the eyes of displaced people, the incremental position must be able, at least minimally, to address their demands and perceptions:

> The demand of the oustees' organisations is simple and perfectly justified. All that they are saying to the government is: if you want us to give up our homes, abandon all we have and move out lock, stock and barrel, at least give us some tangible proof that your promises are sincere and that you mean what you say. Show us...what land you have acquired for us; show us what plans you have formulated as to where and how we are to be rehabilitated, what arrangements you've made to see to it that our relocation to some unfamiliar place will be accomplished reasonably smoothly and without undue hardship.... If you cannot do any of these, you have no right to ask us to move (Engineer 1990: 160–61).

The stress here is on making the displacement–resettlement process accountable and participatory. The orientation is to accept displacement as a necessary evil and to focus on 'damage control' through just resettlement. Both the radical and incremental perceptions have had their influence on the reform and dialogue processes.

Resettlement Reforms at the National Level

The first signs of policy-level reforms are to be seen in the draft national rehabilitation policy circulated by the Ministry of Rural Development in 1995–96. The draft is based on a 1988 document entitled 'National Policy on Development Project Planning: Displacement and Developmental Resettlement' drawn up by an informal body of NGOs called the National Working Group on Displacement. The Working Group included activists of the NBA (which was not yet formed) and was quickly dissolved when the NBA moved towards radical opposition to displacement in the SSP. The political position of the NBA allowed no scope for discussion on resettlement and the 1988 draft document was long suppressed and not 'released for a national debate' until six years later! (NBA 1994, press release, 26 July 1994.)

The policy vacuum on displacement became glaringly obvious during the SSP crisis. The growing awareness of displacement impact in the SSP and elsewhere, as well as the resistance movements around such issues, combined to fill this gap at the national level. The government's draft document, in noting these developments, bears a confessional overtone:

> Today the project-affected people are no longer in a mood to suffer displacement along with its concomitant attributes like occupational degeneration, social disorientation, pauperisation, loss in dignity and often getting cheated on the compensation amount, which serve to make the experience a trauma. This has given rise to protest movements, marked by growing militancy. An interesting feature of the growing protest movement has been the creation of a national awareness of the problem. The press, the activist groups, the social workers and the judiciary have combined together ... to build up a national consciousness. (GoI 1994:1)

A close look at the policy document indicates the 'lessons learnt' from experience and the broad directions in which the state seeks to take the reform process. First, it admits bureaucratic adhocism and the hardships that this causes to the displaced. The prevalent lackadaisical approach is evident from the fact that there are no 'government estimates' of the numbers of people displaced by development projects. The document relies on displacement figures produced by a private research organisation (Fernandes et al. 1989), a straightforward admission that the state has no idea whatsoever of how many people it has displaced in its projects. Second, the document admits that the absence of a national resettlement policy has caused the build-up of national and international pressure, particularly through judicial interventions by courts, and because of heightened international interest in issues of environmental and human rights. Third, the policy document substantially broadens the definition of displacement and the eligibility for compensation from its earlier restricted meaning of land loss. In the new meaning, 'the extent of displacement covers an entire system which is much wider than the loss of land reflected through the process of acquisition', covering 'co-dependents...share-croppers, tenants, landless labourers' among others (ibid.). Fourth, the policy document speaks of the principle of 'total rehabilitation' within which 'social, economic, educational, environmental, physical, occupational and cultural aspects' feature in the programme. While the rehabilitation programme is to be synchronised with implementation of the engineering aspects of the project, its costs are to be included in the total costs of the project. Separate outlay and budgeting for resettlement has been a major recommendation of resettlement experts (Cernea 1999), and it is heartening that the draft policy document incorporates these findings.

The state's preference to focus on resettlement reforms rather than on reforming water and power policies appears to stem from the reluctance of planners to subject large dams to serious critical scrutiny. They hold the general belief that the economy has benefited immensely from the contributions of large dams, notwithstanding the costs, and that such dams are necessary to provide cheap and clean power and water to satisfy growing industrial, agricultural and urban needs. In fact, the *raison d'être* of the draft policy on resettlement is an economic logic that anticipates pressures on land and water resources due to the new economic policies under globalisation, liberalisation and privatisation. This aspect is acknowledged at the very beginning of the draft document:

> With the advent of the New Economic Policy, it is expected that there will be large scale investments both on account of internal generation of capital and increased inflow of foreign investments, thereby creating an enhanced demand for land to be provided within a shorter time-span in an increasingly competitive market ruled economic structure (GoI 1994: 1).

The policy focus on resettlement at the national level runs parallel to developments at the global level in which donor organisations consider displacement and resettlement to be the main problem areas vis-à-vis large dams. In that respect, the 'lesson learning' at the national level matches similar exercises in the World Bank, a major donor of displacement-causing development projects. That the campaign against the SSP had a 'chilling effect on the Bank', according to the then managing director Ernest Stern, has not only led the Bank to be more circumspect with the funding of such projects but has also resulted in substantial refinement of its lending codes and policies governing resettlement (George & Sabelli 1994). In 1993–94, 'prompted by the Narmada experience', the Bank reviewed the resettlement aspects of all projects active during 1986–93. The main conclusions of that review are:

1. Detailed surveys, community consultation and participation, and effective planning are the key to success.
2. Satisfactory resettlement is easier when national policies are supportive.
3. Modest redesign can dramatically lower the numbers of people needing resettlement.
4. Resettlement should be considered right from the outset of project identification (WB 1994a, 1995).

A fuller discussion on the impetus for policy reforms at the World Bank follows later in this chapter. The reforms at the Bank as well as at the national level, however, reflect much-needed sensitivity towards displacement as a social problem. The extent to which resettlement reforms are able to minimise displacement risks in development projects remains to be seen. Furthermore, insofar as governments and donors expect and accept displacement as an integral and inevitable consequence of development, conflicts with those who mobilise for a 'radical approach' seem obvious. This is due to the conviction among advocates of radical approaches that what needs to be changed is not the resettlement policy *per se*, but the policies on water and power that underlie large dam projects such as the SSP.

Outline of Radical Changes: Thinking about Alternatives to the SSP

The Narmada movement has inspired intense debate around water and power policies and the large projects that embody such policies. Given the fact that large dam technologies have addressed hard benefits, i.e. water supply, irrigation, power and flood control, the debate on alternatives that promise such benefits at lower social and environmental costs becomes significant. This is all the more so in the case of the SSP where project authorities have repeatedly claimed that alternatives do not exist (or are less attractive).

In this context, lowering the height of the Navagam dam should not be considered as an 'alternative' but as a practical way of resolving conflict. To the Narmada movement, restructuring the dam height serves only as a strategic option and not as a cherished goal.

In its early days, some people in the movement proposed that 100,000 *small* dams be built instead of the SSP. This gave rise to considerable debate in the late 1980s and early 1990s among planners and irrigation engineers on the merits and demerits of large and small dams (see Dhawan 1990). The juxtaposition of large and small dams in 'either-or' terms led to a foreseeable impasse, with both sides stressing the merits of their respective choice. However, the debate did inspire some planners and experts to think about alternatives to the SSP.

Some of these proposals are discussed here. They are not blueprints, and are not backed by hydrological, financial or environmental surveys. Most of them at best articulate a set of principles and values and identify areas of reform. While some exclude the SSP altogether, others attempt to incorporate its existing structure (albeit with reduced dam height and canal network). Some of the proposals find more favour with the Narmada movement leadership than others. Some of the designs advanced are admittedly 'pre-feasibility' studies, requiring more thorough research and resources. Others rely on significant breakthroughs in techno-economic standards, socio-political institutions and perhaps even community behaviour. Together, however, they give some indication of what technical alternatives are conceivable to avoid the costs of large dam projects while retaining their benefits.

The purpose is not to evaluate these proposals in terms of their feasibility, but to use them to frame the discussion on 'alternative development' in more concrete terms than has so far been undertaken. In referring to the alternative designs we hope to highlight possible new approaches to water and power that emanate from the SSP crisis. The blurred nature of the new knowledge is acknowledged by some of the proponents themselves; their proposals are only *indicative* of what can be done. This is perhaps a serious qualification to the assertion of the movement's leadership, 'that cheaper, quicker, viable, equitable, participatory and sustainable alternatives to the SSP exist' (NBA 1995b: 75).

Eradicating Poverty from Gujarat in 15 Years

One of the first 'alternatives' proposed during the SSP crisis was by Ashvin Shah, a civil engineer from Gujarat based in New York. Shah diagnosed Gujarat's problem to be not drought or the shortage of water, but the shortage of means to store rainfall runoff (Shah 1993: 21, also see Sengupta 1993:14). Shah estimates that, as against the 105 MAF average rainfall in Gujarat, existing reservoir (large, medium and small) capacity in Gujarat is a meagre 8 MAF. 'With *large-scale* implementation of decentralised small rainwater harvesting schemes, it would not be difficult to collect 20% of the rainwater or 21 MAF' (Shah 1993:19, emphasis added). This implies that an additional 13 MAF (21 MAF – 8 MAF = 13 MAF) could be tapped for use. To this, Shah adds 14 MAF from what he calls 'estimated groundwater recharging or resources' and another 5 MAF from water collected in reservoirs outside Gujarat to which it has access through regulated release. In Shah's estimate, 40 MAF (21+14+5) can potentially be used without the Narmada water benefits. To say that the 9 MAF of water available to Gujarat in the SSP looks unimpressive compared to Shah's estimate is an understatement.

Shah refers to three scenarios in which the potential 40 MAF might not be achieved. First, severe conditions of drought could reduce this potential by half. Even then, Shah estimates that this reduced quantity yields a potential 450 gallons (1,680 litres) of water per capita per day to 40 million people, *if* distributed equally. Second, if Gujarat cannot tap rainwater (through small water harvesting schemes) for the potential 13 MAF, it can still have access to 27 MAF (i.e. 40–13). Third, during drought, if this remaining 27 MAF is reduced by half, say to 13.5 MAF, even then this scenario will yield 303 gallons (1,100 litres) of water per capita per day, *when* equally distributed (more than the SSP).

For Shah, 'the best approach to water supply in Gujarat is to promote groundwater recharge schemes with harvested rainwater and imported water (whenever justified) accompanied by groundwater control. Such a major decentralised state-wide effort can only be undertaken by empowering villagers and municipalities to plan, implement, operate and maintain their own schemes with technical and financial assistance provided at their request and to be paid for by the beneficiaries' (ibid.: 26). Shah outlines what he calls 'a sustainable alternative to the SSP' (ibid.: 51) in a three-phase planning for 15 years (see Table 9.1).

Shah's alternative project theoretically removes poverty from Gujarat in 15 years. In contrast, water from the SSP may not even reach the drought-prone areas of Gujarat in the next 15 years. The basic thrust of Shah's proposal is the need to restore the quantity and quality of groundwater. Dubbing groundwater as 'the lost lifeline of Gujarat',[1] Shah argues that groundwater depletion is not because of drought and shortfalls in rain but because of its over-exploitation for water-intensive industries and unsustainable irrigation practices (ibid.: 24–5).

Mainstreaming Innovative Features in the SSP

Nirmal Sengupta, an expert on traditional water resource use, has advanced a set of principles that could be the basis for alternative resource planning for Gujarat. Sengupta (1993) proposes an inversion of the planning processes and strategies featuring in the SSP. He notes the technical and managerial innovations proposed in the SSP, such as efficient water use through computer-regulated supply, augmenting water supply through tank and lift irrigation, and distribution through participatory water distribution cooperatives in a *warabandi* system. Acknowledging these features, Sengupta highlights their unproven character, 'just as unreliable and experimental as any alternative design', hence making for a 'fair comparison with equally untested alternatives' (ibid.: 4–5). Introducing his principles, Sengupta states, 'the SSP project design has used some modern techniques imaginatively, but more as patchwork for developing its need. As alternative

Table 9.1 Phase-wise Elimination of Drought and Poverty in Gujarat

Phase	Objectives	Technical Interventions	Economic Interventions	Social-Political Interventions
Phase 1 First 5 years	Immediate relief of drinking water problem in the 12,000 'no-source' villages. Immediate provision for supplementary irrigation for rain-fed farms	Rainwater harvesting. Small watershed development in Kutch, Saurashtra and North Gujarat.	Changes in pricing, financing and cost-recovery. Incentives for saving water, for waste recycling and energy conservation. Equitable distribution.	User-groups, institution building, organisational development. Policy-level changes. Awareness programmes in sustainable farming and energy generation.
Phase 2 First 10 years	Conservation of water from existing irrigation systems. Drought-proofing the entire state of Gujarat by integrating local and external water supply.	Increasing efficiency of existing Tapi and Mahi projects currently functioning wastefully and sub-optimally. Land restoration from drainage and salinisation problems.	Same as above. Shift in farming practices to integrated farming for crop, horticulture and biomass.	—
Phase 3 2nd year up to 15 years.	Distribution of river waters for restoring groundwater and degraded watershed. Poverty-free Gujarat.	Redesign Narmada project: limited storage reservoir behind the dam, with two-thirds of the capacity stored in the canal command through surface and sub-surface storage. Pumping water to Saurashtra, Kutch and North Gujarat using power from the redesigned SSP.	Same as above.	—

Based on Shah (1993: 51–3).

we are only suggesting *full utilisation of the development poten-tials of these techniques*' (ibid.: 15, emphasis in original). Table 9.2 shows the major features of his proposal.

Table 9.2 An Innovative Approach to an Alternative

Subject	Top-Down (SSP)	Bottom-Up (Alternative)
Source of water	Narmada water	All sources in Gujarat state
Planning region	Potential Command Area of the Project	The entire state
Strategic Initiative	Construction of SSD at Navagam	Collection and management of locally available resources
Secondary Strategy	Construction of distribution network	Integration of small units into watershed and basin plans
Challenges	Covering special areas with ponds, tanks, etc.	Appropriation of Narmada water

Based on Sengupta (1993:23).

The collection of locally available resources has to be under-taken through decentralised surface and sub-surface storage using micro-harvesting methods. Sengupta censures the preva-lent conceptualisation in which 'minor irrigation' is conceived as stand-alone work and in which 'rainwater harvesting' is viewed as a technique to be used only when 'water supplies bet-ter than rainfall are not available'. He argues for a 'bottom-up integrated approach' in which the *scale of intervention is big* and the *technology advanced*. The operating assumptions are:

1. That rainwater harvesting structures are not rudimentary earthworks but require considerable technical expertise. The significant technological advancement in this field in the last few years has to be incorporated.
2. That these structures need to be part of an integrated plan including many micro watersheds and leading to the devel-opment of a whole basin. The spatial integration should facilitate transfer of surplus water across structures, watersheds and basins. The scale of intervention is therefore not minor but major.
3. Ultimately, water from the Narmada can be utilised to aug-ment water resources in the state if we keep an open mind with regard to all technical options, including lowering of the dam height.

4. To achieve these objectives, 'complete reform of the depart-
   mental set-up of irrigation administration in the country is
   an absolute necessity' (ibid.: 16), as the preference for con-
   ventional structures (such as big dams) 'are in-built in the
   current administrative procedure' (ibid.: 7).

A Concrete and Holistic Technical Alternative

A major study on alternatives to the SSP has been undertaken
by a Maharashtra-based group which proposes a comprehensive
restructuring of the SSP and incorporates the ideas developed
by K.R. Datye, a renowned irrigation engineer, who has served
the government in different capacities and is well-known for his
work on water and power generation reforms (Paranjpaye &
Joy 1995). The main features of the proposal—hereafter
called the DPJ proposal, after its proponents—are highlighted
in Table 9.3.

The significance of the features of the DPJ proposal is:

1. It retains Gujarat's allotted share of 9 MAF by the NWDT;
2. it reduces the dam height from FRL 455 ft. to FRL 350 ft.,
   reducing submergence by 70% and displacement by 90%, and
   at the same time incorporating the construction already
   completed of the Sardar Sarovar Dam structure and canal
   network;
3. it promises to increase the service area of the project from
   the current targeted command of 1.8 m. ha. to 4 m. ha.;
4. it delivers as much as 78% of this water to the drought-prone
   areas of Kutch, Saurashtra and north Gujarat (as against
   41% currently allotted to this area from the SSP);
5. the estimated costs of the alternative are equal to (actually a
   little less than) that of the SSP;
6. it is based on the principle of equity, protecting household
   entitlements and laying the foundation for what it terms 'a
   dispersed industrial society'.

The DPJ proposal restructures the SSP into three parts: (*a*)
reorganisation of the central water conveyance system; (*b*) integration of local water resources with the (restructured) central
system; and (*c*) sustainable production of biomass surplus (combined

Table 9.3 Some Salient Features of the DPJ Proposal

Project Features	Objectives
Irrigation through dispersed storage by constructing barrages and canals.	Increase service area from 1.8 m.ha to 4.0 m.ha.
SSP storage system to feed other dispersed local storage.	Make 7.5 MAF water available (through local storage + Narmada water).
Hydroelectric power generation through pump storage reservoir facility at SSD and through a series of micro-units.	Make power available for lift and high land irrigation.
Integrated biomass production by changing cropping pattern and through integrated water harvesting, i.e. rain + local storage + exogenous sources.	Deliver water to Saurashtra and Kutch in 5 years. Every family of five persons to produce 18 tonnes of biomass on 1.5 ha service area (2 tn foodgrains + 2 tn firewood + 5 tn fodder + 3 tn fruit & timber + 6 tn regenerable biomass).

Requirements / Strategies	Impacts
A barrage near the mouth of the river, followed by a series of small barrages both in the command and at the tail end of the main canal.	Submergence of about 26,000 ha in the (new) command area in Gujarat, but reduction in submergence in MP as water requirements at the reservoir will be reduced to 1 MAF. (Reduction of behind-the-dam storage, dam height).
Garland canals in Saurashtra and Kutch, and Kadana high level recharge canal.	
Dredging of monsoon rivers and rivulets. Reclaim wasteland and distribute this land and water supplies among landless.	Supply of water for irrigation upstream of the reservoir through Kadana high level canal.
Formation of village water bodies to regulate water use.	Each household receives basic service (quantum of water needed for livelihood needs) and economic service (over and above basic service).

Sources: Datye (1993), Paranjape & Joy (1995), Ghosh (1995), IRN (1996).

with alternative strategies of sustainable energy use) for a 'dispersed, self-reliant' industrial society.

The proposal emphasises the equitable distribution of water while explicitly stressing cost recovery. It recognises that water is a scarce resource and that its efficient use and development is linked to recovery of operation and maintenance costs as well as capital costs. It thus combines rationing (supply at public cost but nominal cost recovery for basic services) and pricing (supply at higher economic rate for economic services) to limit the burden on the exchequer as well as to ensure the efficient utilisation of water.

The DPJ proposal advocates the development of 'regenerative agriculture' and 'energy self-reliance'. Outlining the significance of these terms in alternative planning of energy and water, Datye notes:

There is a radical difference in the function of external input including water input within regenerative agriculture as opposed to high input agriculture. While the latter depends exclusively on external input, regenerative agriculture requires first the stabilising and enhancement of primary productivity with widely dispersed availability of external inputs in small quantities but in good control over timing and quantum of delivery. Local surface and acquifer storage acquire much more importance in this perspective... There is an associated change in the energy perspective as well. Earlier systems had an implicit principle of avoidance of pumping, resulting thereby in large contiguous, concentrated commands with centralised administration. It has to be recognised that pumping is a necessary concomitant of wide, dispersed and equitable access to water for regenerative agriculture (FMG 1994, Vol: 2: 120).

Along similar lines, self-reliance in energy implies that it is necessary to optimise and limit external inputs, increase the availability of local renewable energy resources and of biomass surpluses in the first instance, and choose energy-saving alternative techniques for infrastructure development for the water and energy sectors (ibid.: 105). Processing bulk biomass such as wood, bamboo and fibre and the operation of biomass-based industries will enhance the coal-replacement value of biomass material and reduce consumption of coal-based products such as cement, steel, metals and petrochemicals.

With the reduction of chemical input for regenerative biomass production system, the enhanced coal replacement value of biomass and renewable energy based production as well as the current surplus of local thermal energy and electricity generation would match the need for import of energy or external energy related inputs. This is the stage at which dependence on external resources will be reduced so that a balance will be attained in the local economy and energy and commodity exchange thereby attaining energy self-reliance. (ibid.)

The DPJ proposal comes closest to what can be termed a blueprint alternative to the SSP. Admittedly, the architects of the model accord it pre-feasibility status and call for more detailed investigation. Unfortunately, the model lacks political backing: the NBA leadership distanced itself from the proposal and instead publicly expressed criticism and scepticism over its claims and benefits.

Power Alternatives to the SSP

The debate on the SSP has also generated alternative proposals to its power component. One advanced by Amulya N. Reddy and associates is significant as it sets out to disprove the official claim that the SSP is the least-cost option for power generation. Moreover, it also develops a critique of the 'energy-intensive' approach to development in general. Like the DPJ proposal that foresees potential increase in water availability and service area at a lower cost than the SSP, the Reddy proposal envisions that, with an equivalent capital outlay, a least-cost mix of alternatives can yield 65% more power benefits than the SSP. In fact, with much less environmental and social costs, the alternatives can yield over 4,000 MW of electricity.

The proposal is based on three questions related to power generation. 'Electricity for whom, electricity for what and electricity how (efficiently)' (Reddy 1993). It characterises the prevalent approach to energy as 'growth-oriented supply-sided consumption biased', i.e. GROSSCON, where the magnitude of energy consumption is deemed to be the indicator of development. This link is outlined in a simple equation:

Development = Economic growth = Energy consumption = Demand projection = Supply increase

Two major crises—financial and environmental—ensue from the current thinking that links energy to growth. On the one hand, a huge 'capital gap' has come into being between what is required and what planners can allocate in the power sector.[2]

On the other hand the strategies of planners (such as the SSP) have serious effects on the environment. Taking its cue from the experiences of some developed economies, the Reddy proposal argues for a reduced coupling of GDP growth and energy consumption, which need not be 'a luxury that can be enjoyed by the industrialised countries' (ibid.: 143). The experiences gained by these countries, i.e. decreases in energy consumption associated with an increased GDP, can not only be translated but also accelerated in India through 'technological leapfrogging' (ibid.: 141). The strategic choice is to adopt a new paradigm for energy, 'development-focused end-use-oriented, service directed', aptly called DEFENDUS. The DEFENDUS model attempts to incorporate almost all established options of demand management and integrates them with some supply-side alterations. Table 9.4 outlines its main features.

In the DEFENDUS model, the SSP in its present shape is not required for the following reasons: (*a*) it is not the least-cost option as the economics of the alternative energy options are markedly superior to that of the SSP; (*b*) these options can achieve power potential many times that of the SSP; and (*c*) the suggested alternatives are socially and environmentally more compatible.

The Proposed Alternatives: Changes and Challenges

It is difficult to evaluate the claims of benefits made in the alternative proposals. Whether the alternative proposals emerging from the SSP crisis will generate political mass in the future and accelerate the reform process in the water and power sectors towards more decentralised techniques and dispersed planning remain to be seen. Together, however, they emphasise the high social and environmental costs of projects such as the SSP.

The proposed alternatives are a radical departure from existing policies and practices as far as water and energy development and management issues are concerned. Not only do they propose technological changes, they also imply institutional and organisational changes. They are also not so antediluvian in orientation as to see virtues only in indigenous and traditional water harvesting systems. True, the proposals require investments in water and energy development in small/dispersed

Table 9.4 DEFENDUS Model for Power Alternatives

Options	Justification	Strategies	Potential Effect / Impact
Demand-side management	Proven large savings from energy efficiency. 1 kWh of electricity saved is more than 1 kWh of electricity generated. Opportunity to surpass the West through technology leapfrogging as wasteful infrastructure still not fully established in India (follies of early industrialisers avoided).	1) Rectification of irrigation pump sets. 2) Lighting improvements 3) Refrigerator efficiency improvements 4) Other strategies: industrial efficiency improvements, power plants tune-ups, reduction of transmission and distribution losses, and load shifting.	1) 539 MW saved. 2) 639 MW saved. 3) 232 MW saved. Efficiency improvements would lead to better energy services.
Supply-side options	Cheaper options than SSP in terms of Rs/KW. Reduce high social and environmental costs of SSP.	1) Modify pump storage scheme of SSP. 2) Modify gas-based power stations. 3) Thermal-solar cogeneration. 4) Wind power	1) Reduce temporary storage and hence submergence 2) 1,600 MW potential (in Western India Grid). 3) 2,500 MW potential (in Gujarat and Maharashtra). 4) 2,000 MW potential (in Gujarat).

Sources: Reddy (1993); Sant (1993).

projects rather than in large, centralised dams. Yet the scale of intervention sought is large and relies on technological advancements. The technological options advanced in the alternative proposals do not rely on a single centre controlling the entire system, but rather on dispersed sub-centres as part of an integrated system. The key factor that determines the need for such technological reorientation is to reduce or avoid displacement and environmental risks, as also to make allowance for more sustainable resource management and equitable distribution of those resources.

It is interesting to note the emphasis that these proposals place on institutional and policy changes. The reliance on such measures suggests that the radical reorientation of technology is a necessary but not sufficient factor to achieve the goals of sustainable, equitable and efficient development. The reforms require a combination of supply-side and demand-side interventions. Thus, groundwater recharges, increased efficiency of existing reservoirs, modification and innovations in renewable energy, may be considered supply-side interventions, whereas pricing policies and cost recovery, formation of user associations, incentives for saving water and power, improvement in service delivery, are demand-side interventions.

It is worth pondering whether the institutional and policy changes implied in the alternative proposals to the SSP could also not be applied in the project to make it more efficient and equitable. The project would no doubt still displace people but policy reforms in resettlement need to be accompanied by changes such as appropriate pricing policies, increased efficiency of existing reservoirs, power plants and service delivery systems, and the formation of user associations.

It is unfortunate that, at the national level, there has been little focused debate on the proposed alternatives, partly because project authorities view this as threatening the very construction of the SSP and similar other large projects, and partly because the Narmada movement has made 'resistance' its central programmatic concern and has not managed to back these alternative proposals politically.

As mentioned earlier, there have been debates on large versus small dams where battlelines have been so sharply drawn that no meaningful dialogue has been possible (see Dhawan 1990; Iyer 1990a, 1990b, 1998). What has not been possible at the local and national levels, however, appears to have been possible at the international level. The dialogue that has been initiated between dam builders and opponents, a process that is a direct outcome of the SSP crisis, is discussed at the end of this chapter.

Globalising Protest: A 'Standard (Anti-) Dam Narrative'

The Narmada movement, while focusing on the SSP and other projects in the Narmada Valley, sees all large projects as sites of struggle. The successful globalisation of protest in the Narmada movement has also meant that the SSP has become an exemplar of 'large dams' gone wrong. For the global alliance, the SSP crisis has meant a stronger struggle against power companies and global funders such as the World Bank. The latter has become a rather obvious target of this global alliance, having provided more than US\$ 50 billion (1992 dollars) for the construction of more than 500 large dams in 92 countries.[3] What is perhaps more significant is the global construction of a 'common script' on large dams that the SSP crisis has inspired and sustained.

As already discussed, the Manibeli Declaration (MD) of July 1994 called for a moratorium on World Bank funding of large dams; 326 NGOs and NGO coalitions in 44 countries endorsed the declaration. Manibeli was the first Maharashtra village to face submergence in the SSP, and the declaration was symbolic of the NBA's leading role in the then emerging global coalition against large dams. The MD clearly articulated what we call the 'standard global narrative' on large dams, which is used today to oppose large dams.

Large Dams: The 'Common Script'

1. *Benefits:* Electricity benefits go to transnational industry and urban elites; irrigation water supply is mainly meant for export-oriented agriculture. Specifically, benefits of procurement contracts go to consultants, manufacturers and contractors based in donor countries.
2. *Costs:* Large dams involve forced displacements. Between 1994 and 1997, 18 large dam projects would have displaced

450,000 people. Dams also cause the destruction of forests, wetlands and fisheries, and increase water-borne diseases. These environmental and social costs 'fall disproportionately on women, indigenous community, tribal peoples and the poorest and most marginal sectors of the population.'

3. *Policy:* The World Bank grossly violates its own policies on resettlement and environmental assessments. It has provided loans in excess of US$ 8 billion (1992 dollar rate) for large dams through IDA, the soft credit window which is meant for poverty alleviation.

4. *Decision-making and Alternatives:* The procedures adopted are secretive and unaccountable. Projects are imposed without meaningful consultation and participation with affected people, and without consideration for more cost-effective and less destructive alternatives. The Bank has consistently ignored alternative energy sources and options which include energy demand management, irrigation rehabilitation, efficiency improvements and rainwater harvesting.

5. *Project Appraisal:* Lessons from past experiences are never incorporated. Delays and cost overruns are systematically underestimated. Appraisals of large dams also make unrealistically optimistic assumptions about project performance. They fail to account for direct and indirect costs of environmental and social impact and costs to cultures.

6. *Project Evaluation and Decommissioning:* There has been no proper assessment of the impact of large dams and no mechanism of measuring the actual long-term costs and benefits. The issue of decommissioning large dams after their useful lifetime has never been addressed by the Bank in policy, research and planning documents.

In considering the SSP as 'a world-wide symbol of destructive development', the Manibeli Declaration (MD) called for a moratorium on Bank funding on large dams until:

(a) adequate reparations have been paid to all those evicted through a fund monitored by an independent agency;
(b) policies and practices are substantially strengthened so as not to force resettlement and to be participatory;
(c) an independent body evaluates all Bank projects and its recommendations are seriously implemented;

(d) the Bank cancels debt owed by large dams which are a failure in terms of costs and benefits; and

(e) project appraisal techniques are suitably modified and large dams are considered integral to 'a locally approved comprehensive river basin management plan'.

The response that the MD elicited from global supporters of large dams is worth noting. First, the International Committee on Large Dams (ICOLD), i.e. the Paris-based dam industry association, in its 14th Congress in 1993 at Durban offered a spirited defence of large dams through its keynote speaker, Nelson Mandela. Second, the Operations Evaluation Department (OED) of the World Bank conducted a desk review of 50 Bank-funded large dams. The review found that in most cases benefits have far outweighed costs, including costs of adequate resettlement programmes, environmental safeguards and other mitigation measures. It also held that the policies of the Bank on resettlement and environment have improved project performances locally and that the projects have had a mixed record in treating displacement and effects on environment. These findings led the OED to conclude that 'the Bank should continue supporting the development of large dams provided that they strictly comply with Bank guidelines and fully incorporate lessons of experiences' (WB 1996:1). It recommended a second phase, field-based evaluation which was accepted by the Committee on Development Effectiveness (CODE) of the Bank's board of executive directors. On its part, CODE urged OED to ensure that the second phase 'reflects the views of civil society, including those of private investors and non-governmental organisations' (ibid.: 5).

Movement against Large Dams:
The 'Common Script'

Three years after the Manibeli Declaration, in March 1997, the first international meeting of people affected by dams was held in Curitiba, Brazil. The Declaration of Curitiba (DC), 'affirming the right to life and livelihood of people affected by dams', endorsed the principles and demands of the MD. More specifically, it reiterated the need for an international independent commission to review all large dams financed and supported by international aid and credit agencies. Also, it demanded that national and regional agencies must also undertake independent comprehensive reviews of each large dam that they funded. These reviews must be subject to the approval and monitoring of representatives of affected peoples' organisations and the international movement of people affected by dams. The DC noted 'the process of privatisation which is being imposed on countries in many parts of the world by multilateral institutions [that] is increasing social, economic and political exclusion and injustice'. This, it held, was not 'a solution to corruption, inefficiency and other problems in the power and water sectors which are under the control of the state'.

If the Manibeli Declaration symbolised the role of the Narmada movement against large dams, the Curitiba Declaration signified the rise to prominence of the anti-dam movement in Brazil. The Brazilian Movement of People Affected by Large Dams (MAB) tells a similar story to that of the NBA on mobilisation over the Machadinho and Ita dams in the Alto Uruguay region.

In Brazil, it is estimated that over 1.5 million people have been displaced to make way for dams, and about 36 new hydroelectric power plants are proposed in an expansion of Brazil's power sector (WB 1994a: 3/3). In the past, particularly in the 1970s, dams such as the Salto Santiago and Salto Osorio in the South, the Passo Real dam in Rio Grande do Sul and the Itaipu dam along the Brazilian-Paraguayan border, have evicted numerous small producers (*atingidos* and *colonos*) with insufficient compensation.

The plans to build the Machadinho and Ita dams on the Uruguai were drawn up in the 1960s. In 1980, after the formation of the state electric company Electrosul (in 1968) and after conducting feasibility studies (1970s), the government planned to build 22 hydroelectric plants over the next two decades. The first dam in the plan, Machadinho, was cleared in 1982 and the second, Ita, in 1983.

Opposition to the dams was immediate. Initial demands for information were made through the Farmers' Commission, which later adopted the acronym CRAB (regional commission of persons to be affected by dams). With no information forthcoming from project authorities, CRAB questioned the need for the dams, criticised the authoritarian imposition of the project, rejected

the monetary compensation, and demanded land for land in southern Brazil. The last demand articulated the fears of the *atingidos*
of possible resettlement in the Brazilian Amazon in the north
(see Navarro 1994). Soon, total opposition to the Alto Urguai
Projects began with the slogan 'no to the dam'. Oppositional
strategies included organising *Romaria*, the Brazilian counterpart
of *yatra*, to project sites, community meetings and assemblies,
and the uprooting of stone-markers for submergence. Inter-state
meetings on dams were organised with the slogan 'water for
life, not for death'. Ideologically, the movement combined 'a vol-
untarist populism of liberation theology, with a polarising ethic
"rich vs. poor", with some notions basic to Marxism' (ibid.: 135).
Politically, it made brief forays into electioneering, campaigning
for candidates for the provincial legislature belonging to the
Workers' Party. More importantly, it sought the immediate reform of national energy policies.

Over the years CRAB achieved significant success; the two
dam projects were suspended because the Brazilian State reduced
investments. In its current incarnation, as MAB, the focus of
the movement seems to be on building international coalition
rather than on local mobilisation. The striking similarities between MAB and the NBA in terms of strategies, politics and
ideology can hardly be ignored. Of particular significance is their
inclination towards global coalitions and networks. As their local
bases dwindle because of the success they achieve in halting/
suspending projects, their predisposition to globalise their protest
bases itself on the strategic thinking that each actor in the coalition benefits from the others. This is evident from the words of
the Executive Coordinator of MAB and one of the organisers of
the Curitiba Conference, Ricardo Luiz Montagner:

[The Curitiba] conference will help raise international
awareness of the struggle for justice of dam-affected people
in Brazil and make Brazilians aware that people all over
the world are making the same demands as MAB (MAB
press release, 11 March 1997).

In demanding an international independent commission to
review large dams, the anti-dam movement perhaps took a step
backward from the radical assertions of the Manibeli Declaration.

Instead of reiterating the 'common script', the movement sought an objective assessment of the projects, as well as the setting up of a dialogical process in which standards could be negotiated and devised.

The World Commission on Dams: 'Learning from the Past, Looking at the Future'

The demand for a global coalition was met with the formation of the World Commission on Dams (WCD), in February 1998, with headquarters in Cape Town. Headed by South Africa's (then) Minister for Water Resources, the WCD has 12 commissioners representing different interest groups and constituencies, grassroots leaders, multinational hydropower companies, international NGOs, state government agencies, international dam-technocrats' associations, academic and corporate consultants and aid agencies.

The mandate of the WCD was to function for two years with the main terms of reference being: to review the development effectiveness of large dams and to develop standards, criteria and guidelines to advise future decision-making.[4] Donor agencies, industrial interest groups, governments, civil society actors including representatives of affected people and experts in relevant fields of knowledge, from water and energy resource development to displacement and resettlement, have a voice in the WCD. Over the next two years, the Commission was to engage in assessing a complex, diverse and controversial set of issues surrounding large dams, ranging from hydrology to cultural diversity, from financial appraisal methods to public policy on resettlement. As a venue for dialogue and communication on dam-related problem areas and policy issues, the WCD was expected to generate policy-relevant research data. Its proposed research activities were to be inclusive, participatory, and based on consensus. Its findings need to be acceptable to all stakeholders and to representatives of over 35,000 large dams around the world.

Perhaps the most interesting aspect of the WCD is that former adversaries meet as commissioners to shape its public profile and research agenda. Major controversial issues concerning water and energy resources planning, civil engineering aspects, river-basin management, environmental assessments and methods, economic and financial appraisals and, last but not least, affected-people's voices, resettlement and social policy issues, have been converted into research problem areas. The preference for an interdisciplinary and participatory approach and multi-criteria methodology have been reiterated. These are welcome first steps.

It would be interesting to observe how the WCD measures 'development effectiveness', given the sharply divided opinions around large dams. As we saw in the SSP, different interest groups have offered different interpretations of issues. Struggles have been waged over studies and their results. Assessment methods and standards have been the subject of controversy. Even independent review findings have been the subject of conflicting interpretations.[5] Underlying these conflicts over standards are conflicts of different interests. In a world body of adversaries, these aspects will come into full play, whether in measuring effectiveness or in proposing standards for the future.

For instance, participating actors in the WCD subscribe to conflicting interests and values. There are those committed to 'economic growth' and others to 'limits to growth'. The mistrust that these conflicting value positions generate towards each other was in full display during the early negotiation phase of the WCD. Negotiations became obscured over representations in the commission as a section of the NGO community threatened to withdraw from the process, alleging that the WCD was becoming a 'pro-dam' outfit. Although the 'balance' was later restored with the civil society groups gaining far better representation in the WCD, it was the turn of Government agencies to dub the WCD an 'anti-dam commission'. It was on this ground that, in September 1998, the Government of India refused the WCD permission to visit the SSP as part of its first public hearing in South Asia.

Similarly, participating actors may have conflicting interests and expectations in the WCD. For example, a powerful impetus for dam-builders to participate in a forum such as the WCD could be to seek the co-option of opponents and to smooth resistance to their pet projects through 'dialogue and research'. Conversely, dam opponents expect results and findings that show dam effectiveness in a poor light in order to set tough criteria for future river harnessing. Proponents of large dams may be expected to seek a softer line on such issues so that future prospects are not enmeshed in a complex set of 'do's and don'ts'.

The need to devise new standards and criteria for the future is expected to be no less contentious. An unreflective accommodation of all positions could only result in a set of complex rules, norms and procedures, perhaps toothless with no authoritative backing. Who is to ensure that these standards are applied? In the SSP crisis there have been conflicts over project yardsticks and norms. NGOs and local communities accuse aid agencies of flouting their own standards. Project-implementing governments, on the other hand, find the standards set by aid agencies far too high. In some instances they have accused donor agencies (from the developed world) of setting standards that are not followed even in their own countries (as in the response to IRM). For the aid agencies, however, the problem lies not so much in devising standards as in their implementation, which is the responsibility of project-implementing agencies.

At a forum where erstwhile opponents meet to negotiate, bargain and debate, these 'struggles' are bound to occur. Yet, for the global coalition against dams, the WCD provides an opportunity not just to devise new standards on dam building but also on reforms related to water and power sectors. An early NGO report on the WCD was optimistic:

The findings on the actual performance of completed dams would show that many promises made by dam builders have not been realised. Recommendations of internal standards for dam construction would provide a set of guidelines, which NGOs could insist be followed by both public and private dam builders. Recommendations on sustainable and equitable land and water management would also be useful campaign tools. Recommendations on reparations for people who have suffered because of dam construction and on ecological restoration could help provide justice to dam-affected people and highlight the costs of dam construction which dam-builders have so far avoided paying (IRN 1997).

The early optimism has been somewhat tempered by the denial of entry into Gujarat of WCD commissioners intending to visit the SSP dam site, affected villages and resettlement sites (*The Hindu*, 4 September 1998). The Gujarat government asked them to stay away from the state, questioned the *locus standi* of the WCD as a global body, and dubbed it an 'anti-SSP outfit'.

These setbacks notwithstanding, it can perhaps be stated that the formation of the WCD reflects the emergence of a global consciousness regarding the impact of large dams as developmental technology and the need for more dialogical approaches to conflict resolutions around them. From the point of view of the global coalition against large dams, these projects can only be constructed when everybody affected is involved in planning, when dams as options are weighed against all feasible alternatives, when all costs are accounted for and mitigated, and when every stakeholder benefits from them. Whether this is an improbable dream or whether there are possibilities for meaningful tradeoffs; and whether large dams will gain a humanist face or whether the WCD marks the beginning of the end for large dams, only time will tell.[6]

Notes

1. Parallel to the project authorities' characterisation of the SSP as Gujarat's lifeline.
2. A recent estimate points to the need for Rs.1,100,000 million investment in the power sector in the next 10 years for the potential generation of 100,000 MW of electricity. The government is in a position to mobilise only one-third of this amount (*The Hindu*, 27 February 2000).
3. Manibeli Declaration 1994.
4. At the time of finalising this book, the WCD was in its last six months of operation.
5. As seen in Chapter 6, the findings of the Morse Commission were welcomed by grassroots groups, rejected by government agencies and subtly challenged by the World Bank management. See IRM (1992).
6. For a theoretical discussion on conflicting perspectives in the context of development, see Friedmann (1992).

10

Development, Crisis and Change: In Conclusion

The Research Question

The central concern in this book is the production of crisis and change in development action, with a view to gain a deeper understanding of when and how state action on resource development generates incompatible experiences, interests and claims as its outcome. In other words, the concern is with the circumstances and processes that cause and sustain the crisis, the ways these forces work and take political form. This question has been pursued by examining the unfurling of the crisis in the well-known Sardar Sarovar Project, thus particularising development action in terms of context, actors and events.

The theoretical significance of this question stemmed from emerging critiques of state-led development action aimed at large resource redistribution and socio-economic transformation. Such critiques held that the crisis was one of distribution and legitimisation and was attributable to the development order, the constellation of actors, institutions and discourses that defined the parameters of development. In order to reveal what circumstances and mechanisms cause crisis and, in turn, how the crisis affects development action, we thought to probe the dynamics of development action against this backdrop and in a given setting.

The broad methodology of the study has integrated political action into questions of resource development, control and use,

placing specific focus on political forms, i.e. struggles, movements, and conflicts among actors, between and within the state
and civil society. We have emphasised how development action
has led to conflicting claims, how particular claims were privileged, institutionalised, and challenged over time.

The question was significant in view of the stalemate between
forces supporting and opposing the SSP. The stalemate served
as a classic illustration of the paradox of development action,
that is, the simultaneous production of crisis, conflict and resistance as its unintended outcome. Rather than subscribing to any
particular claim, we asked how competing and conflicting claims
emerged in the first place, what interests, values and experiences generate those claims, and how different actors mobilise
power and resources to sustain them. The hypothesis was that
irreconcilable interests, experiences and values will reflect a
crisis of standards.

These questions were developed in a wide setting comprised
of general conceptual and analytical tools on the one hand, and
particular circumstances in which the questions were to be pursued, on the other. The theoretical framework involved:

1. theoretical arguments for situating development action in a
   political environment or action field,
2. analytical tools to interconnect structures and actors in the
   making of the action field,
3. methods of anticipating actors in the action field,
4. concept of public action—conflictive and communicative,
5. tools to conceptualise conflicts in the action field, and
6. theories and perspectives on collective actions as particular
   forms of conflict in the action field.

Overall, the framework is an interactive approach that would
be sensitive to the influence of larger structures and relations
governing development action, but which nonetheless views
development as a field in which different actors mobilise power
and resources to influence its outcome. This theoretical framework offers a more nuanced understanding of development crises
as compared to the formulaic structural approaches that gloss
over the domain of action.

The contextualisation provided a vital historically-specific
understanding of development action by analysing views and

debates on conditions and tendencies in state–society relations in India. Our review of these inclined us to suggest that, although democracy conditions relations and actions in development, the tendency of dominant societal actors influencing development is particularly strong. Development action as a consequence was subjected to conflicts and struggles emanating from democratic conditions, as well as dominant economic and power relations over its outcomes in terms of benefits and costs. Thus, the contextual framework led us to approach crisis situations in development as emanating from the economics of dominance and the politics of democracy.

The Findings

The main findings pertain to three aspects of the SSP crisis: its making, involving circumstances and the mechanisms that caused it; the way it was perceived and experienced by various actors; and the political mechanisms and processes that sustained and deepened it. It must be reiterated here that, in methodological terms, these different aspects are deeply interconnected. The findings are therefore not exclusive of each other even while distinct. As Sayer (1992) has argued, to ask for the cause of something is to ask 'what makes it happen' or what 'produces' or 'generates' it. Causality implies mechanics. Furthermore, he stresses that whether a causal mechanism is actually activated or exercised depends on conditions whose presence and configuration are themselves contingent. Depending on the conditions, the same mechanisms may produce different outcomes and, alternatively, different mechanisms may produce the same empirical outcomes. This interconnection forms the backdrop of our findings.

The SSP crisis was conditioned by numerous conflicts and struggles. Development action on the Narmada generated competing interests, struggles over resources, conflicting values, collision of knowledge claims and a crisis of standards. Several actors contended in the crisis. The overall alignment of these actors revealed a complex spreadsheet defying matrices, such as 'state vs. people' or even 'pro- vs. anti-SSP persuasion'. While a broad mapping of actors into pro- and anti-SSP persuasion was useful, it masked the heterogeneity, tensions and conflicts within the respective categories: state actors conflicted with one another; action groups and NGOs clashed over strategic pursuits; interests of different categories of affected people collided; evaluators differed in their assessments.

The SSP generated intense struggles over resources. At one level, the development and redistribution of Narmada waters for irrigation and power brought into opposition riparian states which competed to gain control over the river waters and their potential benefits. Chapter 2 demonstrated how the Narmada waters became the stake in a bitter riparian conflict. Each riparian state sought to protect its regionally-defined interests through its government, parties and legislature. The federal and democratic conditions in the polity fostered the riparian conflicts. Contending states periodically received democratic mandates from their respective electorates to engage in the conflicts. Although the central government tried to mediate in the inter-state conflicts, it was legally inhibited by the federal rules of the Indian Constitution which mandated state governments to be the appropriate decision-makers and law-makers on water use and control. Consequently, as discussed in Chapters 2, 5 and 6, despite administrative mediation, and professional and judicial arbitration at the highest political levels, the logjam over the project's blueprint and shape continued even after 50 years.

In the contest among state governments over the sharing of Narmada water resources, displacement-related aspects and environmental issues did not feature prominently, nor did they find adequate political representation. Successive governments in Madhya Pradesh, however, periodically raised some of these issues out of electoral compulsion and the felt need to thwart Gujarat's efforts at controlling and using Narmada waters. Although there was truth in their claim that the SSP meant development in Gujarat and destruction in MP, their primary interest was to control and command as much of the Narmada waters as they could. It must be said, then, that the SSP became a bargaining chip to get other projects sanctioned and funded in MP.

At another level, the struggles over resources in the SSP involved land, water, forests and other means of livelihood. The conflicts involved the interests of a set of people who feared the loss of resources, livelihoods and lifestyles. However, the local arena appeared divided over risk perception and resistance action. In

Chapters 4 and 7 we have shown that the factors influencing perceptions of losses and gains in the local setting included changes in resettlement entitlements, goals and orientation of action groups in the movement, as well as differences in local economies and environment. People not only perceived displacement risks differently, but their perceptions underwent change over time. Likewise, their orientation towards resistance action also changed over time. Better-off farmers, who were the worst losers, were more inclined to resist. There was evidence to suggest that marginal farmers and landless labourers were less inclined to resist as they perceived some benefit in resettlement. Resettlement entitlements were unevenly distributed, however, raising doubts over their adequate and successful implementation.

Public action ensuing from the struggles to protect livelihoods, resource base and lifestyles in the Valley brought people's rights and entitlements into the arena of struggle. What had until the 1980s been primarily an inter-state conflict over water was transformed into struggles over resources, rights and meanings. The political form of the conflict took the shape of a movement that articulated these struggles at multiple levels—local, national and global—to oppose the SSP. The movement thrived in the opportunity structure provided by the inter-state conflict. Spearheaded by networks of action groups, NGOs and community organisations, the movement sought to protect civil rights and to fulfil the democratic rights of affected people.[1] Violations of rights—arbitrary, illegal and forceful evictions, and police action against the democratic organisation of dissent—were brought into public focus, along with the assertion and creation of rights to information, consultation and participation, just compensation and rehabilitation, and to livelihood and the use of natural resources and the environment.

Like conflicts among state actors, the movement showed conflicting interests and values. Organisations differed as to movement strategies and goals. Some focused on demands for the right to better resettlement. Others like the NBA preferred a radical agenda that widened the struggle to include rights to use and control environmental resources, rights to information and participation, and rights to protect cultural resources. The NBA became a movement for alternative development, seeking the more efficient, equitable and sustainable use of water in and around the Narmada Valley. The movement also indicted the underlying water and power policy through which projects such as the SSP are justified. In the process, the NBA generated a critique of large infrastructural projects in general and large dams in particular. Chapter 8 showed the main features of the anti-dam narrative constructed in the global articulation of the movement. As the anti-dam narrative gained in popularity and appeal at the local, national and global levels, dialogue and debates intensified around project and policy alternatives aimed at more sustainable, equitable and efficient water and power resources development.

The Narmada movement (in particular, its radical stream) strongly reflected some characteristic features of what we theoretically termed New Social Movements. It operated in a multilevel, multi-class network. Its language of resistance reflected struggles over resources as well as over values and visions, and bore a non-negotiable tone. Its political practice showed a tendency to transcend the local and to link up with national and global networks. This network politics enabled the NBA to transform the struggle from one of local peasants resisting displacement to a broad citizens' coalition seeking the demonstration of public purpose in the SSP and other such development projects. The political field was further widened through public actions over information, knowledge, norms and values. The environmental and economic impacts of the SSP, the institutional and methodological determinants and valuations of these impacts, and the remedial measures to prevent and minimise the adverse impacts, came into the ambit of the struggle. As seen in Chapters 4, 5 and 6, the public actions strongly contended the 'public good' in the SSP, and the information and knowledge struggles questioned its viability and desirability as a development project. A spectrum of civil society actors participated in these struggles: environmentalists, professionals, academics, journalists, lawyers, and communities of NGOs and movement activists.

The knowledge struggles in the SSP transformed the conflict over resources into one over how development is planned and perceived. 'Lower-order' interest conflicts over perceived losses and gains were engulfed in a 'higher order' resistance to developmental values and associated truth claims. The environmentalism espoused in the movement trod a scienticised domain, relying heavily on the mobilisation of technological and scientific expertise. At stake were the underlying principles, the rules of the game, and decision-making tools, procedures and processes: the manner in which they were devised and applied, what they revealed and hid, and whom they favoured. Unlike interest conflicts that were limited to the distribution of relative advantage (as in struggles for compensation), the higher-order conflicts in the SSP engulfed the underlying rationality and assumptions of development planning. From the economic uncertainties and assumptions in the project to its social and environmental risks and hazards, to its desirability and appropriateness, the knowledge struggles encompassed the means and ends of development.

The SSP crisis illustrated how planners had underestimated the political callings around the costs and benefits of the project. Their perceptions of development represented in the early project appraisals reflected mainly its economic gains. Fundamental changes accruing to people's lives and livelihoods, their environment and culture, and permanent and drastic alterations to the river ecology, drew the attention of planners only when brought into the ambit of political action. There was some evidence to suggest that project planners were inclined to debate, demonstrate and defend the public interest in the SSP. They also tried to plug gaps in assessment studies and ameliorative measures based on the emerging criticisms. The conflicts forced planners to adapt and reform policy regimes governing the SSP, particularly those concerning resettlement and environment. Nonetheless, these steps were too few and far between to set at rest the serious doubts that were raised regarding the manner in which the SSP was valued and the ways in which it distributed risks and resources.

The application of state-of-the-art planning tools by a battery of planners, financiers and their consultants notwithstanding, uncertainty continued to dog the project's benefits. There was sufficient evidence—both in the politics of the Narmada movement and in the assessments of independent evaluators—to establish that the planners planned the project on a 'minimal information, maximal uncertainty' dictum. Of course, the foreseen economic changes generated political support for the project among potential beneficiaries in Gujarat. Yet, by externalising social and environmental changes (both in planning tools and in political discourses), planners omitted far too many actors from the ambit of the project. Legitimacy of stakes remained unrecognised, liabilities and compensation were not properly negotiated, and information on rights and entitlements did not flow to stakeholders. In fact, there was conclusive evidence to suggest that project authorities preferred to politicise the project as being one for the people of Gujarat. The SSP conflict was projected as between those who favoured the development of Gujarat and 'anti-development' environmentalists. To that end, the authorities contributed to the crisis by closing channels of communication and debate and by showing the SSP to be the only answer to the water problems of Gujarat.

It was not merely a coincidence, therefore, that the SSP crisis set in motion dialogical processes at the national and global levels while the logjam continued at the project level. At the national level, the debate focused on the policy vacuum on development-induced displacement and resettlement. At the global level, the dialogue process yielded a research agenda to measure the effectiveness of large dams, to note best practices, and to develop better standards for the future. At the project level, however, there seems to have been no major breakthrough except that the final verdict of the Supreme Court of India was expected before the monsoon months of the year 2000. Over the past few years, the Court has facilitated the exchange of opinions and views among the conflicting parties. We have mentioned that it would be difficult to predict what the Supreme Court's verdict will be and whether that verdict will resolve the crisis one way or another at the project level. It can be stated with some confidence, however, that the verdict of the Supreme Court will bear most heavily on the capacity of riparian states to successfully resettle those displaced by the project.

Reasons for Research

Before assessing the significance of this study, it should be stated that it shares some common ground with earlier research on the SSP. In the same vein, however, one must also mention that comprehensive research work on the SSP is still at a very early stage. The major parallel effort undertaken during the course of this work has been that of William Fisher's edited volume entitled, *Towards Sustainable Development?* (1995). This is 'not one person's comprehensive view of the Narmada', but consists of 17 contributions that offer conflicting and contradictory viewpoints and values. In acknowledging the numerous institutional and individual actors involved in the Narmada controversy, classified as 'planners, funders, debunkers and resisters' (ibid.: 15), however, it criticises the simplistic projection of the Narmada struggle as being between the powerful development industry and the powerless people in the Valley. Neither the industry nor the non-governmental community is, according to Fisher, 'a homogenous field of thought or action...but]...is multilayered, polyvalent and often internally conflicted' (ibid.: 17).

Fisher frames the SSP as a complex case that could be used to 'understand development and resistance to it in the contemporary world' (Fisher 1995: ix). He also proposes a four-fold classification of opinions and views on the SSP—those planning and supporting the project; those criticising it on grounds of implementation failures; those criticising it for its general rationale and on account of 'fundamental flaws'; and finally, those who view such projects as an integral part of a larger oppressive development process. Clearly this differentiation of actors in terms of the levels of their criticism is useful in categorising the contributors to the volume. Equally importantly, it corroborates the merit of the actor-oriented methodology adopted in the study. Such differentiated treatment of views takes us beyond the simplistic 'big dams are bad' argument, as pointed out by Fisher (ibid.: x). It opens up to analysis how proponents and opponents defend their positions 'with both cost-benefit analysis and grassroots mobilisation' (ibid.: 8).

Fisher's book identifies two major problem areas in the SSP: the burden on development planners to assess costs and benefits accurately, and to compensate fairly for the environmental and social impacts. Somewhat predictably, in all the substantive sections in this book devoted to project benefits, resettlement and rehabilitation, and technical and environmental concerns and alternatives, these problems emerge as significant points of dispute among the contributors. Although the debate between opponents and proponents is set, Fisher seems more inclined towards an *'alternative third approach'* that focuses on 'what actually happens in the development process above and beyond the stated intention and goals of development planners' (ibid.: 39–40). The process of change, Fisher observes, 'doesn't happen only within the context of the goals set by the project, but also in the broader context that includes resistance to and attempts to redirect development intervention towards alternative solutions and alternative means' (ibid.: 40).

The extent to which Fisher's volume contributes to the analysis it generates—the alternative third approach—is open to debate. However, the book seems to be a prisoner of its own structure—it remains a collection of two contending views, i.e. of proponents and opponents of the effectiveness of the SSP as development. The third approach—to unravel development as actually practised—although recognised, remains unrealised. The present book on the Narmada crisis has attempted to advance the debate by developing the alternative approach nascent in Fisher's work. Instead of merely stressing the need for the alternative approach, we attempt to apply it. This is the major significance of our study.

For a start, here is a comprehensive understanding of the SSP crisis. The circumstances and processes that led to, built-up and precipitated the crisis are discussed, and the entire political process within which the crisis rose from an inter-state dispute over resource distribution to the culmination of a social movement is examined. This has revealed a complex interlocking of factors and mechanisms, the scrutiny of which leads to the offered understanding that the SSP crisis is the outcome of a political process that is unable to accommodate and negotiate the conflicts and struggles that it has generated.

To facilitate this understanding, this book adopts a pluralistic, interdisciplinary and interactive approach. We have scrutinised the SSP crisis in terms of its economic, social, environmental and political effects, using techniques such as economic appraisal, social risk assessment and conflict, and social movement analysis. Rather than explaining these efforts in systemic formulaic terms, they have been seen as being played out in the political process involving actor interactions and relations. This approach has shown how actors placed at different levels in the conflicts and struggles have contributed to the making and deepening of the crisis.

Second, considering that the study of development action has by and large remained evaluative, positivist and applied, the attempts in this book to go beyond merely evaluating development action in the Narmada as desirable or undesirable are of significance. Earlier studies, albeit in different degrees, approached the SSP mainly with an evaluative framework and methodology. The same is also true of studies that have focused more generally on large dam projects, the findings of which are limited to the objective of providing corroborative evidence of the badness or goodness of the SSP and other such large dam projects. The debate generated by these studies gradually froze with the periodic recycling of arguments for and against the projects.

This book has evaluated some aspects of the SSP and has looked at its benefits and costs and the proposed manner of their distribution. In the process, the possibility has emerged that the project would have adverse social and environmental outcomes. But rather than merely 'reproducing' this knowledge in the SSP, it shows how knowledge claims are socially and politically constructed and how legitimacy is sought or denied to each knowledge claim. In that sense, this research has not produced a critique of the SSP. Rather, it reflects on how the critique is produced in mechanisms like social movements in order to show that it contributes to the making of the crisis and is integral to its understanding. This is a significant step because it concerns a fundamental yet difficult problem introduced by this research, i.e. the problem of whether we are required to separate concerns about the validity of arguments from concerns of negotiating the overarching validity of mutually exclusive claims in the analysis of development action as in the SSP. That is to say, whether methodological tools of 'critical analysis' need to be different from those of 'crisis analysis'.

In our opinion, critical analysis in development action involves judging the validity of claims, arguments and statements made in relation to the action. It entails using methods with which to evaluate whether claims are true or false, accurate or inaccurate. Methods of seeking validation are many and may range from the use of (single) methods and techniques for the quantitative verification of outcomes to methods that judge the situational validity or relevance of objectives and even their underlying ideologies or social ideals (Apthorpe & Gasper 1996; Fischer 1995).

With regard to the SSP, scholarship has relied almost exclusively on critical analysis to judge whether or not the project is acceptable. It has been used here to evaluate claims as in the cost-benefit calculus or in the resettlement component of the SSP. However, the SSP case also provides evidence of mutually exclusive but valid claims and arguments. In fact, we submit that it is precisely this overarching validity of both claims and counter-claims that sets a crisis apart from a critique (see Kosellek 1987).

A crisis relates to whether or not a sphere of action or, for that matter, a system, ought to continue. It is caused when criticisms or contradictory claims can no longer be accommodated or assimilated. In our view, the methods needed to analyse such situations involve the setting of claims and counter-claims that appear valid (at a technical level) and consistent with preferred ideals and choices but yet difficult to reconcile. At one level, technical validation of some claims becomes difficult because they pertain to outcomes expected to emerge in the future and are in the realm of uncertainty and probability. Here, even 'experts' conflict over the likely outcomes. At another level, the validation or invalidation of such claims from value positions assures a slide-back into the realm of critical analysis. In such circumstances, where claims and counter-claims confront each other, we are led, indeed impelled, to look at factors underlying them: how they are made, by whom, and with the backing of what power and resources.

In the SSP, we look at the critique not simply as a collection of valid, verifiable statements but as a set of claims emerging from new societal values regarding environment, rights and localism which influence the valuation of development action. Although development action in the Narmada produces risks, suffering, curtailment of choices, impoverishment and despoliation, together with opportunities, benefits, expansion of choices and enrichment, these contradictions are necessary but not sufficient conditions for a crisis. It is when new meanings are attributed to such conditions from different value positions that the crisis becomes pronounced. What we have then is a breakdown in the legitimacy of established ways of viewing and doing development action, but without any acceptable and alternative ways available. We are left with claims and counter-claims that crisis analyses bring to the fore.

Third, this book is significant in terms of the understanding it advances of the social-environmental movement around the SSP. Most studies remain focused on the project. A few that have reflected on the movement have understood it as the struggle of local affected people against development-induced displacement. Additionally, it must be noted that, in theoretical terms, such movements have been explained as not just expressing disenchantment with development but as simultaneously posing 'new visions' of alternatives from below.

In our view, the Narmada movement is best understood as a multi-level network that links the local with the global. Through these links, the movement mobilises various resources and discourses in its campaign against the SSP. By seeking alliances at different levels, the movement equips itself with the wherewithal of 'risk politics'. Information, facts, figures and experts have become major resources that are deployed in the campaigns to highlight project risks and uncertainties. Probabilistic assumptions, input-output projections, reliability of performance indicators and methods, become battlegrounds for the movement. It develops as a knowledge producer, generating counter-claims/ critiques to the claims of project planners.

Scholarship on environmental movements in India has so far completely failed to acknowledge this dimension of movements. Due to inadequate understanding on the part of the actors and their politics that constitute such movements, theoretical understanding of environmentalism has remained stunted. At one level the explanations of environmental conflicts have relied on juxtaposing state and people. Variants within this overall dichotomy include the suggested opposition between planners' rationality and peasant rationality, or scientific knowledge vis-à-vis indigenous knowledge. At another level, the understanding of movement politics has relied on structural-functional and Marxian explanations, despite claims of 'new social movement' analysis. The broad understanding has been that political-economic conditions are the main 'causes' that generate social-environmental movements.

This book highlights some of the limitations in existing approaches to social-environmental movements. In terms of actors it shows how the Narmada movement involves a much broader coalition than just the struggles of affected people. Similarly, in terms of ideals, it shows that the movement's environmentalism is based on the contesting of scientific rationality even while it attempts to radicalise it. In the process, the realm of development action (based on scientific rationality) becomes more transparent and accountable. Perhaps it is here that the contemporary significance of the Narmada movement lies. Even without discounting the 'alternative visions' that scholars have bestowed on the Narmada movement, its most significant success has been in the realm of accountable development and strengthening of the public sphere. What needs to be acknowledged is the growing persuasive power of the movement to force planners to debate such projects in the public sphere. The findings of this research certainly point to the increasing accountability of the development process in the Narmada.

Finally, another significance of this work lies in the policy concerns that it addresses. It identifies zones of challenge, change and negotiation which have been informed by the development crisis in the SSP, and it goes on to develop some recommendations regarding problems in these areas. The main challenge to crisis resolution will be the provisioning of institutional avenues of dialogue for negotiating conflicting interests, norms and values ingrained in the different persuasions of the actors involved. We admit that precisely because of its relative durability, this is a much more difficult way out than those proposed by other studies which recommend either the scrapping of large dams and other development projects altogether, or cosmetic changes in their planning and implementation. These opposing views do not take us far in the SSP or comparable debates nor offer any tangible ways of resolving such crises, whether in the short- or the long-run. Insofar as this book attempts to tread the more complex terrain of crisis and change in development action, resisting the temptation of an easy way out, it is both unique and original.

The Way Forward

It is clear then that conflicts in the SSP are multidimensional. Resources, material benefits and costs, rights and entitlements, procedures and norms, knowledge and values, have all been sites of struggle. Conflicts in these aspects have defined the political field and the alignment of actors. Not only that, these dimensions became reflected in the practices of the state and of the Narmada movement.

We have used a three-fold classification of conflicts in development, as already discussed, anticipating that they would be multidimensional: organisational, institutional and cultural.[2] This device is useful not only to analytically distinguish the dimensions, levels and stakes in conflicts, but also to organise the different ways forward, reiterating that there is considerable overlap between the different dimensions when applied to specific empirical cases. Nonetheless, we use the three-dimensional matrix to situate the different SSP conflicts and to delineate the contours of crisis resolution. Our intention here is not to open a discussion on these dimensions to establish priorities and determinants. Rather, using the proposed analytical devices, our effort is to identify possible ways of conflict minimisation in the SSP.

Lower-order Conflicts: Recognition and Representation of Interests

In terms of the organisational dimension of conflicts according to Touraine (1985), the struggles are over the political status of interests. In the SSP this conflict is manifested in the inter-state interest competitions over Narmada's water resources and also in the struggles around displacement, implicating the status and entitlements of adversely-affected people. Resolution of such conflict necessarily implies a process of negotiation and the subsequent settlement of status issues.

Inter-state conflicts involve the interests of riparian states. We have shown that riparian conflicts over the Narmada have a long history. We have also noted the geo-political and political–economic interests that such conflicts generate nationally and internationally. The interests in such conflicts are often articulated within parameters that define the water needs of one state as threatening to the other. At stake are water rights between upstream and downstream states. Whereas lower riparian states claim the benefit of the juridical principle of 'priority of appropriation', the upper riparian states seek the endorsement of another judicial principle, 'equitable distribution' (see RIC 1972a: 340). Even while reiterating that riparian disputes are neither unique nor beyond settlement, the intensity of riparian conflicts in the SSP has overshadowed negotiation and arbitration, calling for a fresh look at institutions governing inter-state rivers. While this demands a shift in gear to look at the political/institutional dimension of conflict (following Touraine), it must be noted that the inter-state conflict in the Narmada has its basis in the political status of interests.

To elaborate, one cannot fail to observe that Gujarat's interests in the SSP have been well defined and protected. In contrast, MP's posturing on the project embodies an attempt to create an interest and a political status for it. Gujarat has assiduously cultivated interest for the project in the benefiting area as well as in the wider body politic. The state has mobilised resources from Gujarati people (from within the state and outside the country) who have invested heavily in the project. In MP, in contrast, although the government and its people seem united in opposing the project, their interests have often conflicted. There are several indicators to this effect: the government's refusal to up-grade its resettlement package; the different standards it has adopted for other projects on the Narmada; and its use of force against its own people. The Gujarat government has also successfully projected an image of struggling with scarcity and supply constraints. Madhya Pradesh, in contrast, emerges as a state that has more but cannot utilise it, i.e. a state with demand constraints. Additionally, the role of politicians and bureaucrats from Gujarat is of great significance. Apart from exercising considerable clout in the water bureaucracy, they have received tacit support from Gujarat courts on SSP matters.

Ultimately, riparian conflict is an issue that rests with the Supreme Court. Noting the intense nature of the conflict of interests, it is difficult to formulate concrete steps forward. It is only possible to outline a probable path. The riparian states could develop a regional cooperation mechanism in consultation with civil society actors for negotiation of interests. While this could initially be limited to the SSP, it may subsequently be extended to cover the Narmada Valley development plan. This could be fruitful not only for breaking the deadlock in the SSP but for paving the way for the resolution of similar conflicts.

This mechanism, however, will also have to recognise and accommodate the political status of the interests of affected people, a complex matter in itself. It is probably correct to claim that the interests of this constituency have lacked political status, despite a history of struggles over displacement. This book has taken us as far back as the Mulshi Satyagraha in the 1920s, but the history of struggles is definitely longer. The fact that the government still does not have a national policy on resettlement illustrates the political status of displaced people. A policy debate to that effect has been initiated only recently, thanks to the Narmada movement. The Land Acquisition Act, under which the state acquires land for public purposes, is 105 years old and has operated on an extremely limited base, both in terms of what and who is compensated.[3] A proposal to revise the Act has been mooted by the Union Ministry of Rural Areas and Employment, yet another accomplishment of the movement. These proposals should be given concrete shape as quickly as possible.

The organisation and collective actions of affected people's communities symbolise first and foremost their recognition as a people 'at risk'. Their interests vary in terms of what they perceive as loss and are compensated for, i.e. land, livelihoods, homes, social capital, etc. We have shown how people act differently on the basis of these different interests. Some prefer to stay and resist. Others seek compensation, and still others move to resettlement sites. We have highlighted several variables that influence people's perceptions. The bottomline, however, is that under conditions of displacement, affected people, by and large, seek both adequate protection against impoverishment risks and opportunities for well-being. Any compensation and resettlement provisions perceived as failing to accommodate these interests reasonably is bound to create conflicts.

In policy terms, this recognition entails, at the very least, formulation of a uniform national resettlement policy with simultaneous changes in the Land Acquisition Act such that it will protect and not impoverish people. In the specific case of the SSP, substantive changes are required in the resettlement policy of the MP government. It must offer affected people in MP the option to settle there instead of compelling them to move to Gujarat. We have indicated how the interests of affected people must find representation in the regional political process in order to resolve the tangle.

Middle-order Conflicts:
Transparency in Decision-making

Since a restriction on interests (material and political) in the SSP would imply a narrow view of things, a question may be recalled here. Will the SSP conflicts end if inter-state aspects and displacement-related issues are resolved? There is more to the conflict than merely competing interests. Touraine (1985) models this higher-level conflict by attributing to it a greater, more integrated political force aimed at changing the 'rules of the game', and therefore different from the competitive pursuit of collective interests. This higher level becomes more apparent when we look at challenges thrown at major decision-making procedures, norms and standards in the SSP.

In Chapter 3 we singled out the economic appraisals of the SSP for closer scrutiny and identified some limitations. We pointed out that, as a planning tool, cost-benefit analysis (CBA) tends to hide more than it reveals. It suppresses uncertainties and risks, ignores 'external' costs to society and the environment at large, and is based on a set of (optimistic) spatial and temporal assumptions. A 'stake and stakeholder analysis' was undertaken in order to bring some of these hidden aspects of the CBA to the fore, especially the central concern in the SSP conflicts: whose costs and whose benefits?

Chapter 5 too noted conflicts around the application and interpretation of other decision-making procedures, mainly environmental and social impact assessments. Our concern has been to demonstrate the struggles to control the 'zones of uncertainty' in the SSP. The political battle over the accuracy and adequacy of social and environmental impact assessments has involved struggles over meanings, methods, actors and time. Thus, what has been assessed, how, by whom and when, become points of contention.

The institutional-level conflicts bring other actors into play: NGOs, social and environmental citizens' groups, research and professional bodies, and rights organisations. The political force generated through their involvement significantly widens the public discourse to include uncertainties and risks in every aspect of the project—economics, finance, environment, hydrology, technology—and not just in displacement and resettlement. Such conflicts signify a challenge to planners' control over the definition and management of uncertainties and risks. Actors in these conflicts contribute to more transparency, a wider information base, and greater participation in the decision-making process.

It is not difficult to see positive outcomes from the institutional-level conflicts in the SSP. Project planners have been pushed towards greater participatory and democratic procedures. For instance, both the environmental and resettlement components of the SSP have monitoring mechanisms (sub-groups) that allow for NGO and civil society participation. The resettlement component also has independent monitoring and evaluation bodies. Project authorities appear to have responded positively to criticisms on gaps in assessments and preventive measures. New ways of involving affected people and NGOs in land allotment have been tried through the organisation of land purchase committees. Similarly, water in the project command area is to be distributed only through user-associations. Further, the system of pricing water in the SSP is advanced compared to other hydro projects in the region.

These advances and innovations, however, will have to be weighed against several other facets of the decision-making structure that as yet remain closed. The SSP case shows that planners were reluctant to open the project to independent professional review. In fact, the Gujarat government tried to stop the Morse review, boycotted the FMG review, and successfully stalled the visit of the World Commission of Dams. Likewise, persistent demands for a more transparent and comprehensive cost-benefit analysis of the project have been ignored with the blunt assertion that the project is a profitable investment.

Overall, these conflicts in the SSP point to gaps and inadequacies in institutional frameworks and appraisal and evaluation methods governing large river valley projects. Restructuring old and outdated frameworks and setting new standards and guidelines through transparent and participatory processes are ways of managing such conflicts. These reforms do not in any way mean that uncertainties and risks in large dam projects will disappear. In any long-term planning, assumptions are, of course, necessary. We are asking for mechanisms that will prevent their conversion into assertions (as planners tend to do). Rather, they should be oriented towards the more forceful mobilisation of information and communication in order to (*a*) reduce uncertainty and risks where possible, (*b*) spread project risks more equally, and (*c*) assign clear liabilities for compensating risks and pricing opportunities.

Thus public domain politics—what we term 'risk politics' in the SSP—is a step towards democratising the institutions that govern large dam projects. New and democratic ways of 'reasoning' are called for, defined as a contested process of constructing logic, whereby practices can be disciplined through prior knowledge of their effects on human beings and nature (Watts & Peet 1996: 261). The task ahead is to strengthen this 'public sphere', to push for further reforms and new institutional frameworks as a means of making development transparent and accountable.

Higher-order Conflicts: Changing the Parameters of Change

The management of organisational- and institutional-level conflicts in the SSP calls for restructuring and reform of the political process so as to recognise and distribute stakes in large dams in a transparent, negotiated and equitable manner. Conflicts of a higher order, in which stakes include *conceptions* of change and development, are not tackled. Touraine (1985) models this higher level as struggles over cultural patterns: over a model of knowledge that represents truth, a type of investment that represents production and a set of ethical principles that represent morality. For Touraine, this conflict is central to the way in which society, through its collectivity, acts upon and changes itself, the main mechanism of society's self-production. It is in this conflict that social movements play the leading role. Their struggle over the control of 'historicity' envisages constructing new social orders, promising new models of development and change, and heralding the emergence of new visions. In other words, movements challenge development as a 'regime of representation' and generate social energy and political resources in attempts to restructure the 'existing political economies of truth' (Escobar 1995: 218). In new social movement (NSM) theory, this order of conflict is seen as not only qualitatively different from interest conflicts, but as a major approach point to analyse (post-industrial and post-colonial) social movements (Cohen 1985; Escobar 1995; Melluci 1992; Wignaraja 1993). What are the so-called 'higher-order' conflicts in the empirical context of the SSP and how does one address them?

Following NSM theory, protest in the Narmada cannot just be explained as the response of people facing the threat of displacement and marginalisation. Neither can it be deemed as the (rational) action of people strategically pursuing their own interests (say, for better compensation). Rather, the movement in the Narmada Valley is to be seen as contributing to the 'historical possibility' of altering the conception of development. Escobar (1995) notes the radical potential of these movements in the Third World which seek 'alternatives to development' rather than development alternatives. Shiva (1991) sees in them new values and rationalities. Our study has dealt with the visions and values espoused in the broad movement around the SSP, especially with those groups associated with the NBA network. In the course of this we have maintained that the unified visions and ideals projected in social movements are partly rhetoric. We also recount, however, how the movement produces both a critique of development as instrument of hegemony and control, and a knowledge system which challenges established ways of doing and seeing. Higher-order conflicts, therefore, imply contests over the power, knowledge and relations that govern resource transformation. In other words, they aim at altering the structural edifice of development itself, at *changing the parameters of change.*

There are several strands in higher-level conflict: technological aspects, including questions of appropriateness; political aspects involving questions of decentralisation, rights and empowerment; cultural aspects involving indigenous cultures and their diversity; and economic aspects of efficiency and distribution. Together they constitute a discourse on 'alternative development' which in some ways articulates alternative methods of 'developing' the Narmada that may be more sustainable, equitable and efficient, and subject to popular local control. The translation of these ideals and values into an 'alternative blueprint' is heavily contested and somewhat elusive. The higher-order conflict in the SSP therefore is not over whether developmental means can be infused with certain values. Rather, in favouring those values, it attempts to demonstrate how they are incompatible with certain means.

Under such circumstances, planners rule out alternatives altogether in their attempt to project the SSP as the only feasible means to fulfil the goals of water and power development in the region. Simultaneously, they attempt in a limited way to ascribe to it the goals of sustainability, equity and decentralisation. The 'non-negotiable' character of this conflict is evident both in the politics of 'no dam' by the movement, and the persistent refusal of planners to renegotiate even the height of the dam to save some areas from submergence. This aspect of the conflict makes it difficult to speak of conflict resolution in the SSP. A resolution seems feasible only when one side wins at the expense of the other: a zero-sum game.

Despite this stand-off, it is possible to negotiate conflicting values and to seek their confluence in development planning. Consider the following. To planners, flowing river water is waste if left untapped. Environmentalists consider it the *dharma* of the river to flow. Underlying these positions are conflicting values. Yet river valley planning must somehow find ways and means of incorporating both these values. Rivers must flow and must be tapped. The way out is not in the reiteration of these cherished ideals, but in finding ways to negotiate how a minimum reserve flow in the river can be maintained while tapping the excess for societal needs.

The higher-order conflict is being played out in a political field much wider than the SSP, as symbolised by the growing political opposition to large dams in general. The SSP's situation is not unique but widespread. Although displacement issues, intra-basin struggles and decision-making aspects feature as points of controversy in other large dams as well, the construction of 'global anti-dam narrative' is worth noting in the opposition to them. This is a powerful tool that will increasingly be put to use to oppose *all* large dams (ongoing and planned) on a global scale. Whether this will or will not culminate in a more serious quest for alternatives remains to be seen. Nevertheless, the anti-dam narrative has made large dams a politically contested terrain and has opened up avenues for wider debate and consultation, as is evident from the proceedings of the World Commission on Dams.

A second sign of the widening field is the activation of social movement networks on issues other than the SSP. In fact, the

Narmada 'struggle' is only one major involvement of the movement networks that are active on other diverse issues, be they multinational corporations, food security and globalisation, protected areas and struggles around forests, nuclear power and peace movements, struggles related to mining and coastal fisheries. This basket of issues shows how seemingly disparate conflict arenas are articulated in movement networks, contributing to the production of a unified discourse on development, displacement and environment. While one could certainly study the strengths and weaknesses of such articulations, we are hinting here at the emergence of a broader struggle around the conception of development, aimed at changing both its strategies and its ideals.

Options for Managing Crisis and Change

It is clear therefore that the conflicts emerging in the SSP can be addressed in three possible ways. The minimum step would be to ensure adequate representation and participation of different interest groups in the distribution of benefits and costs of the project. The central concern is the protection of rights and entitlements of adversely affected people. This option takes the project as given. To that end it addresses only partially the conflicts in the SSP. Yet, inadequacies in resettlement feature as the central concern of project opponents, including the NBA. Moreover, the proceedings of the Supreme Court indicate the primacy that is accorded to this option by the Court. The way forward would be to call for regional consultation in which major actors such as riparian states, project planners, people's organisations and environmental groups could meaningfully and dialogically plan resettlement policies and ensure their proper implementation. This regional consultation should be facilitated by national government through the formulation of national policies, guidelines and legislation on water and resettlement, and should serve as a basis for negotiating other sites of conflict.

The second option is to widen the 'political field' and initiate consultation on institutional frameworks and decision-making aspects governing large dams. The major contribution of the Narmada movement has been precisely to open this hitherto closed field to public scrutiny. In their enthusiasm for advancing the visions of 'alternative development', most contributions on social movements fail to acknowledge what we call their 'contemporary significance' (see Dwivedi 1998). In precise terms, this contemporary significance stems from the ability of social movements to identify gaps and inadequacies in institutions and policies through what we have characterised as 'knowledge struggles'. Such struggles, as has already been shown, are not between indigenous/local/traditional systems and 'modern scientific' ones, but take place within what is understood as the scientific knowledge domain, implicating methods and techniques of appraisal, assessment and evaluation. Furthermore, by exposing knowledge gaps, movements demand their institutional bridging in a democratic and accountable manner.

The exercise of this option would entail reforming a body of policies and practices that govern large dam projects. In the context of the SSP, this option would require at least a panel of experts with adequate participation of local affected people to conduct a fresh, independent cost-benefit appraisal of the project, incorporating social and environmental impacts. Within the economic analysis itself issues pertaining to externalities, uncertainties, and risk and time factors need to be addressed. The economic bias in appraisal must be corrected with a multiple criteria approach for informed decision-making. The multiple criteria must incorporate the distributional impact of large dams through stakeholder analysis and participation. Assessments should be carried out by independent researchers, and disseminated and debated in the public sphere. Depending on the results of these analyses, the SSP could be suitably restructured.

The third option is a radical one inspired by the 'alternative development' movement. It involves a moratorium on large dam projects like the SSP and a radical reorientation of water resource policy towards the formulation and implementation of decentralised systems of water harvesting and hydropower projects. This would entail mainstreaming and replicating 'model alternatives' that are being tried in different parts of the country and elsewhere. The focus here shifts to identifying technological alternatives for flood control, power and water supply. Chapter 8 gives some indication of this effect in the context of the SSP. Exercising this option would also mean reorienting the goals of development.

Here, the criteria for development goals are not (unlimited) growth and accumulation but basic needs, distributive justice and environmental sustainability. Water resources planning in this option shifts from a project-based, technocrat-driven framework to a basin-based, community-oriented framework. A parallel thread running through this option is the restoration of diverse traditional water-harvesting and management systems that have fallen into disuse.

This option would also mean that decentralised water harvesting and development projects become the major components of water resources planning. The assumption is that while planners hold large and small projects as necessary to integral planning, they prefer the former and pay only lip service to alternatives. The radical option is based on the assumption that cumulative outcomes of local efforts can substitute for large projects; therefore, their possibilities need to be examined seriously.

All three options address the SSP conflicts. Linked as they are to the wider aspect of resource development policies, each option has its advantages and disadvantages. Although they appear to cancel each other out in the context of the SSP, the SSP provides the opportunity to draw on the positive aspects of all of them. Our argument is based on our findings, that the SSP conflicts could substantially be resolved even if only the first option were properly implemented. The second option allows for a more inclusive discussion on the mechanisms and institutions governing large dams in general. A fair debate would further minimise dam-related conflicts. More importantly, it could provide sufficient political impetus for exercising the third option in resource planning and management.

Conclusion

The nature and scope of this research has generated its due share of limitations. Its strength lies in its width, in its comprehensive understanding of the SSP. The emphasis on width, however, is at some cost to the in-depth understanding of issues. It may be said that while a more in-depth study on displacement-resettlement problems, or on the Narmada movement or, for that matter, on the SSP, was feasible, such a sectoral approach to the problem on hand would barely have grasped the crisis.

Similarly, a deeper understanding of people's perceptions and views would have been feasible had an ethnographic approach been adopted and a particular set of people observed over a length of time. Rather, the coverage in this case was extensive rather than intensive. It should be mentioned that the emotions, views and perceptions of the people in this study rose and fell with the unfolding of events in the Narmada Valley. We have attempted to circumvent this limitation by showing how and when their views and perceptions changed.

Last but not least, the study is of a single case. The felt necessity to deal exhaustively with the range of issues involved in the SSP left us with limited time and resources to link it more forcefully with the large dam debates. While other large dam projects are covered and compared with the SSP, this can only be illustrative of the debate, and although we draw heavily upon it, it is not a significant focus of the study due to obvious constraints of time and space.

Notes

1. Civil rights or liberties are the rights of citizens against the state, whereas democratic rights are the articulation of expectations of entitlements to just governance, insisting on fulfilment of positive obligations of the state towards the people. Whereas civil rights are codified and therefore exist, democratic rights could represent claims, which are entitlements-in-making (Baxi 1998: 338). According to Ram (1986), the struggle for democratic rights is the struggle to assert rights already guaranteed but not ensured in practice. They are needed for those struggling for social justice (Hargopal & Balagopal 1998).
2. The terms are used loosely here and seem to correspond to what Gadgil and Guha (1995) call the material, political and ideological dimensions of environmental mobilisations in India.
3. The compensation is asset-based and not livelihood-based, and often goes to the family patriarch.

References

Action Committee for National Rally Against Destructive Development (1989) 'We want development, not destruction: to Harsud! To
Narmada! Badwani, 28 September.

Adams, J. (1984) 'Infrastructure and economic development: the evidence from central India', *Journal of Asian and African Studies*, 19
(1–2): 36–48.

Adams, W.M. (1993) 'Development's deaf ear: downstream users and
water releases from the Bakolori dam, Nigeria', *World Development*,
21(9): 1405–16.

Aditjondro, G. and D. Kowaleski (1994) 'Damning the dams in Indonesia: a test of competing perspectives', *Asian Survey*, 34 (4): 381–95.

Agarwal, B. (1994) *A Field of One's Own: Gender and Land Rights in
South Asia*. Cambridge: Cambridge University Press.

——— (1997) 'Environmental action, gender equity and women's participation', *Development and Change*, 28 (1): 1–44.

Alagh, Y.K. (1991) 'Planners approach to Sardar Sarovar developments',
All About Narmada. Gandhinagar: Government of Gujarat.

Alagh, Y.K., M. Pathak & D.T. Buch (1995) *Narmada and Environment: An Assessment*. New Delhi: Har-Anand Publishers.

Alvares, C. & R. Billory (1988) *Damming the Narmada*. Penang: Third
World Network/APPEN.

Ambasta, A. (1998) 'Capitalist restructuring and formation of Adivasi
proletarians: agrarian transition in Thane District (Western India)
c. 1817–1990'. The Hague: Institute of Social Studies.

Amte, B. (1990) 'Narmada project: the case against and an alternative
perspective', *Economic and Political Weekly*, 25 (16): 811–18.

Appa, G. (1992) 'Narmada Projects Without World Bank Backing', *Economic and Political Weekly*, 28 November: 2577–80.

Apthorpe, R. & D. Gasper (1996) 'Introduction: discourse analysis and
policy discourse', in R. Apthorpe & D. Gasper (eds) *Arguing Development
Policy: Frames and Discourses*, pp. 1–15. London and Portland, Or:
Frank Cass.

ARCH-Vahini (1988) 'Sardar Sarovar oustees, which way to go: activists' dilemma', mimeo, Mangrol: ARCH-Vahini.

ARCH-Vahini (1988a) 'The 14th May 1988 Vadagam Convention: before and after', mimeo, Mangrol: ARCH-Vahini.

————— (1991) 'Rehabilitation and resettlement in Sardar Sarovar project: are the critics right', paper presented in seminar on Major Issues on Sardar Sarovar Project, Vallabh Vidyanagar: Agro-economic Research Centre, 7–8 March.

Areeparampil, M. (1987) 'The impact of Subarnarekha multipurpose project on the indigenous people of Singhbhum', in People and Dams, workshop proceedings, pp. 83–105. New Delhi: PRIA.

ASG (Alternative Survey Group) (1994) *Alternative Economic Survey 1994–95*. New Delhi: New Age International Ltd. and Wiley Eastern Ltd.

Attwood, D.W. (1992) *Raising Cane: The Political Economy of Sugar in Western India*. Boulder, CO: Westview Press

Baboo, B.G. (1991) 'State policies and people's response: lessons from the Hirakud dam', *Economic and Political Weekly*, 26 (41): 2373–79.

Banaji, J. (1995) 'The farmers' movements: a critique of conservative rural coalitions', in T. Brass (ed.) *New Farmers Movement in India*, pp. 228–45. Ilford: Frank Cass.

Bardhan, P. (1984) *The Political Economy of Development in India*. Oxford and New York: Basil Blackwell.

Barnett, T. (1981) 'Evaluating the Gezira Scheme: black box or Pandora's box?' in J. Weyer, P. Roberts & G. Williams (eds) *Rural Development in Tropical Africa*, pp. 306–24. Macmillan Press.

Bates, R. (1981) *Markets and States in Tropical Africa: The Political Basis of Agricultural Policies*. Berkeley: University of California Press.

————— (1989) *Beyond the Miracle of the Market: The Political Economy of Agrarian Development in Kenya*. Cambridge: Cambridge University Press.

Baviskar, A. (1995) *In the Belly of the River: Tribal Conflicts Over Development in the Narmada Valley*. Delhi: Oxford University Press.

Baxi, U. (1998) 'The state and human rights movements in India', in M. Mohanty & P.N. Mukherji (eds) *People's Rights: Social Movements and the State in the Third World*, pp. 335–52. New Delhi: Sage Publications.

Beck, U. (1992) *Risk Society: Towards a New Modernity*. London, Thousand Oaks and New Delhi: Sage Publications.

————— (1995) *Ecological Politics in an Age of Risk*. Cambridge: Polity Press.

Bhalla, G.S. & Y.K. Alagh (1979) *Performance of Indian Agriculture— A District Study*. Delhi: Sterling.

Bharadwaj, K. (1995) 'Regional differentiation in India', in T.V. Sathyamurthy (ed.) *Industry and Agriculture in India Since Independence*, vol. 2, pp. 189–218. New Delhi: Oxford University Press.

Bharati, I. (1991) 'Bihar's dams, tribals' woes', *Economic and Political Weekly*, 26 (22 and 23): 1385–88.

Blackwelder, B. & P. Carlson (1986) *Disasters in International Water Development. Fact Sheets on International Water Development Projects.* Washington, DC: Environmental Policy Institute.

Blinkhorn, T.A. & W.T. Smith (1995) 'India's Narmada: river of hope', in W. Fisher (ed.) *Towards Sustainable Development, Struggling Over India's Narmada River,* pp. 89–112. Armonk and London: M.E. Sharpe.

Booth, D. (1985) 'Marxism and development sociology: interpreting the impasse', *World Development*, 13 (7).

————— (1994) *Rethinking Social Development: Theory, Research and Practice.* Harlow: Longman.

Brass, P.R. (1990) *The Politics of India Since Independence.* Cambridge: Cambridge University Press.

Brass, T. (ed.) (1995) *New Farmers Movement in India.* Ilford: Frank Cass.

Breman, J. (1985) *Of Peasants, Migrants and Paupers: Rural Labour Circulation and Capitalist Production in West India.* Delhi: Oxford University Press.

Bridger, G.A. & J.T. Winpenny (1983) *Planning Development Projects: A Practical Guide to the Choice and Appraisal of Public Sector Investments.* London: HMSO.

Bryant, R. (1992) 'Political ecology', *Political Geography*, 11: 2–36.

Buttel, F. & P. Taylor (1994) 'Environmental sociology and global environmental change: a critical assessment' in M. Redclift & T. Benton (eds) *Social Theory and the Global Environment*, pp. 228–55. London and New York: Routledge.

Byres, T.J. (1979) 'Of neo-populist pipe-dreams: daedalus in the third world and the myth of urban bias', *Journal of Peasant Studies*, 6 (2).

————— (1991) 'The agrarian question and differing forms of capitalist agrarian transition: an essay with reference to Asia', in J. Breman & S. Mundle (eds) *Rural Transformation in Asia*, pp. 3–76. Delhi: Oxford University Press.

————— (1995) Preface, in T. Brass (ed.) *New Farmers Movement in India,* pp. 1–2. Ilford: Frank Cass.

Byres, T.J., B. Crow & M.W. Ho (1983) *The Green Revolution in India.* Milton Keynes: Open University Press.

Capra, F. (1982) *The Turning Point. Science, Society and the Rising Culture.* Flamingo.

Cassell, P. (ed.) (1993) *The Giddens Reader.* Stanford, California: Stanford University Press.

Castells, M. (1983) *The City and the Grassroots: A Cross-cultural Theory of Urban Social Movements.* Berkeley: University of California Press.

Castells, M. (1997) 'The power of identity', *The Information Age: Economy, Society and Culture*, vol. II. Oxford: Blackwell.

CBIP (1989) *Modern Temples of India: Selected Speeches of Jawaharlal Nehru at Irrigation and Power Projects and Various Technical Meetings of Engineers and Scientists*. New Delhi: Central Board of Irrigation and Power.

Cernea, M.M. (ed.) (1985) *Putting People First: Sociological Variables in Rural Development*. New York: Oxford University Press.

———— (1990) 'Poverty risks from population displacement in water resources development', Development Discussion Paper No. 355. Harvard Institute for International Development, Harvard University.

———— (1995) 'Understanding and preventing impoverishment from displacement: reflections on the state of knowledge', *Journal of Refugee Studies*, 8 (3): 245–64.

———— (1996) 'Impoverishment Risks and Livelihood Reconstruction: A Model for Resettling Displaced Populations', draft mimeo. Washington, DC: World Bank, Environment Department.

———— (1997) 'The risks and reconstruction model for resettling displaced populations', *World Development*, 25 (10): 1569–87.

———— (1999) *The Economics of Involuntary Resettlement: Questions and Challenges*. Washington: World Bank.

Chakravarty, S. (1987) *Development Planning: The Indian Experience*. Oxford: Clarendon Press.

Chatterjee, P. (1984) 'Gandhi and the critique of civil society', in R. Guha (ed.) *Subaltern Studies I: Writings on South Asian History and Society*. Delhi: Oxford University Press.

Chattopadhyay, P. (1992) 'India's capitalist industrialisation: an introductory outline', in B. Berberoglu (ed.) *Class, State and Development in India*, pp. 141–56. New Delhi: Sage Publications.

Chaudhuri, P. (1995) 'Economic planning in India', in T.V. Sathyamurthy (ed.) *Industry and Agriculture in India Since Independence*, vol. 2, pp. 94–114. New Delhi: Oxford University Press.

Choksey, R.D. (1958) *Economic Life in the Bombay Gujarat 1800–1939*. Bombay: Asia Publishing House.

Cholchester, M. (1986) 'An end to laughter? The Bhopalpatnam and the Godavari projects', in E. Goldsmith and N. Hildyard (eds.) *The Social and Environmental Effects of Large Dams*, vol. 2, case studies, pp. 245–54. Wadebridge, UK: Wadebridge Press.

CMIE (1992) *Basic Statistics Relating to the Indian Economy*. Bombay: Centre for Monitoring Indian Economy.

Cohen, J.L. (1985) 'Strategy and identity: new theoretical paradigms and contemporary social movements', *Social Research*, 52: 663–716.

Conklin, Beth and Laura Graham (1995) 'The shifting middle ground: Amazonian Indians and eco-politics', *American Anthropologist* 97(4): 695–710.

Corbridge, S. (1990) 'Post-Marxism and development studies: beyond the impasse', *World Development*, 18(5).

Cotgrove, S.F. (1982) *Catastrophe or Cornucopia: The Environment, Politics and the Future*. Chicester, Sussex: John Wiley.

Cox, B.S. (1987) 'Thailand's Nam Choan dam: a disaster in the making', *Ecologist*, 17(6): 212–19.

CSE (1982) *India: The State of Environment 1982: A Citizens' Report*. New Delhi: Centre for Science and Environment.

CSE (1985) *India: The State of Environment 1984–85: The Second Citizens' Report*. New Delhi: Centre for Science and Environment.

CSS (various years) *Resettlement and Rehabilitation: Sardar Sarovar Project on the Narmada* (Report 15, 16, 19, 20). Surat: Centre for Social Studies.

CWC (1991) *Assessment of Utilisable Annual Flow Volumes in Narmada Basin*. New Delhi: Central Water Commission.

D'Souza, R., P. Mukhopadhyay & A. Kothari (1994) *Watery Dreams and Unfulfilled Promises: How Beneficial are Large-scale Irrigation Projects?* New Delhi: Kalpavriksh.

Dalal, L. (1991) 'Namami Devi Narmade', in *All About Narmada*. Gandhinagar: Government of Gujarat, Directorate of Information.

Dansie, J. (1994) 'Don't deny us our dams!' ICOLD Congress, Durban, *International Water, Power and Dam Construction*, December, pp. 14–15.

Das Gupta, A. (1975) *Indian Merchants and the Decline of Surat*. Wiesbandon: Franz Stenier Vering.

Das, V. (1996) 'Dislocation and rehabilitation: defining a field', *Economic and Political Weekly*, 31(24): 1509–14.

Datye, K.R. (1993) 'Narmada: outline of an alternative', notes for presentation to the Review Committee, Bombay.

Desai, B. (1995) 'Narmada dam will not help save Kutch', *The Telegraph*. New Delhi, 19 May.

Dey, I. (1993) *Qualitative Data Analysis: A User-friendly Guide for Social Scientists*. London and New York: Routledge.

Dhanagare, D.N. (1987) 'Green revolution and social inequalities in rural India', *Economic and Political Weekly*, 22 (19, 20 & 21): AN 137–44.

Dhanagare, D.N. (1995) 'The class character and politics of the farmers' movement in Maharashtra during the 1980s', in T. Brass (ed.) *New Farmers Movement in India*, pp. 72–94. Ilford: Frank Cass.

Dharamadhikary, S. (1993) 'Hydropower from Sardar Sarovar: need, justification and viability,' *Economic and Political Weekly*, 28 (48): 2584–88.

Dhawan, B.D. (ed.) (1990) *Big Dams: Claims, Counterclaims*. New Delhi: Commonwealth Publishers.

Diani, M. & R. Eyerman (eds) (1992) *Studying Collective Action*. London: Sage Publications.

Doornbos, M. & P. Tehral (1993) 'The limits of independent policy research: analysing the EEC-India dairy aid nexus', in A. Hurskainen & M. Salih (eds) *Social Science and Conflict Analysis*, pp. 183–98. Uppsala and Helsinki: Human Life in African Arid Lands and Department of Asian and African Studies, University of Helsinki.

Douglas, M. (1992) *Risk and Blame: Essays in Cultural Theory*. London and New York: Routledge.

Dowie, J. & P. Lefrere (1980) *Risk and Chance*. Milton Keynes: Open University Press.

Dreze, J. & A. Sen (1995) 'Public action and social inequality', in *India: Economic Development and Social Opportunity*, pp. 87–108. Oxford: Oxford University Press.

Dreze, J., M. Samson & S. Singh (eds) (1997) *The Dam and The Nation: Displacement and Resettlement in the Narmada Valley*. Delhi: Oxford University Press.

Drucker, C. (1985) 'Dam the Chico. Hydropower development and tribal resistance', *Ecologist*, 15 (4): 149–57.

Dwivedi, R. (1997a) 'Parks, people and protest: the mediating role of environmental action groups', *Sociological Bulletin*, 46 (2) 209–43.

———— (1997b) *People's Movements in Environmental Politics: A Critical Analysis of the Narmada Bachao Andolan in India*. The Hague: Institute of Social Studies.

———— (1998) 'Resisting dams and "development": contemporary significance of the campaign against the Narmada projects', *European Journal of Development Research*, 10 (2): 135–83.

———— (1999) 'Displacement, risks and resistance: local perceptions and actions in the Sardar Sarovar', *Development and Change*, 30 (1): 43–78.

Eder, K. (1995) 'Does social class matter in the study of social movements: a theory of middle-class radicalism', in L. Maheu (ed.) *Social Movements and Social Classes: The Future of Collective Action*, pp. 21–54. London: Sage Studies in International Sociology.

Elster, J. (1986) 'The theory of combined and uneven development: a critique', in J. Roemer (ed.) *Analytical Marxism*, pp. 54–63. Cambridge: Cambridge University Press.

Engineer, R. (1990) 'The Sardar Sarovar controversy', in B.D. Dhawan (ed.) *Big Dams: Claims and Counterclaims*, pp. 155–74. New Delhi: Commonwealth Publishers.

EP (1992) Resolution on the Narmada Dam (India), No. B3-1012/92, European Parliament.

Escobar, A. (1992) 'Reflections on "development": grassroots approaches and alternative politics in the third world', *Futures* (June): 411–36.

———— (1995) *Encountering Development: The Making and Unmaking of the Third World*. Princeton, NJ: Princeton University Press.

———— (1996) 'Constructing nature-elements for a poststructural political ecology', in R. Peets & M. Watts (eds) *Liberation Ecologies: Environment, Development, Social Movements*, pp. 46–68. London and New York: Routledge.

Evers, T. (1985) 'Identity: the hidden side of movement in Latin America', in D. Slater (ed.) *New Social Movements and the State in Latin America*, pp. 43–71. Amsterdam: CEDLA.

EW (1964) 'Weekly notes', *Economic Weekly*, 16 (27): 1078–81.

Eyerman, R. & A. Jamison (1991) *Social Movements: A Cognitive Approach*. Cambridge: Polity Press.

Ferguson, James (1990) *The Anti-Politics Machine: 'Development', Depoliticization and Bureaucratic Power in Lesotho*. Cambridge: University of Cambridge Press.

Fernandes, W & E. Ganguly-Thukral (eds) (1989) *Development, Displacement and Rehabilitation*. New Delhi: Indian Social Institute.

Fernandes, W., J.C. Das & S. Rao (1989) *Displacement and Rehabilitation—An Estimate of Extent and Prospects*. New Delhi: Indian Social Institute.

FFC (1986) 'The Srisailam resettlement experience: the untold story', in E. Goldsmith & N. Hildyard (eds) *The Social and Environmental Effects of Large Dams*, vol. 2, case studies, pp. 255–60. The Fact-finding Committee on the Srisailam Project, Wadebridge, UK: Wadebridge Press.

Fiagoy, G.F. (1987) 'Chico river dam and beyond: people's struggle for self-determination', in *People and Dams*, workshop proceedings, pp. 158–65. New Delhi: PRIA.

Fischer, F. (1995) *Evaluating Public Policy*. Chicago: Nelson-Hall Publishers.

Fisher, W. (ed.) (1995) *Towards Sustainable Development, Struggling Over India's Narmada River*. Armonk and London: M.E. Sharpe.

FMG (1994) *Report of the Five Member Group on Sardar Sarovar Project*, Vols. 1 and 2. New Delhi.

FMG (1995) *Supplementary Report of the FMG on Certain Issues Relating to the Sardar Sarovar Project*. New Delhi.

Foucault, Michel (1991) 'Governmentality', in G. Burchell, C. Gordon and P. Miller (eds) *The Foucault Effect: Studies in Governmentality*. Chicago. University of Chicago Press.

Frankel, F.R. (1990) 'Decline of a social order', in F.R. Frankel & M.S.A. Rao (eds) *Dominance and State Power in Modern India: Decline of a Social Order*, vol. II, pp. 482–517, Delhi: Oxford University Press.

Friedmann, J. (1992) *Empowerment: The Politics of Alternative Development*. Cambridge and Oxford: Blackwell.

Friedmann, J. & H. Rangan (eds) (1993) *In Defense of Livelihood: Comparative Studies on Environmental Action*. West Hartford, CT: Kumarian Press.

Gadgil, M. & R. Guha (1995) *Ecology and Equity: The Use and Abuse of Nature in Contemporary India*. London and New York: Routledge.

Gandhi, M.K. (1951) *Towards Non-violent Socialism*, B. Kumarappa (ed.). Ahmedabad: Navajivan Publishing House.

Gasper, D. (1987) 'Motivations and manipulations: some practices of project appraisal and evaluation', *Manchester Papers on Development*, 3 (1): 24–70.

George, S. & F. Sabelli (1994) *Faith and Credit: The World Bank's Secular Empire*. Penguin Books.

Ghosh, A.S. (1995) 'Narmada: waters of despair, waters of hope', *Economic and Political Weekly*, 16 March: 643–46.

Giddens, A. (1984) *The Constitution of Society: Outline of the Theory of Structuration*. Cambridge: Polity Press.

Giddens, A. (1990) *The Consequences of Modernity*. Cambridge: Polity Press.

Gill, M.S. (1995) 'Resettlement and rehabilitation in Maharashtra for the Sardar Sarovar Narmada Project', in W.F. Fisher (ed.) *Toward Sustainable Development: Struggling over India's Narmada River*, pp. 231–64. Armonk and London: M.E Sharpe.

GoG (1983) 'The Gujarat economy', *Sardar Sarovar (Narmada) Project Development Plan*, vol. I. Gandhinagar: Government of Gujarat, Irrigation Department, Narmada Planning Group.

———— (1989) 'Planning for prosperity', *Sardar Sarovar Development Plan*. Gandhinagar: Government of Gujarat, Narmada Development Department.

———— (1991) *All about Narmada*. Gandhinagar: Government of Gujarat, Directorate of Information.

GoI (1979) *The Gazette of India*. Ministry of Agriculture and Irrigation, Department of Irrigation, Notification, Part II-sec 3(ii), 12 December 1979, 1429–39.

———— (1994) *Draft National Policy for Rehabilitation of Persons Displaced as a Consequence of Acquisition of Land*. New Delhi: Government of India, Ministry of Rural Development.

Goldsmith, E. & N. Hildyard (1984) *The Social and Environmental Effects of Large Dams*, vol. 1, overview. Wadebridge, UK: Wadebridge Press.

Goldsmith, E. & N. Hildyard (1986) *The Social and Environmental Effects of Large Dams*, vol. 2, case studies. Wadebridge, UK: Wadebridge Press.

———— (1992) *The Social and Environmental Effects of Large Dams*, vol. 3,. Wadebridge, UK: Wadebridge Press.

GoMP (1970) *Jalsindhi Project Report*, July 1970. Bhopal: Government of Madhya Pradesh.

———— (1975) *Revised Project Report on the Harinfal Dam*. Bhopal: Government of Madhya Pradesh.

———— (1994) Written Submission on Behalf of the Government of Madhya Pradesh Before the FMG, 18 April 1994. Bhopal: GoMP.

Gopal, S. (1975) *Jawaharlal Nehru: A Biography*, vol. I, 1889–1947. London: Jonathan Cape.

Gopal, S. (ed.) (1980) *Jawaharlal Nehru: An Anthology*. Delhi: Oxford University Press.

Gorter, P. (1989) 'Canal irrigation and agrarian transformation', *Economic and Political Weekly*, 24 (39): A94–A105.

Grillo, R.D. & R.L. Stirrat (eds) (1997) *Discourses of Development: Anthropological Perspectives*. Oxford: Berg.

Guha, R. & J. Martinez-Alier (1997) *Varieties of Environmentalism: Essays North and South*. London: Earthscan.

Guha, R. (1988) 'Ideological trends in Indian environmentalism', *Economic and Political Weekly*, 23 (49): 2578–81.

———— (1989) *The Unquiet Woods: Ecological Changes and Peasant Resistance in the Himalaya*. New Delhi: Oxford University Press.

Habermas, J. (1978) *Knowledge and Human Interest*. London: Heinemann.

Hakim, R. (1996) 'Vasava identity in transition: some theoretical issues', *Economic and Political Weekly,* 31(24): 1492–99.

Hannigan, J.A. (1995) *Environmental sociology: a social constructionist perspective*. London: Routledge.

Hargopal, G. & K. Balagopal (1998) 'Civil liberties movement and the state in India', in M. Mohanty & P.N. Mukherji (eds) *People's Rights: Social Movements and the State in the Third World*, pp. 353–71. New Delhi: Sage Publications.

Harris, J. (1994) 'Between economism and post-modernism: reflections on research on "agrarian change" in India', in D. Booth (ed.) *Rethinking Social Development: Theory, Research and Practice,* pp. 172–201. Harlow: Longman.

Hart, H.C. (1956) *New India's Rivers*. Bombay: Orient Longmans.

Hashim, S.R. (1991) 'SSP: issues considered in its planning', in *All About Narmada*. Gandhinagar: Government of Gujarat.

Hassan, Z. (1995) 'Shifting grounds: Hindutva politics and the farmers' movement in Uttar Pradesh', in T. Brass (ed.) *New Farmers Movement in India,* pp. 165–94. Ilford: Frank Cass.

Heimer, C. (1988) 'Social structure, psychology and the estimation of risk', *Annual Review of Sociology*, 14: 491–519.

Helmers, F.L.C.H. (1979) *Project Planning and Income Distribution*. Martinus Nijhoff.

Hilhorst, D. (1997) 'Discourse formation in social movements: issues of collective action', in H. de Haan & N. Long (eds) *Images and Realities of Rural Life: Wageningen Perspectives on Rural Transformation*, pp. 121–49. The Netherlands: Van Gorcum.

Hirschman, A.O. (1967) *Development Projects Observed*. Washington, D.C.: The Brookings Institution.

Hirsh, P (1987) Social implications of Nam Choan dam, Thailand, in *People and Dams*, Workshop Proceedings, pp. 119–24. New Delhi: PRIA.

Hobart, M. (1993) 'Introduction: the growth of ignorance?' in M. Hobart (ed.) *An Anthropological Critique of Development: The Growth of Ignorance*. London and New York: Routledge.

House, E. (1980) 'The major approaches', in *Evaluating with Validity*, pp. 21–43. Newbury Park, Calif. Sage.

Hulme, D. (1988) 'Project planning and project identification: rational, political or somewhere in between,' *Manchester Papers on Development*, 4 (2): 272–93.

Hulme, D. & M.M. Turner (1990) *Sociology and Development: Theories, Policies and Practices*. New York, NY: Harvester Wheatsheaf.

Humphrey, C.R. & F.H. Buttel (1982) 'Exploring environmental sociology: from environment, energy and society', in M. Redclift & G. Woodgate (eds) (1995) *The Sociology of the Environment*, vol. III. Aldershot and Brookfield: Edward Elgar.

IIM (1991) *Compensation to Canal Affected People of Sardar Sarovar Narmada Project*. Ahmedabad: Indian Institute of Management, Centre for Management in Agriculture.

IRM (1992) *Sardar Sarovar: Report of the Independent Review*, Bradford Morse (Chairman). Ottawa: Resource Future International Inc.

IRN (1996) *A Critique of 'the World Bank's Experience with Large Dams: A Preliminary Review of Impacts'*. Berkeley: International Rivers Network.

————— (1997) 'An NGO report on the April 1997 World Bank-IUCN dams workshop and on the proposal for an independent international dam review commission. http://www.irn.org: International Rivers Network.

IIPA (1988) '*Study on large dams in India',* mimeo, New Delhi: Indian Institute of Public Administration.

IUCN & World Bank (1997) *Large Dams: Learning from the Past, Looking at the Future*. Washington, DC: IUCN-World Conservation Union and World Bank.

Iyer, R.R. (1994) 'Federalism and water resources', *Economic and Political Weekly*, 29 (13): 733–36.

————— (1990a) 'Large dams: the right perspective', in B.D. Dhawan (ed.) *Big Dams: Claims, Counterclaims,* pp. 61–88. New Delhi: Commonwealth Publishers.

_____ (1990b) 'The large dam debate: a response to an intervention', in B.D. Dhawan (ed.) *Big Dams: Claims, Counterclaims*, pp. 95–100. New Delhi: Commonwealth Publishers.

_____ (1998) 'Water resource planning: changing perspectives', *Economic and Political Weekly*, 33 (50): 3198–205.

Jasanoff, S. (1997) 'NGOs and the environment: from knowledge to action', *Central Asian Survey*, 18 (5): 579–94.

Jazairy, I., M. Alamgir & T. Panuccio (1992) *The State of World Rural Poverty: An Inquiry into its Causes and Consequences*. New York: New York University Press for the International Fund for Agricultural Development (IFAD).

Jodha, N.S. (1994) 'Common property resources and the rural poor', in R. Guha (ed.) *Social Ecology*, pp. 150–89. New Delhi: Oxford University Press.

Jones, D.E. & R.W. Jones (1976) 'Urban upheaval in India: the 1974 Nav Nirman riots in Gujarat', *Asian Survey*, 16 (11): 1012–33.

Joshi, V. (1987) *Submerging Villages*. New Delhi: Ajanta Publishers.

_____ (1991) *Rehabilitation, A Promise to Keep—A Case of SSP.* Ahmedabad: Tax Publishers.

_____ (1997) 'Rehabilitation in the Narmada Valley: human rights and national policy issues', in J. Dreze, M. Samson & S. Singh (eds) *The Dam and the Nation: Displacement and Resettlement in the Narmada Valley*, pp. 168–183. Delhi: Oxford University Press.

Jouhar, A.J. (ed.) (1984) *Risk in Society*. London: Libbey.

Kabeer, N. (1994) '"And no-one could complain at that": claims and silences in social cost-benefit analysis', *Reversed Realities*, pp. 163–86. London and New York: Verso.

Kalpavriksh & The Hindu College Nature Club (1986) 'The Narmada Valley Project: development or destruction', in E. Goldsmith & N. Hildyard (eds) *The Social and Environmental Effects of Large Dams*, vol. 2, case studies. Camelford: Wadebridge Ecological Centre.

Kaviraj, S. (1997) 'The modern state in India', in M. Doornbos & S. Kaviraj (eds) *Dynamics of State Formation: India and Europe Compared*, pp. 225–50. New Delhi: Sage Publications.

Kiely, R. (1998) 'Introduction: globalisation, (post-) modernity and the Third World', in R. Kiely & P. Marfleet (eds) *Globalisation and the Third World*, pp. 1–22. London and New York: Routledge.

King, J.A., L.L. Morris & C.T. Fitz-Gibbon (1987) 'Planning for data collection', *How to Assess Program Implementation*, pp. 42–59. Newbury Park, CA: Sage.

Knight, J. (1994) *Institutions and Social Conflicts*. Cambridge: Cambridge University Press.

Kohli, A. (1984) 'Communist reformers in West Bengal: origins, features and relations with New Delhi', in J.R.Wood (ed.) *State Politics in Contemporary India: Crisis or Continuity?* pp. 81–102. Boulder and London: Westview Press.

Kohli, A. (1989) *The State and Poverty in India: The Politics of Reform.* Cambridge: Cambridge University Press.

———— (1990) *Democracy and Discontent: India's Growing Crisis of Governability.* Cambridge: Cambridge University Press.

Korten, D. (1989) 'Social science in the service of social transformation', in C.C. Veneracion (ed.) *A Decade of Process Documentation Research: Reflections and Synthesis,* pp. 5–20. Quezon City: Institute of Philippine Culture, Ateneo de Manila University.

Kosellek, R. (1987) *Critique and Crisis.* Leamington Spa: Berg.

Kothari, R. (1984) 'Party and state in our times: the rise of non-party political formations', *Alternatives,* 9 (Spring): 541–64.

———— (1990) *State Against Democracy: In Search of Human Governance.* London: Aspect Publications.

Kothari, S. (1995) 'Damming the Narmada and the Politics of Environment', in W. Fisher (ed.) *Towards Sustainable Development, Struggling Over India's Narmada River,* pp. 421–44. Armonk and London: M.E. Sharpe.

Kriesi, H. (1989) 'New social movements and the new class in the Netherlands', *American Journal of Sociology,* 94 (5): 1078–116.

Laclau, E. (1985) 'New social movements and the plurality of the social', in D. Slater (ed.) *New Social Movements and the State in Latin America,* pp. 27–42. Amsterdam: CEDLA.

———— (1990) *New Reflections on the Revolution of Our Time.* London: Verso.

Laclau, E. & C. Mauffe (1985) *Hegemony and the Socialist Strategy: Towards a Radical Democratic Politics.* London: New Left Books.

Lenin, V.I. (1965) *Collected Works.* Moscow: Progress Publishers.

Lindberg, S. (1995) 'New farmers movement in India as structural response and collective identity formation: the cases of Shetkari Sangathana and BKU', in T. Brass (ed.) *New Farmers' Movement in India,* pp. 95–125. Ilford: Frank Cass.

Lipton, M. (1978) *Why Poor People Stay Poor: A Study of Urban Bias in World Development.* London: Temple Smith.

Long, N. & J.D. van der Ploeg (1994) 'Heterogeniety, actor and structure: towards a reconstitution of the concept of structure', in D. Booth (ed.) *Rethinking Social Development: Theory, Research and Practice,* pp. 62–89. Harlow: Longman.

Maheu, L. (ed.) (1995) *Social Movements and Social Classes: The Future of Collective Action.* London: Sage Publications.

MARG (1986) *Sardar Sarovar Oustees in Madhya Pradesh: What Do They Know?* vol. 1, Tehsil-Alirajpur, District Jhabua. New Delhi: Multiple Action Research Group.

MARG (1987–88) *Sardar Sarovar Oustees in Madhya Pradesh: What Do They Know?* vols. 2–5, Districts Khargone and Dhar. New Delhi: Multiple Action Research Group.

Marshall, G. (ed.) (1994) *Oxford Concise Dictionary of Sociology*.

McCully, P. (1996) *Silenced Rivers: The Ecology and Politics of Large Dams*. London: Zed Books.

McDonald, M. (1993) 'Dams, displacement and development: a resistance movement in southern Brazil', in J. Friedmann & H. Rangan (eds) *In Defence of Livelihood: Comparative Studies on Environmental Action*, pp. 79–105. West Hartford: UNRISD and Kumarian Press.

Mehta, L. (1992) 'Tribal women facing submergence: the Sardar Sarovar (Narmada) Project in India and the social consequences of involuntary resettlement', M.A. Dissertation, University of Wein.

––––––– (1997) *Water, Difference and Power: Kutch and the Sardar Sarovar (Narmada) Project*. IDS Working Paper No. 54. Sussex: IDS.

Melucci, A. (1988) 'Getting involved: identity and mobilisation in social movements', in H. Kriesi, S. Tarrow & B. Vui (eds) *International Social Movements Research*, vol. 1. London: JAI Press.

––––––– (1989) 'Nomads of the present: social movements and individual needs', in J. Keane & P. Mier (eds) *Contemporary Society*. London: Radius.

––––––– (1992) 'Frontier land: collective action between action and systems', in M. Diani & R. Eyerman (eds) *Studying Collective Action*, pp. 238–58. London: Sage Publications.

Merton, R.K. (1957) *Social Theory and Social Structure*. New York: Free Press.

Miles, M.B. & A.M. Huberman (1994) *Qualitative Data Analysis: An Expanded Sourcebook*, 2nd. ed. Thousand Oaks, CA: Sage Publications.

Mishra, A. (1986) 'The Tawa dam: an irrigation project that has reduced farm production', in E. Goldsmith & N. Hildyard (eds) *The Social and Environmental Effects of Large Dams*, vol. 2, case studies. Wadebridge, UK: Wadebridge Press.

Mistry, J.F. (1991) 'Theme paper on large dams vs. small dams', Symposium on large dams vs. small dams organised by the Central Board of Irrigation and Power and Sardar Sarovar Narmada Nigam Limited, New Delhi, 2 December 1991.

Mitchell, Timothy (2002) *Rule of Experts: Egypt, Techno-Politics, Modernity*. Berkeley: University of California Press.

Moore, D.B. (1995) 'Development discourse as hegemony: towards an ideological history, 1945–1995', in D.B. Moore & G.J. Schmitz (eds) *Debating Development Discourse: Institutional and Popular Perspectives*, pp.1-53. New York: Macmillan, St. Martin's Press.

Moore, D.B. and G.J. Schmitz (eds) (1995) *Debating Development Discourse: Institutional and Popular Perspectives*. New York, Macmillan, St. Martin's Press.

Mosse, D. (1998) 'Process-oriented approaches to development practice and social research', in D. Mosse, J. Farrington & A. Rew (eds) *Development as Process: Concepts and Methods for Working with Complexity*, pp. 3–30. London: Routledge.

Mosse, David (2005) *Cultivating Development: The Ethnography of Aid Policy and Practice*. London: Pluto Press.

Mouzelis, N. (1988) 'Sociology of development; reflections on the present crisis', *Sociology*, 22 (1).

———— (1990) *Post-Marxist Alternatives: The Construction of Social Orders*. London: Macmillan.

Mukta, P. (1995) 'Wresting riches, marginalising the poor, criminalising dissent: the Narmada Dam in western India', *South Asia Bulletin*, 15 (2): 99–108.

NAPM (1995) *Strengthening Struggle against Globalisation and Displacement*, Declaration of the National Alliance for Peoples' Movements, New Delhi.

NAPM (1996) *For Just and Sustainable Development*, Campaign Leaflet of the National Alliance for Peoples' Movements, New Delhi.

Narmada (1989–93) *Narmada: A Campaign Newsletter* (Nos. 1–13), New Delhi: Narmada Bachao Andolan.

Narmada Samachar (1993–96) 'Action alert and update' (various issues). Baroda, Badwani: Narmada Bachao Andolan.

Navarro, Z. (1994) 'Democracy, citizenship and representation: rural social movements in southern Brazil, 1978–1990', *Bulletin of Latin American Research*, 13 (2): 129–54.

NBA (1992) *Towards Sustainable and Just Development: The Peoples' Struggle in the Narmada Valley*. Narmada Bachao Andolan, October.

———— (1993) 'Sardar Sarovar Project: environmental impacts, costs and clearances', mimeo, Narmada Bachao Andolan.

———— (1993a) *Assessment of Utilisable Annual Flow Volumes in Narmada Basin*. Baroda: Narmada Bachao Andolan.

———— (1994) *Campaign Letter*, 10 November 1994.

———— (1995a) *A Note on the Height of the Sardar Sarovar Dam*, submission to the FMG.

———— (1995a) *International Update*, 18 July 1995.

———— (1995b) Writ Petition [civil] No. 319 of 1994, Supreme Court of India: NBA vs Union of India and Others.

———— (1996) *International Update*, June 1996.

———— (undated) *Sardar Sarovar Project, Displacement and Resettlement in Madhya Pradesh: A Critical Assessment from the Point of View of Affected People*. Baroda, Badwani, Manibeli: Narmada Bachao Andolan.

NCA (1990) *Need for Hydro-electric Power Development at Sardar Sarovar Project*. NCA publication No.1/90. New Delhi: Narmada Control Authority.

_____ (1991) *Drinking Water from the Sardar Sarovar Project*. Indore: Narmada Control Authority.

_____ (1992) *Benefits to Saurashtra and Kutch Areas in Gujarat*. Indore: Narmada Control Authority.

_____ (1995) *Master Plan for Resettlement and Rehabilitation (R&R): Sardar Sarovar (Narmada) Project*. Indore: Narmada Control Authority.

NCAER (1967) *Cropping Pattern in Madhya Pradesh*. New Delhi: National Council of Applied Economic Research.

NDS (undated) *Sardar Sarovar on the Narmada. Why Do We Oppose?* Narmada Dharanagrasta Samiti.

Nehru, J.N. (1956) *The Discovery of India*. London: Meridian Books Ltd.

NIHRP (1992) *Interim Report*, 7 October, Prepared by C.A. Wold. Narmada International Human Rights Panel.

Niyogi, S.G. (1982) 'Chattisgarh and the national question', in *Nationality Question in India*. Seminar papers, pp. 107–18. Hyderabad: Andhra Pradesh Radical Students' Union.

NWDT (1978a) *The Report of the Narmada Water Disputes Tribunal with its Decision*, vol. I. New Delhi: NWDT.

_____ (1978b) *The Report of the Narmada Water Disputes Tribunal*, vol. II. New Delhi: NWDT.

_____ (1979) *The Report of the Narmada Water Disputes Tribunal*, vol. III. New Delhi: NWDT.

NWRDC (1965) *Report of the Narmada Water Resources Development Committee, Government of India, Ministry of Irrigation and Power, September*. New Delhi: Government of India, NWRDC.

O'Connor, J. (1989) 'Uneven and combined development and ecological crisis: a theoretical introduction', *Race and Class*, 30 (3): 1–11.

_____ (1992) 'A political strategy for ecological movements', *Capitalism, Nature and Socialism*, 3 (1): 1–6.

OECD (1991) *Guidelines for Aid Agencies on Involuntary Displacement and Resettlement in Development Projects*. Paris: OECD.

Offe, C. (1985) 'New social movements: challenging the boundaries of institutional politics', *Social Research*, 52 (4): 817–68.

Ohlsson, L. (ed.) (1995) *Hydropolitics: Conflicts Over Water as a Development Constraint*. London: Zed Books.

Oliver-Smith, A. (1991) 'Involuntary resettlement, resistance and political empowerment', *Journal of Refugee Studies*, 4 (2): 132–49.

Omvedt, G. (1992) 'Capitalist agriculture and rural classes in India', in B. Berberoglu (ed.) *Class, State and Development in India*, pp. 82–138. New Delhi: Sage Publications.

_____ (1993) *Reinventing Revolution: New Social Movements and the Socialist Tradition in India*. Armonk: M.E. Sharpe.

_____ (1995) '"We want the return for our sweat": the new peasant movement in India and the formation of a national agricultural policy', in T. Brass (ed.) *New Farmers' Movement in India,* pp. 126–64. Ilford: Frank Cass.

ORG (1982) *Regionalisation of Narmada Command.* Baroda: Operations Research Group.

Padaki, V. and V.R. Prasad (1995) 'The development context', in V. Padaki (ed.) *Development Intervention and Programme Evaluation—Concepts and Cases*, pp. 26–50. New Delhi: Sage Publications.

Palmlund, I. (1992) 'Social drama and risk evaluation', in S. Krimsky & D. Golding (eds) *Social Theory of Risk*. Westport, CT: Praeger.

Parajuli, P. (1991) 'Power and knowledge in development discourse: new social movements and the state in India', *International Social Science Journal*, February: 173–90.

Paranjape, S. & K.J. Joy (1995) *Sustainable Technology: Making the Sardar Sarovar Project Viable.* A comprehensive proposal to modify the project for greater equity and ecological sustainability. CEE.

Paranjpye V. (1990) *High Dams on the Narmada: A Holistic Analysis of the River Valley Projects*. New Delhi: INTACH.

Parasuraman, S. (1993) *The Anti-Narmada Project Movement in India: Can the Resettlement and Rehabilitation Policy Gains be Translated into a National Policy?* Working Paper Series No.161. The Hague: Institute of Social Studies.

Parekh, B. (1989) *Gandhi's Political Philosophy*. London: Macmillan.

———— (1995) 'Jawaharlal Nehru and the crisis of modernisation', in U. Baxi & B. Parekh (eds) *Crisis and Change in Contemporary India*, pp. 21–56. New Delhi: Sage Publications.

Patel, A. (1995) 'What do the Narmada valley tribals want?', in W.F. Fisher (ed.) *Towards Sustainable Development, Struggling Over India's Narmada River,* pp. 179–200. Armonk and London: M.E. Sharpe.

Patel, C.C. (1992) 'Narmada Project: what is wrong with the IIM study', *The Times of India*. Ahmedabad, 11 February.

———— (1991) 'SSP: is the project justified?', in *All About Narmada*. Gandhinagar: Government of Gujarat.

———— (1991a) *Sardar Sarovar Project—India; What it is and What it is not*. Gandhinagar: Sardar Sarovar Narmada Nigam Ltd.

———— (1995) 'The Saradar Sarovar Project: a victim of time', in W. Fisher (ed.) *Towards Sustainable Development: Struggling Over India's Narmada River,* pp. 71–88. Armonk and London: M.E. Sharpe.

Patnaik, U. (1986) *The Agrarian Question and the Development of Capitalism in India*. New Delhi: Oxford University Press.

———— (1992) 'A perspective on the recent phase of India's economic development', in B. Berberoglu (ed.) *Class, State and Development in India*, pp. 185–206. New Delhi: Sage Publications.

Patnaik, U. & Z. Hasan (1995) 'Aspects of the farmers' movement in Uttar Pradesh in the context of uneven capitalist development in

Indian agriculture' in T.V. Sathyamurthy (ed.) *Industry and Agriculture in India Since Independence*, vol. 2, pp. 274–300. New Delhi: Oxford University Press.

Pattanaik, S.K., B. Das & A. Mishra (1987) 'Hirakud dam project: expectations and realities', *People and Dams*, workshop proceedings, pp. 47–59. New Delhi: PRIA.

Patton, M.Q. (1987) *How to Use Qualitative Methods in Evaluation*. Newbury Park, CA: Sage Publications.

———— (1997) *Utilization-focused Evaluation: The New Century Text*. Thousand Oaks, CA: Sage Publications.

Pearce, D.W. & C.A. Nash (1981) *The Social Appraisal of Projects: A Text in Cost-benefit Analysis*. Basingstoke: Macmillan.

Peet, R. & M. Watts (1996) *Liberation Ecologies: Environment, Development, Social Movements* (2nd. ed. 2004). London: Routledge.

Porter, D., B. Allen & G. Thompson (1991) *Development in Practice: Paved with Good Intentions*. London: Routledge.

Preston, P.W. (1994) *Discourses of Development: State, Market and Polity in the Analysis of Complex Change*. Aldershot: Avebury.

Ram, M. (1986) 'Civil rights situation in India', in A.R. Desai (ed.) *Violation of Democratic Rights in India*. Bombay: Popular Prakashan.

Ram, R.N. (1992) 'Drinking water from the SSP, doubtful claims', *Economic and Political Weekly*, 25 July: 1589–93.

———— (1993) *Muddy Waters: A Critical Assessment of the Benefits of the Sardar Sarovar Project*. New Delhi: Kalpavriksha.

Rangan, H. (1993) 'Romancing the environment: popular environmental action in the Garhwal Himalayas', in J. Friedmann & H. Rangan (eds) *In Defence of Livelihood: Comparative Studies on Environmental Action*, pp. 155–81. West Hartford: UNRISD and Kumarian Press.

———— (1996) 'From Chipko to Uttaranchal: development, environment and social protest in the Gharwal Himalayas, India', in R. Peet & M. Watts (eds) *Liberation Ecologies: Environment, Development, Social Movements*, pp. 205–26. London and New York: Routledge.

Redclift, M. (1987) *Sustainable Development: Exploring the Contradictions*. London: Methuen.

Reddy, A.K.N. (1993) 'The electrical part of the Sardar Sarovar Project. Part I: Introduction and main conclusions', in *FMG*, vol. 2, pp. 138–47. New Delhi.

Renard, R. & L. Berlage (1992) 'The rise and fall of cost-benefit analysis', in L. Berlage & O. Stokke (eds) *Evaluating Development Assistance: Approaches and Methods*, pp. 33–55. London: Frank Cass.

Renn, O. (1992) 'Concepts of risk: a classification', in S. Krimsky & D. Golding (eds) *Social Theories of Risk*. Westport, CT: Praeger.

Repetto, R. (1986) *Skimming the Water: Rent-seeking and the Performance of Public Irrigation Systems*. Washington, DC: World Resources Institute.

RIC (1972a) *Report of the Irrigation Commission,* vol. I. New Delhi: Government of India.

————— (1972b) *Report of the Irrigation Commission,* vol. II. New Delhi: Government of India.

RLA (1991) '1991 Right Livelihood Award Sends Strong Message to 1992 Earth Summit', Press Release. Stockholm: Right Livelihood Award Stiftelsen, 2 October.

Rodrigues, L. (1984) 'Rural protest and politics: a study of peasant movements in western Maharashtra 1875–1947'. D.Phil thesis. London: London School of Economics and Political Science.

Roemer, J. (1986) '"Rational choice" Marxism: some issues of method and substance', in J. Roemer (ed.) *Analytical Marxism,* pp. 191–201. Cambridge: Cambridge University Press.

Rossi, P.H. & H.E. Freeman (1993) 'Strategies for impact assessment', in *Evaluation: A Systematic Approach,* pp. 221–45. Newbury Park, CA: Sage Publications.

Roy, S.K. (1987) 'Tehri dam', *People and Dams,* workshop proceedings, pp. 60–76. New Delhi: PRIA.

Rudolph, L.I. & S.D. Rudolph (1987) *In Pursuit of Lakshmi: The Political Economy of the Indian State.* Chicago, IL: University of Chicago Press.

Ryle, M. (1988) 'Socialism in an ecological perspective', in *Ecology and Socialism,* pp. 17–40 and 104–10. London: Radius Hutchinson.

Saith, A. (1992) *The Rural Non-farm Economy: Processes and Policies.* Geneva: ILO.

Salman, T. (1990) 'Between orthodoxy and euphoria research strategies on social movements: a comparative perspective', in A. Williem, G. Burgwal & T. Salman (eds) *Structures of Power, Movements of Resistance: An Introduction to the Theories of Urban Social Movements in Latin America,* pp. 99–162. Amsterdam: CEDLA.

Sant, G. (1993) 'The electrical part of the Sardar Sarovar Project, Part II: A comparison with other low-cost options', in *FMG* Vol. 2, pp. 148–70. New Delhi.

Sayer, A. (1992) *Method in Social Science: A Realist Approach.* London: Routledge.

Schaffer, B.B. (1984) 'Towards responsibility: public policy in concept and practice', in E.J. Clay & B.B. Schaffer (eds) *Room for Manoeuvre: An Exploration of Public Policy in Agricultural and Rural Development,* pp. 142–90. London: Methuen.

Schuurman, F.J. (1994) *Current Issues in Development Studies: Global Aspects of Agency and Structure.* Saarbrücken: Verlag für Entwicklungspolitik Breitenbach.

Scott, J. (1990) *Domination and the Arts of Resistance: Hidden Transcripts.* New Haven and London: Yale University Press.

SCSS (1987) 'Bakun dam: educational and organising efforts', in *People and Dams*, workshop proceedings, pp. 149–52. New Delhi: PRIA (Society of Christian Service, Sibu, Sarawek).

Scudder, T. (1989) *Supervisory Report on the Resettlement and Rehabilitation (R&R) Component of the Sardar Sarovar Project (SSP)*, 22 May. Washington, DC: World Bank.

——— (1991) 'Development-induced relocation and refugee studies: 37 years of change and continuity among Zambia's Gwembe Tonga', *Journal of Refugee Studies*, 6.

Sengupta, N. (1993) 'Issues in Project Design: The Sardar Sarovar Project and its Alternatives', paper prepared for the Narmada Forum. New Delhi: Centre for Development Economics and Institute of Economic Growth.

Seth, D.L. (1983) 'Grassroots stirrings and the future of politics', *Alternatives*, 9 (Spring): 1–24.

Sethi, H. (1993a) 'Survival and democracy: ecological struggles in India', in P. Wignaraja (ed.) *New Social Movements in the South: Empowering the People*, pp. 122–48. London: Zed Books.

——— (1993b) 'Action groups in the new politics', in P. Wignaraja (ed.) *New Social Movements in the South: Empowering the People*, pp. 230–55. London: Zed Books.

Shadish, W., T. Cook & L. Leviton (1991) 'Good theory for social program evaluation', in *Foundations of Program Evaluation: Theories of Practice*, pp. 36–60. Newbury Park, CA: Sage Publications.

Shah, A. (1993) *Water for Gujarat—An Alternative. Technical Overview of the Flawed Sardar Sarovar Project and a Proposal for a Sustainable Alternative*. Vishakhapatnam: Jan Vikas Andolan and others.

Shah, A.M. (1964) 'Political system in eighteenth century Gujarat', *Enquiry*, 1(Spring).

Shah, G. (1974) 'The upsurge in Gujarat', *Economic and Political Weekly*, 9 (32–34): 1429–54.

——— (1977) *Protest Movements in two Indian States*. New Delhi: Ajanta Publishers.

——— (1990) 'Caste sentiments, class formation and dominance in Gujarat', in F.R. Frankel & M.S.A. Rao (eds) *Dominance and State Power in Modern India: Decline of a Social Order*, vol. II, pp. 59–114. Delhi: Oxford University Press.

Shah, U. (1994) 'Lowering height of Sardar Sarovar Dam: what purpose does it serve', *Economic and Political Weekly*, 29 (12): 667–68.

Sharma, L. & T.R. Sharma (1981) *Major Dams: A Second Look*. New Delhi and Sirsi: Gandhi Peace Foundation and Totgars Co-operative Sale Society Ltd.

Sheth, P.N. (1994) *Narmada Project: Politics of Eco-development*. New Delhi: Har-Anand Publications.

Sheth, P.N. (1995) 'The Sardar Sarovar project: ecopolitics of development', in U. Baxi & B. Parekh (eds) *Crisis and Change in Contemporary India*, pp. 400–1. New Delhi: Sage Publications.

Shiiu-Hung, L. & J. Whitney (eds) (1993) *Mega Project: A Case study of China's Three Gorges Project*. Armonk, New York: M.E. Sharpe.

Shiva, V. (1988) *Staying Alive: Women, Ecology and Development*. London: Zed Books.

———— (1991) *Ecology and the Politics of Survival*. New Delhi: United Nations University and Sage Publications.

Shiva, V. & J. Bandhopadhyay (1989) 'Political economy of ecology movements', *IFDA Dossier*, 71, pp. 37–60.

Singh, S.K. (1990) 'Evaluating large dams in India', *Economic and Political Weekly*, 25 (11): 561–74.

———— (1997) *Taming the River: Political Economy of Large Dams in India*. New Delhi: Oxford University Press.

Slater, D. (1992) 'Theories of development and politics of the post-modern—exploring a border zone', *Development and Change*, 23 (3): 283–319.

———— (ed.) (1985) *New Social Movements and the State in Latin America*. Amsterdam: CEDLA.

Social Action (1988) 'Action against big dams', *Social Action*, 38: 297.

Soni, T. (1993) 'Sardar Sarovar Project: downstream impact—The voiceless tragedy', unpublished mimeo.

Sovani, N.V. & N. Rath (1960) *Economics of Multi-purpose River Dam: Report of an Inquiry into the Economic Benefits of the Hirakud Dam*. Bombay: Gokhale Institute of Politics & Economics.

Srinivasan, B. (1993) 'Repression in the Narmada valley', *Economic and Political Weekly*, 28 (49): 2640–41.

———— (1994) 'Dissent and democratic practice: attack on NBA office' *Economic and Political Weekly*, 29 (18): 1058–59.

Srinivasan, B., R. Prajapati & W. D'Costa (1989) 'Dam workers on strike', *Economic and Political Weekly*, 24 (7): 339–41.

Srinivasan, S. (1997) 'Engendering the World Bank: an enquiry into the mainstreaming of gender concerns in development aid' MA research paper. The Hague: ISS.

Srivastava, R. (1995) 'India's uneven development and its implications for political processes: an analysis of some recent trends', in T.V. Sathyamurthy (ed.) *Industry and Agriculture in India Since Independence*, vol. 2, pp. 219–47. New Delhi: Oxford University Press.

SSNNL (1994a) *Facts*. Gandhinagar: Sardar Sarovar Narmada Nigam Ltd.

———— (1994b) *Facts Updated*. Gandhinagar: Sardar Sarovar Narmada Nigam Ltd.

Stewart, F. (1975) 'A note on social cost-benefit analysis and class conflict in LDCs', *World Development*, 3 (1).

Sunkel, O. (1980) 'The interaction between styles of development and the environment in Latin America', *CEPAL Review*, 12.

Swami, P. (1995) 'Height in question: debate on Narmada dam changes course', *Frontline*, 10 March, Chennai.

Tarrow, S. (1994) 'National politics and collective action: recent theory and research in Western Europe and the USA', *Annual Review of Sociology*, 14: 421–40.

TECS (1983) *Economic Appraisal of Sardar Sarovar Project*. Report prepared for Narmada Planning Group. Bombay: Tata Economic Consultancy Services.

Thakar, H. (1993) 'Can the Sardar Sarovar ever be financed', *Economic and Political Weekly*, 28 (42): 2262–66.

Thakar, U. & M. Kulkarni (1992) 'Environment and development: the case of the Sardar Sarovar Project', *South Asia Bulletin*, 12 (2): 96–103.

Thukral, E.G. (ed.) (1992) *Big Dams, Displaced People. Rivers of Sorrow, Rivers of Change*. New Delhi: Sage Publications.

TISS (1993) 'Sardar Sarovar Project: review of resettlement and rehabilitation in Maharashtra', *Economic and Political Weekly*, 28 (34): 1705–14.

Touraine, A. (1985) 'An introduction to the study of social movements', *Social Research*, 52 (4): 749–88.

Toye, J. (1993) *Dilemmas of Development: Reflections on the Counter-revolution in Development Economics*. Oxford: Blackwell.

Udall, L. (1995) 'The international Narmada campaign: a case of sustained advocacy', in W. Fisher (ed.) *Towards Sustainable Development: Struggling Over India's Narmada River*, pp. 201–27. Armonk and London: M.E. Sharpe.

Uphoff, N. (1992) *Learning from Gal Oya: Possibilities for Participatory Development and Post-Newtonian Social Science*. Ithaca, NY: Cornell University Press.

Usher, A.D. (1997) *Dams as Aid: A Political Anatomy of Nordic Development Thinking*. London: Routledge.

Vaidyanathan, A. (1995) 'The political economy of the evolution of anti-poverty programmes', in T.V. Sathyamurthy (ed.) *Industry and Agriculture in India Since Independence*, vol. 2, pp. 329–48. New Delhi: Oxford University Press.

Wade, R. (1975) 'Administration and the distribution of irrigation benefits', *Economic and Political Weekly*, 10 (44 & 45): 1743–47.

——— (1976) 'Performance of irrigation projects', *Economic and Political Weekly*, 11 (3): 63–66.

Watts, M. & R. Peet (1996) 'Towards a theory of liberation ecology', in R. Peet & M. Watts (eds) *Liberation Ecologies: Environment, Development and Social Movements*, pp. 260–69. New York and London: Routledge.

WB (1985a) Report and Recommendation of the President of the IDA to the Executive Directors on a Proposed IDA Credit of SDR 149.5 Million to India for the Narmada River Development (Gujarat) Dam and Power. Washington, DC: World Bank.

———— (1985b) Report and Recommendation of the President of the IDA to the Executive Directors on a Proposed IDA Credit of SDR 149.5 Million to India for the Narmada River Development (Gujarat) Water Delivery and Drainage Project. Washington, DC: World Bank.

———— (1985c) Staff Appraisal Report, Narmada River Development—Gujarat, Sardar Sarovar Dam and Power Project. Washington, DC: World Bank, South Asia Projects Department, Irrigation II Division.

———— (1985d) Staff Appraisal Report, Narmada River Development—Gujarat, Sardar Sarovar Project: Water Delivery and Drainage. Washington, DC: World Bank, South Asia Projects Department, Irrigation II Division.

———— (1985e) Staff Appraisal Report, India, Narmada River Development—Gujarat. Supplementary Data Volume, Part I. Washington, DC: World Bank, South Asia Projects Department, Irrigation II Division.

———— (1985f) Staff Appraisal Report, India, Narmada River Development—Gujarat. Supplementary Data Volume, Part II. Washington, DC: World Bank, South Asia Projects Department, Irrigation II Division.

———— (1990) *Review of Sardar Sarovar (Narmada) Projects Economic Analysis*. Washington, DC: World Bank, India Agriculture Operations Division.

———— (1991a) *India Irrigation Sector Review*, vol. I, main report. Washington, DC: World Bank, Agriculture Operations Division, India Department, Asia Region.

———— (1991b) Letter and Enclosure of 18 January 1991 of Jan Wijnand, World Bank, to Bruce Rich and Lori Udal.

———— (1992) INDIA: Sardar Sarovar Projects—Mission Briefing to the Executive Directors, 5 August.

———— (1994a) *Resettlement and Development: The Bank-wide Review of Projects Involving Involuntary Resettlement 1986–1993*. Washington, DC: World Bank, Environment Department.

———— (1994b) *World Development Report 1994: Infrastructure for Development*. Oxford: Oxford University Press for the World Bank.

———— (1995) *Learning from Narmada*. Washington, DC: World Bank,

Operations Evaluation Department.

_____ (1995a) Project Completion Report: Narmada River Development—Gujarat, Sardar Sarovar Dam and Power Project: Environmental Impacts and Management. Washington, DC: World Bank, India Department.

_____ (1995b) Project Completion Report: Narmada River Development—Gujarat, Sardar Sarovar Dam and Power Project: Resettlement and Rehabilitation. Washington, DC: World Bank, India Department.

_____ (1995c) Project Completion Report: Narmada River Development—Gujarat, Sardar Sarovar Dam and Power Project: Revised Cost-benefit Analysis. Washington, DC: World Bank, India Department.

_____ (1995d) Project Completion Report: Narmada River Development—Gujarat, Sardar Sarovar Dam and Power Project: Water Delivery and Drainage Project. Washington, DC: World Bank, India Department.

_____ (1996) *World Bank Lending for Large Dams: A Preliminary Review of Impacts*. Washington, DC: World Bank, Operations Evaluation Department.

Weber, T. (1987) *Hugging the Trees: Story of the Chipko Movement*. New Delhi: Viking Press.

Wignaraja, P. (1993) 'Introduction', in P. Wignaraja (ed.) *New Social Movements in the South: Empowering the People*, pp. 4–35. London: Zed Books.

Williams, R. 'Socialism and ecology', London Socialist Environmental Resources Association, pp. 3–20, in M. Redclift & G. Woodgate (eds) (1995) *The Sociology of the Environment*, vol. I, pp. 504–21. Aldershot and Brookfield: Edward Elgar.

Wood, J.R. (1986) 'Congress restored? The "Kham" strategy and Congress (I) recruitment in Gujarat', in J.R. Wood (ed.) *State Politics in Contemporary India: Crisis or Continuity?* pp. 197–228. Boulder and London: Westview Press.

World Commission on Dams (2000) *Dams and Development: A New Framework for Decision-making*. London: Earthscan Publishers.

Index

Member Group 240–1, 260–5;
260–5
Indian political economy 48
industrial development and
growth of working class 55n
Integrated Rural Development
Programme (IRDP) 46
International Committee on
Large Dams (ICOLD) 336
inter-state river water disputes
64–6

Jalsindhi dam 71–2, 73–4, 78,
81, 84–6, and Maharashtra
and Rajasthan states'
interest 71–2; project and
Madya Pradesh's interest 71
Jan Mukti Morcha (Peoples'
Liberation Front) 175
Jan Vikas Andolan (JVA-
Movement for People's
Development) 166–7, 174
Jan Vikas Sangharsh Yatra
(Struggle March for People's
Development) 166
Japanese government, with-
drawal of loan to Sardar
Sarovar Project 225

Kalpavriksh and Hindu College
Nature Club study on
Narmada Valley 154
Kerala Sastra Sahitya Parishad
and Silent Valley project 148
Khedut Mazdoor Chetna
Sangath (KMCS) 156, 175,
and resettlement of 281–2;
and Sardar Sarovar Project
306
Khosla Committee 67, 69;
Gujarat, Madhya Pradesh
and Maharashtra 77;
recommendations of 74–7; *vs*

Narmada Water Dispute
Tribunal 90, 92
'knowledge struggles' 366;
Narmada movement and
Sardar Sarovar Project 228–
32; about other projects
286–90; that represents
truth model 362
Kohli 56n; and state-society
relations 46–7

Land Acquisition Act (1894)
178n, 182n, 280, 311n
Land Ceiling Act (1972)
opposition to 95n
Land Ceiling Act (1974) 96n;
amendment and opposition
to 79
landless agricultural labourers,
and compensation to 313n;
action plan of Gujarat and
Madhya Pradesh (MP)
governments and landless
labourers 290; risk of 290–5;
vs Narmada Bachao Andolan
294–5
'landlord capitalism' 55n
large projects 36–8; and 'hiding
hand principle' 194–5; and
Nehruvian doubt 67–9;
movement against 336–9;
Nehru on 38
Levy Act (1973) 96n; opposition
to 79
Liberation Ecology, by Peet and
Watts 32; and 'regional
discursive formation' 32

Madhya Pradesh (MP) 74–8;
formation of 94n; its chief
minister for Prime Minister's
intervention on Sardar
Sarovar Project height
reduction 186n, 188n